Radioisotope Gauges for Industrial Process Measurements

Radioisotope Gauges for Industrial Process Measurements

Geir Anton Johansen
University of Bergen, Norway

Peter Jackson
Tracerco, Cleveland, UK

John Wiley & Sons, Ltd

Telephone (+44) 1243 779777

Email (for orders and customer service enquiries): cs-books@wiley.co.uk
Visit our Home Page on www.wileyeurope.com or www.wiley.com

Other Wiley Editorial Offices

John Wiley & Sons Inc., 111 River Street, Hoboken, NJ 07030, USA

Jossey-Bass, 989 Market Street, San Francisco, CA 94103-1741, USA

Wiley-VCH Verlag GmbH, Boschstr. 12, D-69469 Weinheim, Germany

John Wiley & Sons Australia Ltd, 33 Park Road, Milton, Queensland 4064, Australia

John Wiley & Sons (Asia) Pte Ltd, 2 Clementi Loop #02-01, Jin Xing Distripark, Singapore 129809

John Wiley & Sons Canada Ltd, 22 Worcester Road, Etobicoke, Ontario, Canada M9W 1L1

Wiley also publishes its books in a variety of electronic formats. Some content that appears in print may not be available in electronic books.

British Library Cataloguing in Publication Data

A catalogue record for this book is available from the British Library

ISBN 0-471-48999-9

Typeset in 10.5/13pt. Times by TechBooks Electronic Services, New Delhi, India

This book is printed on acid-free paper responsibly manufactured from sustainable forestry in which at least two trees are planted for each one used for paper production.

Contents

Everything should be made as simple as possible,
but not simpler.

Albert Einstein, 1879–1955

Preface

The beginning of this book was university lecture notes and training material for process engineers. Although it now is a lot more comprehensive, it has been our intention to maintain a mixed academic and industrial approach to the various subjects. Our motivation for writing this book has been the need for a text covering the full range from the underlying physics to the process applications of radioisotope gauges. We could not deal with all subjects in detail; however, we have included references to many excellent books and articles where further details can be found.

We wish to acknowledge help and support from many colleagues and friends: Prof. Richard Thorn at the University of Derby, Dr. Ken James and Dr. Dave Couzens at Tracerco, Prof. Erling Hammer and Prof. Jan Vaagen at the University of Bergen, Dr. Paul Schotanus at Scionix, Prof. Robin Pierce Gardner at North Carolina State University, Dr. Jaafar Abdullah at the Malaysian Institute for Nuclear Technology Research, Dr. Stein-Arild Tjugum at Roxar Flow Measurement, and Mr. Truls Roar Søvde at the Norwegian Metrology and Accreditation Service.

Last but not least, loads of patience and support from our wives, Kari Anne and Marilyn, are highly appreciated, and likewise the patience of Anne, Peter, Bendik and Victor, all accepting their father's absence in writing when they had other activities in mind.

Bergen/Billingham
February 2004

Geir Anton Johansen
Peter Jackson

Symbols, Units and Abbreviations

For names of elements, such as ^{226}Ra and ^{137}Cs, see Appendix A.3. Several symbols are listed without their subscripts; these are often listed separately. Also note that some symbols have multiple meanings.

1D	One-dimensional
2D	Two-dimensional
3D	Three-dimensional
A	Activity or decay rate of radioisotope
A	Mass number or atomic weight ($A = N + Z$, in terms of u)
A	Amplifier gain
AC	Alternating current
ADC	Analogue to digital converter
ALARA	As low as reasonably achievable
A_2^M	Throat cross-section area in Δp metres
ANS	American National Standard
APD	Avalanche photodiode
$B(\mu, x)$	Build-up factor, also denoted B
BGO	$Bi_4Ge_3O_{12}$ scintillation crystal
BIPM	Bureau International des Poids et Mesures
BLR	Baseline restorer
Bq	Becquerel (SI unit of activity, 1 Bq = 1 disintegration per second)
C	Discharge coefficient (of Δp metres)
c	Speed of light in vacuum ($=2.9979992458 \times 10^8$ m/s)
CCD	Charged coupled device
C_f	Feedback capacitance
Ci	Curie (old unit of activity, 1 Ci = 3.7×10^{10} Bq)
c_i	Sensitivity coefficient
cps	counts per second (often denoted c/s)
CR^2–RC^n	Bipolar shaping network
CR–RC^n	Semi-Gaussian unipolar shaping network
CSDA	Continuously slowing down approximation
CT	Computerised tomography
CWO	$CdWO_4$ scintillation crystal

D	Absorbed dose
d	Distance from radiation (point) source
d	(Pipe) diameter
d	Cathode–anode separation
DAC	Digital to analogue converter
DC	Direct current
DDL-RC	Double delay line shaping network
$\mathrm{d}E/\mathrm{d}x$	Stopping power (charged particle energy deposition per unit path length)
DET	Dual energy transmission
DL-RC	Delay line shaping network
DSP	Digital signal processor
e	Electron
e	Elementary charge ($=1.602176462 \times 10^{-19}$ C)
E	Energy of nuclear radiation, normally expressed in terms of eV
E	Electric field strength
E	Effective dose
E	Velocity approach factor in Δp metres
E_A	Preamplifier and biasing network noise
E_{bi}	Electron binding energy of the ith atomic shell
ECT	Electrical capacitance tomography
E_D	Detector (diode) noise
E_{det}	Energy deposited in radiation detector
E_E	Total electronic noise
E_{kin}	Kinetic energy
EMI	Electromagnetic interference
ENC	Equivalent noise charge
eV	Electron Volt unit of energy (1 eV $= 1.6 \times 10^{-19}$ J)
E_γ	γ-Ray energy (other subscripts also used for other radiations, e.g. α, β and X)
F	Fano factor
FET	Field effect transistor
FWHM	Full width at half maximum
G	Gain
g	Gas
g_m	Transconductance
GMT	Geiger–Müller tube
GSO	$Gd_2SiO_5(Ce)$ scintillation crystal
GVF	Gas volume fraction (void fraction) $= \alpha_g$
Gy	Gray (SI unit of absorbed dose, H)
h	holes
h	Planck's constant ($=6.62606876 \times 10^{-34}$ Js)
HPD	Hybrid photon detector, also known as the hybrid PMT (HPMT)
H_T	Equivalent dose
HV	High voltage (sometimes denoted HT – High Tension)
I	Radiation beam intensity
I	Mean excitation energy of absorber

I_0	Initial or incident radiation beam intensity
IAEA	International Atomic Energy Agency
IATA	International Air Transport Association
IC	Internal conversion
IC	Integrated circuit
ICRP	International Commission on Radiological Protection
ID	Inner diameter
I_E	Beam intensity with empty pipe (or vessel)
I_l	Leakage current (also known as dark current)
ISO	International Standards Organisation
IT	Isomeric transition
k	Boltzmann's constant ($=1.3806503 \times 10^{-23}$ J/K $= 0.8617 \times 10^{-4}$ eV/K)
k	Confidence coverage factor
K_α	Characteristic X-ray emission from the L to the K atomic shell
K_β	Characteristic X-ray emission from the M to the K atomic shell
l	Liquid
L	Loss fraction of the scintillation light
LCD	Liquid crystal display
LED	Light emitting diode
LET	Linear energy transfer
LLD	Lower level discriminator
LLD	Lower limit of detection
LSA	Low specific activity
LSO	$Lu_2SiO_5(Ce)$ scintillation crystal
m	Particle mass
M	Sometimes used for atomic weight instead of A
MC	Monte Carlo
MCA	Multi-channel analyser
MCP	Micro channel plate
m_e	Electron rest mass ($=9.1093818872 \times 10^{-31}$ kg)
MRI	Magnetic resonance imaging
MSM	Metal–semiconductor–metal
MTBF	Mean time between failure
n	Neutron
n	Count-rate
N	Number of neutrons in the atom's nucleus
N	Number of atoms per unit volume
N	Number of charge carriers
N_0	Number of radioactive atoms present at a time $t = 0$
NBS	National Bureau of Standards
n_C	Number of counts
NCS	Nucleonic control systems
NDT	Non-destructive testing
NIM	Nuclear instrument modules
NORM	Naturally occurring radioactive material
$\langle N_\Delta^2 \rangle$	Delta noise coefficient of shaping network

$\langle N_S^2 \rangle$	Step noise coefficient of shaping network
o	Oil
OD	Outer diameter
p	Pressure
p	Probability
p	Proton
Δp	Differential pressure (measurements)
PADC	Poly-allyl diglycol carbonate
PEEK	Polyetheretherketone
PET	Positron emission tomography
PGNAA	Prompt γ-ray neutron activated analysis
PHA	Pulse height analyser
PIN	p-type–intrinsic–n-type semiconductor material
PMT	Photomultiplier tube
PSD	Position sensitive detector
PUR	Pile-up rejection
PZC	Pole-zero cancellation
q	Volumetric flow rate
Q_C	Scintillation efficiency
Q_E	Quantum efficiency of light detectors
R	Energy resolution
R	Resistance
R	Range of particles is in absorbers
$R(\lambda)$	Radiant sensitivity
r_A	Anode radius
rad	Radiation absorbed dose (old unit of absorbed dose, 1 rad $= 10^{-2}$ Gy)
r_C	Cathode radius
rem	Röntgen Equivalent Man (old unit of equivalent and effective dose, 1 rem $= 10^{-2}$ Sv)
R_f	Feedback resistance
rms	root mean square
RPA	Radiological protection advisor
RPS	Radiological protection supervisor
s	Solid
S_0	Isotropic emission intensity of isotopic source
S20	Trialkali or multialkali photocathode (NaKSbCs)
SCA	Single-channel analyser
SCO	Surface contaminated object
SNR	Signal-to-noise ratio
SPECT	Single particle emission computed tomography
Sv	Sievert (SI unit of equivalent and effective dose)
t	Time
$T_{1/2}$	Half-life of a radioactive isotope
TEC	Thermoelectric cooler
TEGRA	Triple energy γ-ray absorption

TLD	Thermoluminescent dosimeter
u	The unified atomic mass constant ($=1.66053873 \times 10^{-27}$ kg)
ULD	Higher level discriminator
UN	United Nations
UNSCEAR	United Nations Scientific Committee on the Effect of Atomic Radiation
UV	Ultraviolet radiation
v	Particle velocity
v	Charge carrier drift velocity
V	Volt
V	High voltage or bias
w	Average energy required to create one charge carrier pair in an absorber
w	water
w_i	Weight fraction of the ith component in a mixture
w_R	Radiation weighting factor (dimensionless)
w_T	Tissue weighting factor (dimensionless)
X	Exposure unit
x	Thickness of absorber
$x_{1/2}$	Average half-thickness, where the radiation beam intensity is half of its initial value
XRF	X-ray fluorescence analysis
YAP	$YAlO_3(Ce)$ scintillation crystal
Z	Atomic number (number of protons in the atom's nucleus)
Z_{eff}	Effective atomic number of a mixture or chemical compound
α	Alpha particle; energetic helium nucleus (^4He) originating from radioactive decay
α_g	Gas volume fraction (GVF, void fraction)
α_i	Volume fraction of the ith component in a mixture
(α, n)	Nuclear reaction, here initiated by an alpha particle with the emission of a neutron
β^-	Negative beta particle or fast electron originating from radioactive decay
β^+	Positive beta particle
β^+	Fast positron originating from radioactive decay
Γ	Specific γ-ray dose rate constant (SGRDC)
ε	Expansibility coefficient of the fluid in Δp metres
ε_r	Relative dielectric constant (permittivity)
θ	Scattering angle of Compton scattered photon
κ	Interaction cross section of pair production
λ	Decay constant of a radioactive isotope
λ	Wavelength of electromagnetic waves
λ	Mean free path of ionising photons
μ	Linear attenuation coefficient of ionising photons
μ	Mobility of charge carriers
μ_β	Linear absorption coefficient of beta particles
μ_{eff}	Effective linear attenuation coefficient
μ_M	Mass attenuation coefficient of ionising photons

μ_{mix}	Linear attenuation coefficient of a mixture of components
v	Neutrino
$\bar{\mathrm{v}}$	Antineutrino
ν	Wavelength of electromagnetic waves, normally expressed in nm
ρ	Density
Σ	Attenuation coefficient or macroscopic cross section of neutrons
σ	Interaction cross section of Compton scattering
σ	Standard deviation
σ_{R}	Interaction cross section of Rayleigh scattering
σ_{TOT}	Total interaction cross section
τ	Interaction cross section of the photoelectric effect
τ	Time constant
τ_0	Peaking time of shaping amplifier
τ_{C}	Charge collection time
τ_{C}	Noise corner; filter time constant for optimal noise performance
τ_{D}	Scintillator light decay constant
τ_{I}	Counting time
Φ	Flux
Ω	Solid angle
ω	Fluorescence yield of characteristic X-rays
Ø	Diameter

1

Introduction

Many people fear radioactivity; they associate it with the fallout of atomic bombs or disasters such as the explosion at the Chernobyl nuclear power station. It is, however, a natural process happening constantly all around us. It occurs in our homes, in the food we eat; even our bodies are radioactive. Today radioactive materials and radiations are used in medicine, industry, agriculture, pollution control, energy production and research [1–3]. In this book we will study how even low activities of ionising radiation can be used as a powerful tool in solving difficult industrial measurement problems. We shall also see that the risks involved in applying radioactivity this way are very small. This is ensured through strict recommendations and legislation to which typical radioisotope gauges comply with good margin.

1.1 IONISING RADIATION

Radiation with sufficient energy to ionise atoms in matter is called *ionising radiation*. This includes both electromagnetic radiation such as γ-rays and X-rays, and energetic particles such as α- and β-particles, and neutrons, which although not directly ionising, produce secondary ionising radiation. Ionising radiation is often named after the origin of the radiation. Radiation emitted when an unstable nucleus in an element, a *radioisotope*, disintegrates is called *nuclear radiation*.

1.2 INDUSTRIAL NUCLEONIC MEASUREMENT SYSTEMS

The foundation of all industrial nucleonic measurement systems is the combination of one or several ionising radiation sources and one or several radiation detection units. Important process or system parameters are then derived from the measurement of interactions between the ionising radiation and the process or system under investigation. This type of industrial instrumentation has been boosted by research and development in the nuclear power reactor industry where the radioisotope in many ways is a by-product. Additionally, high-energy physics research has played an important role in the development of new and improved detector principles.

Nucleonic methods are frequently used in modern industrial measurement systems because for several reasons, they are robust and reliable:

Radioisotope Gauges for Industrial Process Measurements. Geir Anton Johansen and Peter Jackson.
 ISBN 0-471-48999-9

- Ionising radiation responds to the fundamental physical properties of matter; the density of elementary particles.
- Nucleonic measurement methods are non-contacting, a very attractive measurement system property, which often allows 'clamp-on' installation and operation.
- The interaction of ionising radiation can be detected and measured with high sensitivity.

The drawback with many industrial radioisotope (nucleonic) measurement systems is relatively high costs. This is partly due to material costs for the radiation source and detector system, but in some cases also indirect costs for preparations and paperwork necessary to comply with the legislation for transport and operation of ionising radiation sources. The latter is particularly important for first time use of a new instrument, or for first time installation in an application or at a site. It is less significant once efficient and adequate routines are established. This is of course to ensure safe transport and operation of the equipment with less risks than other risks encountered at industrial plants. Correct handling of nucleonic measurement systems is thus not hazardous, although many people unfortunately have the opposite apprehension. Such attitudes are often based on lack of knowledge and erroneous preconceptions, and are in most cases dealt with by bringing facts to the table.

The focus of this book will be on permanently installed gauges. This implies that γ-ray methods are the main theme of the book, mainly because for a variety of reasons radioisotope sources are most suitable for permanent installation. Industrial measurement systems based on ionising electromagnetic radiation involve a large diversity of methods and principles. To facilitate matters it is useful to categorise these in a few ways. Firstly, the different measurement systems may be regarded with respect to the type of source used:

- *Natural occurring radioactive materials* (*NORM*): In this case, the measurement system comprises the detector system only. A typical example is lithology where γ-ray emission analysis is used to distinguish between different sedimentary layers in ground boreholes.
- *Sealed sources* and *X-ray tubes*: This involves a well-defined geometry where a process or a system is exposed to radiation from one or several sources and the response is measured by one or several detectors. Here it is handy to introduce three sub-categories:
  - Transmission measurements, where the source and detector are placed on opposite sides of the process. Here a typical example is measurement of process density.
  - Scatter measurements, where the source and detector are placed closer to each other, and often side-by-side (backscatter). This is often used for density measurements on large process vessels.
  - Measurement of secondary emissions such as X-ray fluorescence. This is most often used for element composition and concentration analysis.
- *Tracers*: Here small amounts of a short-lived radioisotope are added to a substance of interest in a process or a system. The pathways of these radioisotopes into a complex system are then followed by detecting γ-ray or annihilation emissions as they appear at different locations. The substance may be in either gas, liquid or solid state. This is often used to measure process dynamics and element residence times in vessels, the degree of mixing or separation of process elements, leakages, etc.

Secondly, and particularly in the context of this book, it is useful to categorize the measurement methods in a different way, according to their mode or nature of operation:

I. *Laboratory instrumentation*, where process samples are taken and brought in for off-line analysis in specialised facilities. The instruments may be characterised as complex and sophisticated, yielding high performance measurements and advanced data analysis. Nuclear reactors, accelerators and X-ray machines are used as radiation sources, and the measurements are often carried out with cryogenic high-resolution radiation detectors. The samples being investigated are in some cases placed in vacuum chambers to allow the use of particle radiation such as α-particles and protons, which otherwise have very short range. For industrial purposes such facilities are in some cases used for periodic process samples; however, these are more often used for research and development on processes, process models and instrumentation.

II. *Process diagnostics instrumentation*, which is brought to the plant and used by specialised personnel. Data are normally recorded for subsequent off-line analysis. Typical applications are scanning of process columns and reactors, tracer measurements and non-destructive testing of equipment and plant. The instrumentation needs to be portable and rugged, suitable for operation in rough environment. Radioisotope sources are used for the majority of these examinations. Various logging and NDT (non-destructive testing) applications also fall in this category.

III. *Permanently installed gauges*, also known as *nucleonic control systems (NCS)*, provide real-time measurements and online analysis, and in some cases are used for closed-loop control. Here, speed of response able to cope with the process dynamics is often of primary importance in comparison to what is the case with the two previous categories. Sealed radioisotope sources are used for most permanently installed gauges, although there are a few examples of automatic injection tracer installations and systems using X-ray tubes and neutron generators.

In cases where only sporadic measurements are required, end-users of nucleonic instrumentation often prefer category II solutions to category III solutions. This is because the process diagnostics company will then be responsible for the use of the ionising radiation and all the paperwork related to it.

1.3 HISTORICAL PERSPECTIVE

The line of discovery that leads to today's nucleonic instrumentation includes many of the most eminent scientists of the past 200 years. Three inventions made possible the discovery and investigation of ionising radiation and led to the development of the radiation measuring instruments and radioactive sources that we use today. In this chapter we will explore the discoveries and try to show a glimpse of the people involved and the workings of the inventive process. The three key inventions are the electroscope, the photography and the cathode ray tube, and all were invented long before anyone imagined that we are surrounded by natural radioactive materials.

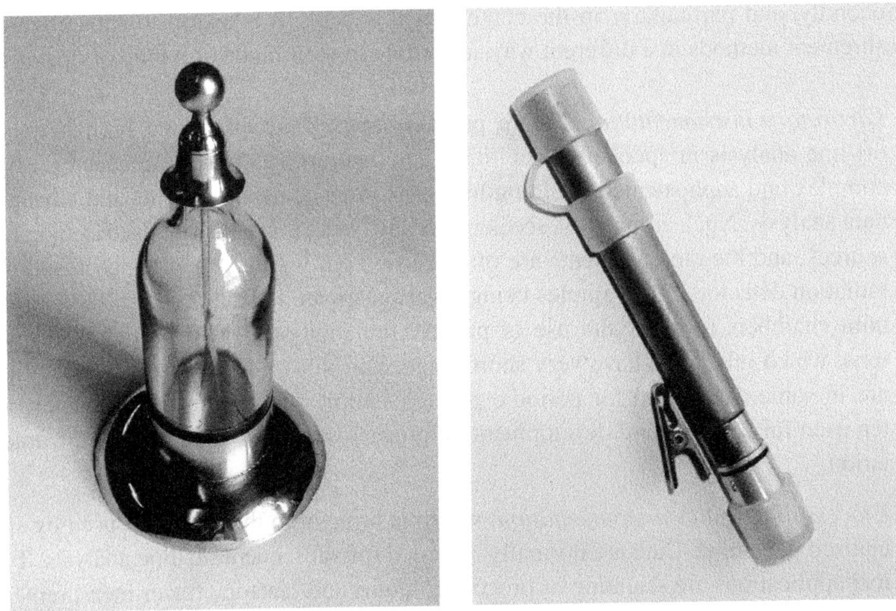

Figure 1.1 The old electroscope (left). The modern quartz fibre electroscope dosimeter (right)

1.3.1 *The Electroscope* was invented around 1748 by the French clergyman *Abbe Jean Antoine Nollet*. The electroscope is a device for detecting and measuring electric charge by using the deflection caused by repelling electric charges.

The first electroscopes consisted of a glass jar in which were suspended two plates or balls which could be charged by applying a voltage to the common suspension point (see Figure 1.1). When the plates become charged they repel each other and move apart on very light pivots. When discharged, either by touching the grounded side of the jar or by the grounding of the pivot point, the two plates fall back together. Thus the plates can be made to flap up and down at a rate that is proportional to the applied charge. For small charges a graduated scale on the side of the jar could be used to measure the angle of separation of the plates. The electroscope was around for about 150 years before it was used as an ionising radiation detector but is still in use as a radiation dosimeter (see Figure 1.1).

1.3.2 The second invention of significance to our saga is *Photography*. The first known photograph, called by its inventor the 'heliograph', was produced in France in the summer of 1827 by *Joseph Nicephore Niepce* (see Figure 1.2). Niepce was born in 1765 in France and lived in Chalon-sur-Saone. Many other researchers were working simultaneously on various photographic processes, all driven by the desire to freeze the images produced by camera obscura, which were commonly used to project images that could then be traced by hand. Niepce collaborated with Louis Jacques Monde Daguerre who 4 years after Niepce's death discovered the means to fix photographs and introduced the Daguerrotype. By now the potential of photography was becoming evident but it seems unlikely that anyone could have dreamed just how important and popular it would become, with nearly every family owning its own camera. The French government bought the patents for Daguerre's process and waived their rights to royalties so as to make the process freely available to all. Meanwhile others claimed the invention: an English inventor and Member

Figure 1.2 The first photograph taken in Paris in 1827
Credit: National Museum of Photography, Film and Television/Science and Society Picture Library, UK

of Parliament, William Henry Fox Talbot, patented his process in England and Wales and another French photographer Hippolyte Bayard claimed prior art but was too late with his claim. In summary, photography played an important part in the discovery of ionising radiation and is still paramount in medical and industrial radiography.

1.3.3 The third of the path-finding discoveries of importance to this narrative was the *Crookes tube* (to be known later as the cathode ray tube). *Sir William Crookes* (1832–1919) was a typical English Victorian scientist with very wide ranging interests and a mainly experimental approach to his science. He was born in London and was educated at Chippenham Grammar School and then the Royal College of Chemistry, Hanover Square, London. Crookes' most important discovery was that of the element thallium in 1861, and his most entertaining for physicists is the Crookes radiometer or lightmill, which ensures endless discussions as to how it works. The lightmill (see Figure 1.3) consists of an evacuated glass bulb inside of which is suspended a rotor of vanes that are blackened on one side and silvered on the other. The vane rotates under incident light. By 1880 Crookes had his own private laboratory at his home in Kensington Gardens, London. Here he began experimenting with electrical discharges in rarefied gases. He noticed that rays emanating from the electrode caused some substances to fluoresce and he observed that the rays travelled in straight lines. Crookes called the rays 'radiant matter' and later J.J. Thompson showed that they were in fact electrons. The cathode ray tube of course became the main component of televisions, oscilloscopes and computer monitors, which keep us entertained when we are not arguing about how the radiometer works. We will leave Sir William Crookes for a while but we will come across his ingenuity again later.

1.3.4 *William Conrad Röntgen* was born in 1845 at Lennep in Germany but moved to Apeldoorn in the Netherlands when he was 3 years old. Röntgen enrolled at the University of Utrecht in 1865 to study Physics. He then sat an entrance exam for Zurich Polytechnic and passed. Here he studied under Kundt and Clausius and attained his doctorate in 1869 from the University of Zurich. By 1875 he became a professor at the academy of

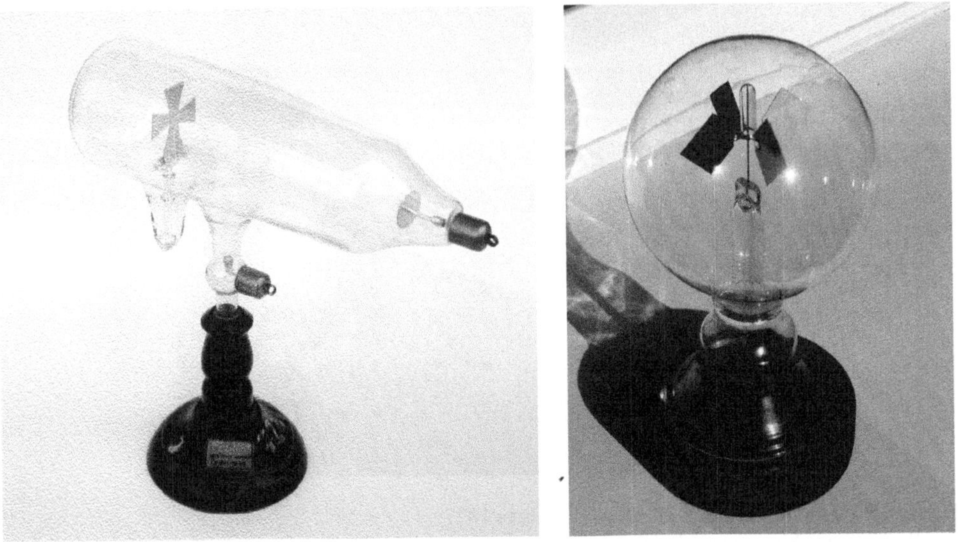

Figure 1.3 Crookes tube (left) and radiometer (right)
Source: Left-hand-side figure – Reproduced by permission of the Oak Ridge Associated Universities, USA

Agriculture at Hohenheim and then became Professor of Physics at Strasbourg (1876), Giessen (1879) and the University of Wursburg (1888), finally ending up at the University of Munich in 1900. In 1895, while at the University of Wursburg, Röntgen began studying the effect of electrical discharges in rarefied gases with the use of the Crookes tube. He was researching what were by now called cathode rays and noticed that some rays seemed to emanate from the tube in spite of it being screened with thick black card. He found that the rays made screens fluoresce, that they affected photographic plates and furthermore, that items of varying thicknesses or densities were more or less transparent to the rays. Soon Röntgen had produced a photograph of his wife's hand, showing the bones and her gold ring (see Figure 1.4). Further investigations by Röntgen showed that the rays were produced by the impact of the cathode rays on a target and since he did not know their nature he called them *X-rays*. The importance of Röntgen's discovery was immediately recognised by the scientific and medical communities. Within 4 months of Röntgen's discovery a team from University of Manchester rushed its X-ray equipment from the University to the hospital to search for a bullet in a shooting victim. Röntgen received the first Nobel Prize for Physics in 1901 in recognition of his discovery of X-rays. Researchers all over the world began reproducing Röntgen's work and studying the new phenomenon. Among these was the man who discovered naturally occurring radioactive material (NORM).

Natural radioactivity has always been around and the means for its detection were around for about a century before the actual discovery. So it seems strange that the discovery of a man-made source of radiation was the stimulus that led to the discovery of natural radioactive materials.

1.3.5 *Antoine Henri Becquerel*, son of a French Professor of Physics, was born in Paris in 1852. He followed in his father's footsteps to become Professor of Applied Physics

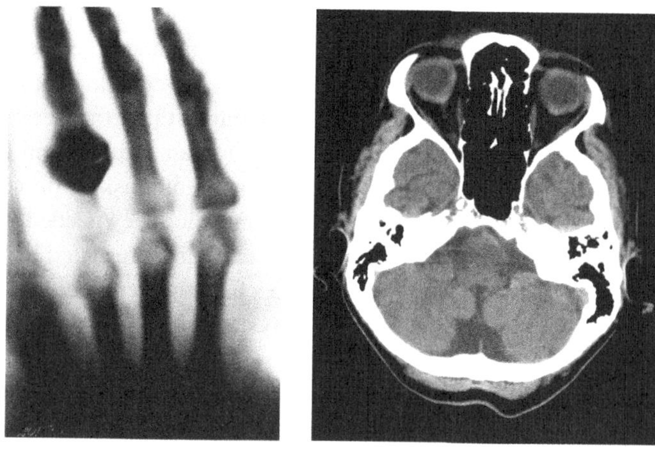

Figure 1.4 X-ray photograph of the hand of Röntgen's wife taken in 1895 (left). A modern axial head image produced by computerised X-ray tomography (right)
Source: Left-hand-side figure – *Credit*: Science Museum/Science and Society Picture Library, UK
Right-hand-side figure – Courtesy of Haukeland University Hospital, Norway

at the Conservatoir des Arts et Métiers. In 1896, Becquerel was investigating the X-rays that Röntgen had discovered the previous year. Becquerel was particularly interested in the fluorescence that is observed around the cathode in the cathode ray tube and thought it might be related to the fluorescence caused by light on uranium salts. When the days were not bright enough for good experiments Becquerel stored his uranium salts in the same drawer as his photographic plates. He soon noticed that his plates were fogged where they had been in contact with the uranium salts, and resolved to find out what had caused the exposure of the plates. Becquerel found that all the salts of uranium had the same effect and concluded that rays must be emanating from uranium. He further discovered that unlike X-rays the new rays could be deflected by electric and magnetic fields and that the rays made gases conduct electricity. Becquerel's discovery was of course a lot less spectacular than Röntgen's without the pretty pictures, and other researchers were slow to pick up on this interesting line of research.

1.3.6 One researcher *Marie Curie* was looking for a subject for research for her doctorate and decided to further investigate the Becquerel radiation. Marie Sklodowska was born in Warsaw in Poland in 1867; her mother and father were both schoolteachers and Marie was very bright. Marie became a governess and funded her elder sister's medical studies at the Sorbonne in Paris, and then at 24 started her own studies at the Sorbonne with financial assistance from her sister, who was by now a doctor. Marie met and married *Pierre Curie* while she was studying at the Sorbonne. Pierre was a successful scientist responsible for discovering the piezoelectric effect and the effect of heat on magnets. This effect was known as the *Curie point*.

By testing all the known elements she could, Marie found that thorium also emitted the same rays that Becquerel had detected from uranium. She ascertained that the intensity of the rays was related only to the amount of uranium or thorium present in any compound, and so concluded that the rays were from the atoms of thorium or uranium. In order to find a supply of raw materials to process for thorium and uranium, Marie started to test natural ores and found that pitchblende was more active than its uranium and thorium content suggested it should be. Marie thought that there must be some other active element in the pitchblende and set out to isolate this more active ingredient. By the end of 1898 the Curies, who were by now working together, announced that they believed that they had discovered two new metallic elements, which they named polonium (after Marie's homeland) and radium. The Curies now began the mammoth task of processing tons of pitchblende to extract sufficient polonium and radium, which they needed to confirm that they were indeed newly discovered metallic elements. Marie presented her work for her doctorate in 1903 and it was received with much acclaim and a Nobel Prize, definitely a first for a doctorate thesis. The Nobel Prize for Physics in 1903 was presented jointly to Marie and Pierre Curie and Antoine Henri Becquerel.

1.3.7 In 1900 another French Physicist *Paul V. Villard* was investigating the Becquerel rays from uranium when he observed some radiations that resembled X-rays but were more penetrating. He called them γ-rays to fit in with Rutherford's α, β nomenclature.

1.3.8 *Ernest Rutherford* was born in 1871 in New Zealand. He was 1 of 12 children and his parents were by no means well off. Ernest was a bright student who gained a first class education by winning scholarships, which led, eventually, to first class honours in mathematics and physics from Canterbury College, Christchurch, UK. It was here that he started his research into radio waves, which led to him being awarded a scholarship that enabled him to travel to England to continue his studies at Trinity College, Cambridge, UK. Rutherford became a research student under J.J. Thompson at the Cavendish Laboratory where he invented a detector for electromagnetic waves and in 1895 studied the behaviour of ions produced in gases by X-rays, which had just been discovered by Röntgen. In 1898 Rutherford left Cambridge to take up the Chair of Physics at McGill University, Montreal, Canada. While at McGill University, Rutherford developed his theory of radioactive disintegration and discovered a number of new radioactive isotopes.

Rutherford was visiting Paris in 1903 when Marie Curie was celebrating the acclaim with which her doctoral thesis had been accepted. Rutherford was invited to the party where no doubt he was able to thank the Curies in person for the radioactive preparation they had sent him to aid his own experiments at McGill University. At the celebration Pierre Curie performed his spectacular party trick, which involved a vial of radium solution coated with zinc sulphide and kept hidden in his pocket. When he displayed the vial it glowed brightly, greatly impressing the assembled guests. Rutherford noticed that Pierre's hand was looking damaged and burnt and it is probable that by now both Pierre and Marie Curie were suffering from the effects of radiation poisoning. Pierre Curie's total dedication is demonstrated by the fact that he deliberately inflicted a radiation burn on his arm in order to study the slow healing process. This experiment suggested to Pierre that radiation may be used to destroy cancerous growths.

In 1907 Rutherford became Professor of Physics at the University of Manchester, where he was to continue the work he had begun at McGill on α-rays and β-rays.

All the radiation measurements up to 1908 were carried out on equipment based on Nollet's electroscope, which is capable of measuring only the rate of interactions and was not capable of resolving individual particles. In 1908 Sir William Crookes made a reappearance with an ingenious little invention that he called the *spinthariscope* (see Figure 1.5). This was the first scintillation counter and consisted of a lens that viewed the back of a thin zinc sulphide screen. When an ionising radiation particle impinged on the screen

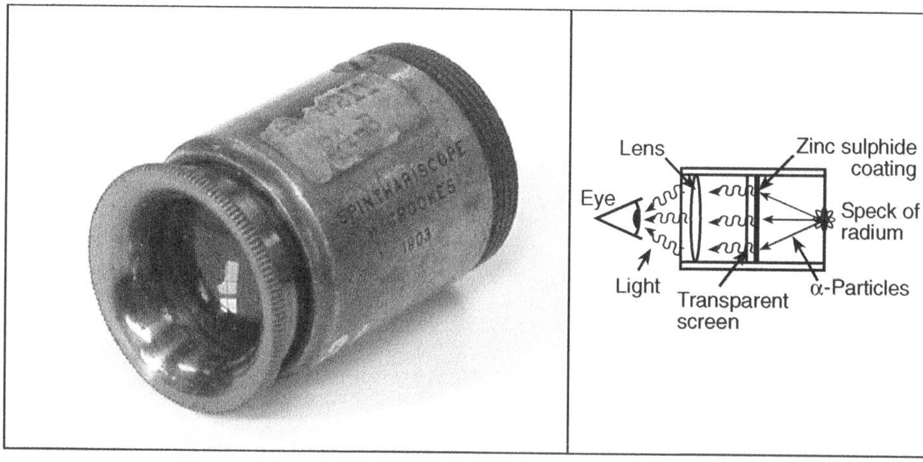

Figure 1.5 The spinthariscope with diagram
Source: Reproduced by permission of the Oak Ridge Associated Universities, USA

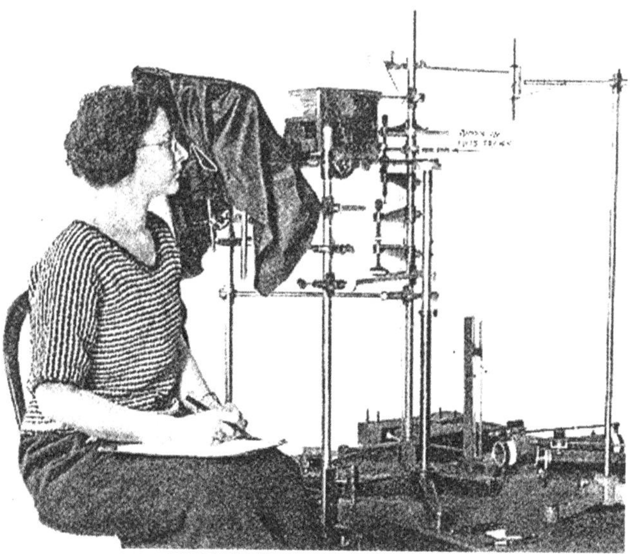

Figure 1.6 A scintillation counter (the lady from the National Bureau of Standards). Actually she is using an electroscope to measure radium activity, but the principle is the same
Source: Reproduced by permission of the US National Institute of Standards and Technology, Photographic Collection

an observer looking through the lens could see the flash of light from the screen. Now for the first time each individual particle could be observed and counted. Visual scintillation counting soon became more sophisticated and became a popular counting method (see Figure 1.6). The method was limited to highly ionising particles and the maximum count-rate is about 60 counts/min.

1.3.9 At the University of Manchester, Rutherford had a research assistant named *Hans Geiger* (see the picture in Figure 1.7). In order to confirm that all α-particles caused flashes

Figure 1.7 Geiger (left) and Rutherford in the laboratory at University of Manchester
Credit: Science Museum/Science and Society Picture Library, UK

in the visual scintillation counter Rutherford and Geiger devised an ionisation detector that was the forerunner of the *Geiger–Müller tube*. In this device each particle produced ionisation in a gas-filled tube at low pressure. The tube had a central electrode connected to earth via an electrometer and a high resistance; the outer case of the tube was held at a high potential (see Figure 1.8). Each particle causing ionisation in the gas produced a flick of the electrometer. Note that the pulses still had to be counted manually as no electronic counting means yet existed.

The ionisation counter agreed exactly with visual scintillation counting and both methods were used thereafter. Visual scintillation counting was used right through until the 1930s in a lot of experiments, which led to the revealation of atomic structure.

Rutherford, by 1908, had explained that α-rays were helium nuclei and that β-rays were high-energy electrons. He had measured the polarity of the charge on both and demonstrated that α-rays carried a positive charge and that electrons carried a negative charge. Rutherford was awarded the Nobel Prize for Chemistry in 1908 for his work on the disintegration of the natural radioactive elements radium, thorium and actinium. Why Chemistry? Because until Rutherford demonstrated that one element could transmute into another, chemists had believed that the atom was invincible and indivisible. Now a whole new understanding of chemical reactions at the atomic level was possible.

1.3.10 Let us leave the physicists in their laboratories for a while and step back a couple of years. What did the public make of the recent discoveries of X-, α- and β-rays, stuff that glowed in the dark and gave off heat endlessly? They were in fact fascinated and excited at the possibilities. Within months of Röntgen's discovery of X-rays, the British army fighting in Sudan was using X-rays to find shrapnel in wounded soldiers. After the discovery of polonium and radium the Curies became famous throughout the world and found it difficult to get any work done because of the constant press and public interest.

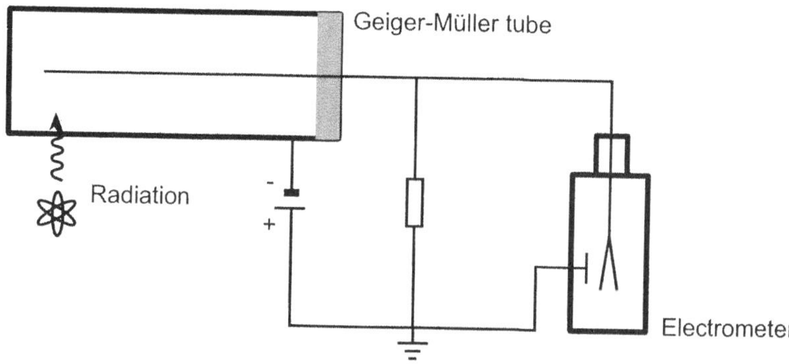

Figure 1.8 Diagram of Geiger–Müller tube counting arrangement

Tragically in 1906 Pierre Curie was knocked down and killed by a horse drawn cart in Paris, leaving Marie with two young children. Undaunted Marie carried on working, taking over Pierre's post at the Sorbonne and in 1911 she was awarded a second Nobel Prize, this time for Chemistry. When in 1914 the world was at war, mobile X-ray units were soon in action, helping the medics to treat the wounded. Marie Curie drove one such vehicle and later in the war her daughter Irene, who was by then 18 years old, joined her.

By now medical treatments for cancerous tumours were common and effective. In the United States, radium production was now industrialised and sources were being produced to the international standard established in 1911. The International Radium Standards Committee, which included among its members Marie Curie and Earnest Rutherford, agreed on a standard measure of radioactivity in Brussels. The standard agreed was the Curie, named in honour of Pierre Curie, and was the quantity of ^{222}Rn in equilibrium with 1 g of the parent isotope ^{226}Ra.

With the medical successes of radioactive material came a bizarre crop of weird patent medicines and curative treatments, which without exception made unproven claims of benefits and which were at best useless and at worst positively harmful. The worst of these devices encouraged the ingestion of radioactive material and included devices for dissolving radon gas in drinking water and pills and suppositories containing radium. Almost as bad were pads containing radium or thorium that were applied as compresses to relieve aches and pains caused by rheumatism or arthritis. The pad pictured in Figure 1.9, the Q-ray, was confiscated from a colleague's grandmother and probably worked because in addition to being radioactive it had a heating element that could have alleviated aches and pains.

Meanwhile, back in the physics laboratories of the world, scientists were identifying the properties of the radiation and their interactions with the elements. In 1919 Rutherford performed possibly the most important experiment in nuclear physics. He bombarded nitrogen with α-particles and discovered that hydrogen particles or protons were produced in the interaction. In popular terms Rutherford had split the atom. This was the first example of artificial nuclear fission, which demonstrated that a proton is a hydrogen atom without its electron, thus giving it a positive charge.

In tandem with the discoveries relating to the radioactive emanations there were advances in detector technology. The interactions of the less ionising rays such as X-rays, γ-rays and β-particles are so small that their impact on a zinc sulphide screen is impossible

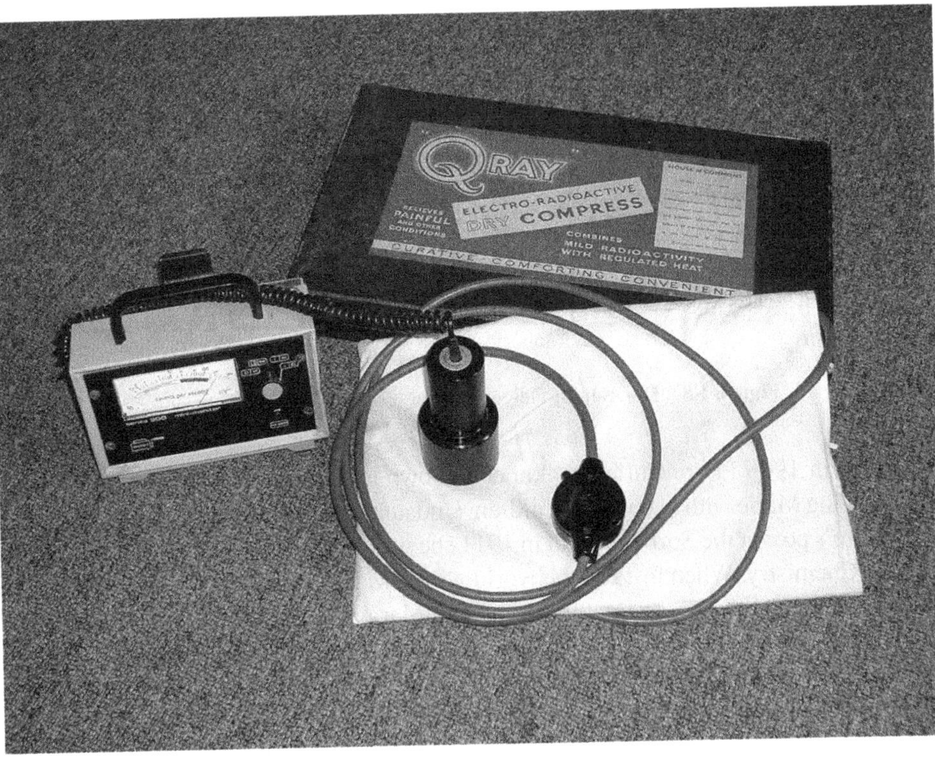

Figure 1.9 Picture of grandmother's Q-ray

to see using visual scintillation counting. In order to measure these radiations the gas ionisation counter of Rutherford and Geiger underwent many developments, resulting in the Geiger–Müller counter and the gas proportional counter, both of which are in common use today (see Section 4.4.5). By the 1930s with the development of electronic amplifiers and counting circuits, electronic pulse counting became possible and the visual scintillation counter was finally obsolete.

1.3.11 In 1932 *Sir James Chadwick* discovered the *neutron*, the last of the fundamental particles of the atom. The neutrons were produced by bombarding the metal beryllium with α-particles; this is how today's neutron sources are constructed. Chadwick was a student of Rutherford at the University of Manchester and later worked with Rutherford at the Cavendish Laboratories in Cambridge. In 1920 Rutherford postulated that in order to build heavy elements there would have to be a heavy particle about the same weight as the proton but with no charge. This neutral particle would then be able to enter the nucleus without being repelled by the positive charge on the nucleus; how else could heavy elements be built?

The first experimental indication that this may be true came from an experiment carried out by Irene Joliot–Curie and her husband Jean Frederick Joliot. Irene, who we last mentioned as driving a radiography truck in the war, was by now married and she and her husband were carrying on the family business of being eminent physicists.

They were also investigating the 'beryllium radiations', which were produced by bombarding the light element beryllium with α-particles. They noticed that when the radiation interacted with paraffin wax, protons were ejected from the wax. This result led Chadwick to the conclusion that the beryllium radiation must be a large particle. Further experiments with absorbers showed that the radiation could pass easily through 20 cm of lead whereas protons are very easy to stop. Thus Chadwick concluded that the particles must have no charge and a mass of one – the neutron.

In 1935 James Chadwick was awarded the Nobel Prize for Physics and the Joliot–Curies were awarded the Nobel Prize for Chemistry.

1.3.12 *Nuclear power*: As usual when a new particle is discovered, researchers in nuclear physics start to bombard everything they can lay their hands on with the new particle in order to see what happens. Bombarding elements with neutrons produced numerous radioactive isotopes, with the most interesting reactions being observed when the heaviest natural element uranium was bombarded. A team led by *Enrico Fermi* working in Rome produced several radioactive isotopes from uranium but were unable to unravel the complex reactions. They were expecting the introduced neutrons to add to the nucleus and produce only transuranic elements, i.e. those heavier than uranium. What in fact was happening was revealed by Otto Hahn and Lise Meitner in 1938 in Berlin. They showed that the uranium was being split into two parts, one of which they recognised as barium, and that the process of splitting resulted in a loss of mass and a release of energy. Amazingly this was the first experimental demonstration of Einstein's theory of the equivalence of mass and energy, which was published in 1905.

In 1939 war broke out in Europe and the legitimate flow of information from country to country ceased. Scientists all over the world demonstrated that when uranium is split there is not only a release of energy but also production of excess neutrons, which can produce a chain reaction, splitting more uranium and releasing even more neutrons and energy.

The most successful program was driven in Britain by a committee code-named the Maud Committee whose aim was to study the feasibility of the atomic bomb and the atomic boiler. (The latter was envisaged as being particularly useful in submarines.) The committee was established after two refugee physicists, Otto Frich and Rudolf Peierls, sent a letter to the British authorities stating their conviction that an atomic bomb was a realistic possibility. The committee was headed by Sir Henry Tizzard and established the principles of fission bomb design and uranium enrichment with the help of scientists at the universities of Liverpool, Bristol, Birmingham, Oxford and Cambridge. In addition, industrial expertise was recruited to study the problems of uranium enrichment. Dr Philip Baxter of Imperial Chemical Industries (ICI) produced the first sample of uranium hexafluoride for use in research at Liverpool University headed by Sir James Chadwick. ICI then went on to build a production unit under the code name 'Tube Alloys Project' and produced all of the uranium used in the British bomb project. The Maude Committee produced two reports in July 1941 that confirmed that both the atomic bomb and the nuclear boiler could be achieved in a realistic time span, and Winston Churchill urged that the bomb project should be urgently pursued. The reports were shared with Canada and the United States. In the United States, the power-producing aspects of the technology were receiving most interest and there was little pressure to produce the bomb until in December 1941 Pearl Harbour was attacked and the American attitude toward developing the bomb changed overnight. With American industrial might concentrated on the project, things started to move fast.

Within a year of the Pearl Harbour attack Fermi had built the world's first nuclear reactor at the University of Chicago. Code-named Metallurgical Laboratory the reactor CP-1 went critical on December 2, 1942. The US program to produce the bomb began on December 18, 1941, and was headed by Arthur H. Compton. The Chicago reactor project was started and programs to produce fissile materials were initiated. By June 1942 it became obvious that the bomb project was going to be a massive undertaking and would need a huge organisation to control it. The military took over and in August 1942 an organisation known as the Manhattan Engineering District was formed. Under this suitably confusing code name and under the direction of Colonel (soon to become Brigadier General) Leslie Richard Groves the Manhattan project started serious procurement including from sites at Hanford in Washington, Oak Ridge and Los Alamos. Hanford and Oak Ridge were to produce and process plutonium and Los Alamos was to be the Central Laboratories. As they say the rest is history but not the history of instrumentation that we are interested in. Out of the project to produce the bomb came nuclear reactors, which were capable of creating a whole range of radioactive isotopes useful for instrumentation and medicine, such as ^{137}Cs, ^{60}Co and ^{192}Ir and secondly the most versatile radiation detector was developed at Los Alamos Laborator.

Gas ionisation counters were by now the detector of choice and in 1941 Krebs used a light-sensitive Geiger–Müller tube to count the scintillations from a phosphor. Such counters could be made sufficiently sensitive to count the weak scintillations from β- and γ-radiations, which were invisible with the visual scintillation counter. At Los Alamos amazingly complex and elegant gas ionisation chambers were developed to measure α-, β-, X-rays, and neutrons, and in 1944 Curran and Baker at Los Alamos Laboratories made a most significant development in detectors. They placed a zinc sulphide screen in front of an RCA IP 21 photomultiplier tube and the most versatile of all radiation detectors was born. The photomultiplier tube was developed as an amplifier tube for cinema projectors many years before and was used by John Logie Baird in his early experiments with television picture cameras in the 1920s.

1.3.13 *The scintillation counter*, as the combination of phosphor and photomultiplier tube is known, is adaptable to measure all types of radiation. Furthermore the energy deposited by each scintillation can be measured as it is related directly to the size of the output pulse from the photomultiplier (see Section 4.6.3).

All of the components for a modern nucleonic level gauge or density gauge are now in place; in fact the tools to produce a gauge of sorts were available as soon as the first radium source was refined. At this time the level in a tank could probably have been measured with a radium source on one side of the vessel and an electrometer on the other, but of course it was easier to just look in the top or dip it with a stick. Only when remote control of plant became possible and fully closed pressure vessels became necessary, it was useful to have a remote level measurement system.

1.3.14 In 1948 A.P. Schreiber published a description of the *First Nucleonic Level Gauge*. The gauge consisted of a Geiger detector on top of a tank and a source on a float inside of the tank. As the level moves the source/detector distance alters and the level can be deduced from the count-rate using the inverse square law: count-rate $= 1/\text{distance}^2$.

When the Second World War ended, Britain, Canada, the United States and the Soviet Union all had nuclear establishments and soon all of them had nuclear reactors. By 1951 about 600 new isotopes had been produced and many of these had obvious commercial uses. The most obvious were Radium replacements for medical and industrial radiography, such as ^{137}Cs, ^{60}Co and ^{192}Ir. The situation now was one of a new technology and quite a large industry looking for applications, giving rise to the Atoms for Peace programs. By 1961 some 21 countries had a total of 15,000 nucleonic gauges installed, about half of these were in the United States and most of the other half were in Canada, France, Britain and Germany.

Since the development of the scintillation counter in the 1940s up to the present day, the greatest changes in nucleonic equipment has not been in the detector systems but in the associated electronic equipment. As in most fields, huge racks of equipment with thermionic valves and high power consumption can be compressed into pocket size and microprocessors are commonplace in detectors and counting systems.

1.3.15 The development of the transistor led to the last new family of detectors, the *solid-state detectors*. These detectors were developed in the early 1960s and use transistor style junctions as the detector element. Such detectors as lithium-drifted germanium and cadmium zinc telluride have a better energy resolution than do scintillation detectors but physical limitations in the doping process make the detection elements small and therefore they are most suitable for low-energy applications. As yet they have not made a great impact on the industrial nucleonic gauges, but that is not history, perhaps it is the future.

1.4 THE OBJECTIVE OF THIS BOOK

There is a trend to transfer and develop laboratory methods for use in permanently installed gauges. This is made possible by new and improved detectors and detector electronics combined with compact, efficient and online computing systems for the implementation of demanding models. The driving force is the requirement for more information to facilitate increasingly complex processes running at smaller margins. Key issues are improved process control, process utilisation and process yields, ultimately brought forward by cost-effectiveness, quality assurance, environmental and safety demands.

Because of this the intention of this book is not only to explain the mode of operation of permanently installed gauges, but also to present nucleonic methods in general to enable the readers to evaluate these for measurement problems in their applications of interest. For this reason typical laboratory methods also will be explained in some detail, particularly those based on electromagnetic radiation, but with references to more extensive coverage of the methods. Furthermore, possible methods utilising other radiation types, such as neutrons and β-particles, are also presented, again with references to more extensive coverage of the subjects.

To achieve this it is necessary to see both the advantages and the limitations of the different methods. Many gauges developed some 40 years ago are still being used with minor changes only. This is often satisfactory, but there are clearly unused potentials in many nucleonic methods. Generally in measurement science one seeks to better exploit the inherent information content of different measurement principles. Very often it is about using multiple simultaneous measurements instead of just one, for instance multiple energies, multiple sensors or multiple modalities, all to provide complementary, but in some cases also redundant, information. Altogether this requires some insight into the underlying physics, and this will be the subject of the next chapter. We shall take a look at atomic and nuclear physics and the interaction of radiation with matter, all from a measurement science point of view. However, it is worthwhile to note that in most cases knowledge of the process or system being investigated is just as important as the physics behind the measurement principle.

2

Radiation Sources

2.1 A PRIMER ON ATOMIC AND NUCLEAR PHYSICS

The atom consists of two parts, the nucleus and the electrons orbiting it. The diameter of the atom is about 10^{-10} m, whereas that of the nucleus is about 10^{-14} m. Nevertheless the nucleus accounts for more than 99.9% of the total mass of the atom. Considering that the protons and the neutrons in the nucleus, collectively called the *nucleons*, each have mass of about $1u$,* the density of the nucleus is about 10^{14} g/cm^3. This gives us an indication that the forces holding the particles in the nucleus together are truly enormous – much bigger than forces found elsewhere.

By definition, electrons and protons have unit negative ($-e$) and positive ($+e$) charge, respectively, whereas neutrons are electrically neutral. Since the atom as a whole is electrically neutral, the number of protons in the nucleus is equal to the number of electrons surrounding it. This number is the *atomic number* Z of the atom. Atoms with the same atomic number are atoms of the same element, and the physical and chemical properties of an atom are fixed primarily by the atomic number. The atom has shells of electrons and those in the outermost shell are the valence electrons, which take part in chemical combination. The innermost, most tightly bound shell is called the K-shell, and can be occupied by no more than 2 electrons. If this is full the next 8 electrons can occupy the next (outer) shell. This is the less tightly bound L-shell. If this in turn is full the M- and N-shells can be occupied by 18 and 32 electrons, respectively, and so on for the O-, P- and Q-shells. Electrons within each shell occupy levels, the energies of which are sharply defined and lie close together. These energies are distinct for atoms of the same element.

Atoms of the same element but with different number of neutrons, N, in the nucleus are called *isotopes* of that element. The number of neutrons affects the nuclear properties of the atom. The *mass number* or *atomic weight* A of an atom is the sum of the numbers of protons and neutrons in the atom: $A = Z + N$. The protons and neutrons occupy discrete nuclear energy levels analogous to those occupied by the electrons. A *nuclide* is the name of any isotope of any element, and the internationally accepted nomenclature of indicating the characteristics of a nuclide is ${}^{A}_{Z}\mathrm{X}_{N}$, or ${}^{A}_{Z}\mathrm{X}$, or simply ${}^{A}\mathrm{X}$, A-X or X-A (e.g., ^{137}Cs, 137-Cs or Cs-137), where X is the name of the element in question. The latter, which is the shorthand notation, is unambiguous because the number of protons is given by the name of the element.

* The unified atomic mass constant $u = 1.66053873 \times 10^{-27}$ kg.

Radioisotope Gauges for Industrial Process Measurements. Geir Anton Johansen and Peter Jackson.
 ISBN 0-471-48999-9

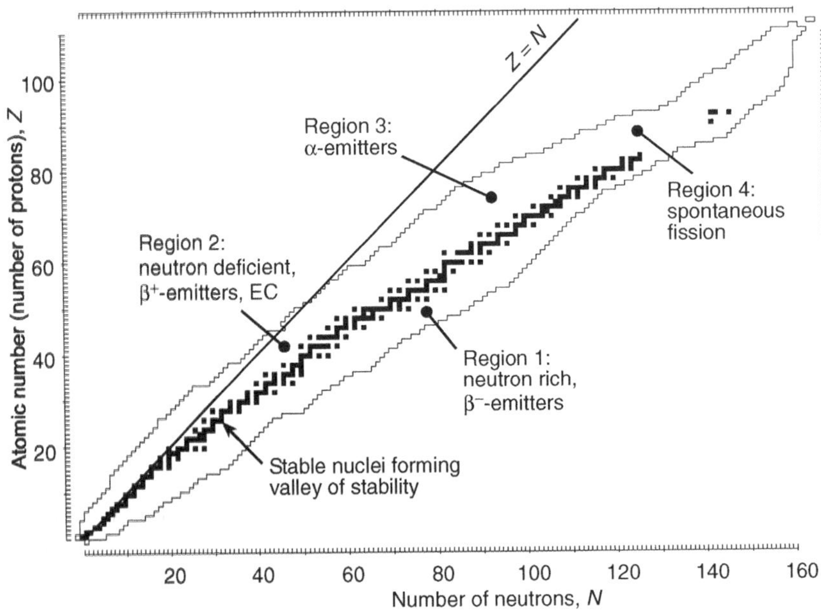

Figure 2.1 Map of nuclides, with stable nuclei (solid squares) and unstable nuclei enclosed by the outer envelope, plotted according to proton (Z) and neutron (N) numbers. Note that the region indications are meant only to show where the different modes of decay are mainly found. In all regions (particularly 3 and 4) there will be competitive modes of decay, including rare ones not been mentioned in this text [227]. A comprehensive presentation is given in colour-coded versions of this map, such as one in Reference [4] from where these data have been taken

2.1.1 Radioactive Decay

Radioactive decay, also referred to as *disintegration*, is a spontaneous change within the nucleus of an atom, that results in the emission of particles and electromagnetic radiation. It is always exoergic; the mass of the product, the *daughter*, is always less than the mass of the original nuclide, the *parent*. It is beyond the scope of this book to explain what makes a nuclide radioactive (*unstable*) [226]. For each isotope this is basically determined by the ratio of Z to N. This is plotted in the map of nuclides shown in Figure 2.1. The filled squares denote stable and long-lived naturally occurring nuclides and are commonly referred to as the *valley of stability*. Neighbouring nuclides are known as *unstable nuclides, radionuclides* or *radioisotopes*. For low-Z elements the stable nuclides are found at $Z = N$, whereas for higher Z values N becomes appreciably greater than Z.

2.1.2 Modes of Decay

If an atom contains too many neutrons, one of the neutrons (n) will, sooner or later, undergo a spontaneous transformation into a proton (p^+). This happens through a *negative beta decay*, with the emission of a *negative beta particle* (β^-) and an *antineutrino* ($\bar{\nu}$):

$$n \rightarrow p^+ + \beta^- + \bar{\nu} \tag{2.1}$$

The proton remains in the nucleus whereas the β-particle, which is a fast electron, carries away some or all of the energy involved in the mass change as kinetic energy. Its energy ranges from zero up to the maximum energy represented by the mass loss. The remaining portion of energy is carried away by the antineutrino. This has negligible mass, no charge and a velocity near that of light. As an example ^{137}Cs is a β^--emitter that disintegrates to ^{137}Ba:

$$^{137}_{55}\text{Cs} \rightarrow {}^{137}_{56}\text{Ba}^+ + \beta^- + \bar{\nu} \tag{2.2}$$

In the map of nuclides the β^--emitters are found on the low-Z side of the valley of stability (region 1 in Figure 2.1). Nuclides on the high-Z side of the valley (region 2) have too few neutrons in the nucleus. There are two possible mechanisms by which a proton in these is transformed into a neutron. One of these is *positive beta decay*, which involves the emission of a *positive beta particle* (β^+) and a *neutrino* (ν):

$$\text{p}^+ \rightarrow \text{n} + \beta^+ + \nu \tag{2.3}$$

The β^+-particles from this decay, called *positrons*, are emitted in a continuum of energies from zero up to some characteristic maximum energy, as in the case of β^--decay. When the kinetic energy of the positron has been expended, it combines with an electron, and the pair is *annihilated*; the positron and electron disappear. In this process their mass (energy) is converted into two photons that are emitted in nearly opposite directions. Each photon has energy very close to the rest mass of each particle, that is 511 keV. This *annihilation radiation* is characteristic of β^+-decay. However, this also means that for this decay mode to take place, the mass loss energy resulting from the β^+-decay needs to be more than 1022 keV. As an example ^{22}Na is a β^+-emitter that disintegrates to ^{22}Ne:

$$^{22}_{11}\text{Na} \rightarrow {}^{22}_{10}\text{Ne}^+ + \beta^+ + \nu \tag{2.4}$$

Actually, this is not the only way ^{22}Na may disintegrate into ^{22}Ne. It may also happen through *electron capture* (EC). This is the second mechanism by which protons in neutron-deficient atoms are transformed into neutrons. This is the only possibility in cases where the mass loss energy resulting from the decay is less than 1022 keV. In this process an electron, most likely from the K-shell (which is closest to the nucleus), is captured by a proton in the nucleus and transformed into a neutron:

$$\text{p}^+ + \text{e}^- \rightarrow \text{n} + \nu \tag{2.5}$$

The energy released by EC is carried away by a neutrino. The EC often results in characteristic X-ray emission, and this will be explained further in Section 2.1.5.

A decay mode most often found in high-Z nuclides is *alpha decay*. These nuclides are found in region 3 in the map of nuclides shown in Figure 2.1. The α-particle, denoted α or ^{4}He, is emitted at a discrete energy, not at a continuum of energies like the β-particle. One example is ^{226}Ra, which disintegrates to ^{222}Rd:

$$^{226}_{88}\text{Ra} \rightarrow {}^{222}_{86}\text{Rd} + \alpha \tag{2.6}$$

Since the α-particle is a ^{4}He nucleus the parent nucleus looses 4 mass units and 2 charge units.

For the sake of completeness *spontaneous fission* needs to be mentioned. Some transuranic elements break up with the production of lighter elements (fission products) and the emission of neutrons. An example is

$$^{252}_{98}\mathrm{Cf} \rightarrow {}^{140}_{54}\mathrm{Xe} + {}^{108}_{44}\mathrm{Ru} + 4\mathrm{n} \qquad (2.7)$$

where four neutrons are produced. The spontaneous fission nuclides are found in region 4 in the map of nuclides shown in Figure 2.1. *Neutron-induced fission* may happen when nuclides such as ^{235}U and ^{239}Pu absorb a neutron. More neutrons are emitted, energy is released and under certain circumstances a chain reaction is started. This is the basis of nuclear power production and some nuclear weapons.

2.1.3 γ-Rays

The majority of decay modes encountered are not single-step disintegrations. For all of the presented decay modes, the daughter nucleus often has some residual energy and is left in an excited state. The change to the ground state often involves the emission of a *γ-ray* with a discrete energy. This may be regarded as a massless particle, a *photon*, with energy equal to the difference in the energy of the excited and stable levels of the nucleus. In some cases, there are two or more excited levels with the consequence that two or more γ-photons of discrete energies are emitted in cascade. Energy analysis of the γ-ray emission may be used to identify the daughter isotope.

The γ-emission usually happens about 10^{-13} s after the primary disintegration. Sometimes, however, for some nuclei this may be considerably longer. If it exceeds about 1 μs, the excited (metastable) state is called an *isomer* of that nucleus. The subsequent decay of an isomer by a γ-emission is called *isomeric transition* (IT). An isomer is indicated by placing a lower case m after the atomic weight in the isotope symbol, for instance ^{137m}Ba, the daughter of ^{137}Cs β^--decay.

As an alternative to γ-ray emission, a nucleus may become de-excited by transferring the energy to an extranuclear electron which is ejected. This is called *internal conversion* (IC). The electron is most likely ejected from one of the shells closest to the nucleus, and its energy is equal to the transition energy minus the electronic binding energy and a small nuclear recoil energy. IC is most likely for low-energy transitions in heavy nuclei. As with EC, IC may be followed by characteristic X-ray emission.

2.1.4 Competitive Modes of Disintegration

Some radionuclides have competitive modes of disintegration, for instance ^{22}Na, which disintegrates to ^{22}Ne either through β^--emission or through EC, and different combinations of the same or different disintegration modes may also exist. The latter is best illustrated graphically by disintegration schemes as shown in Figure 2.2. In these schemes the losses of energy and changes in Z are conveniently represented. Vertical distances

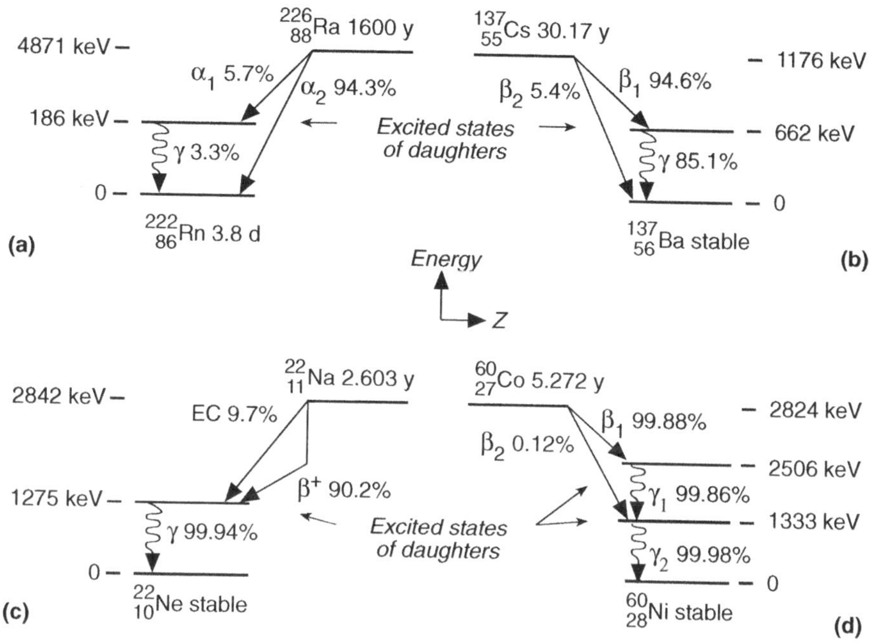

Figure 2.2 Radionuclide decay schemes of (a) ^{226}Ra to ^{222}Rn, (b) ^{137}Cs (to ^{137m}Ba) to ^{137}Ba, (c) ^{22}Na to ^{22}Ne and (d) ^{60}Co to ^{60}Ni. Data are taken from Reference [4]
Source: Reproduced by permission of John Wiley & Sons, Inc.

represent energy, movement to the right represents a gain of positive charge (that is transmutation to an element of higher atomic number) and movement to the left indicates a loss of positive charge. Note that any one atom can disintegrate in only one particular way, and so the disintegration scheme is a statistical description of the decay of a particular radioisotope.

2.1.5 Characteristic X-rays

Following IC and EC, the daughter atom is left in an excited or unstable state with a shell vacancy. There are two processes by which it can revert to its original state. Firstly, the vacancy is filled by an electron dropping in from a higher, less tightly bound shell. The energy released in this process often appears as a *characteristic X-ray*. The emission of one X-ray may very well be followed by others of lower energy as electrons cascade down from shell to shell towards greater stability. The X-ray nomenclature is such that transitions from the L-, M- and N-shells to the K-shell are labelled K_α, K_β and K_γ, respectively. Likewise, transitions from the M- and N-shells to the L-shell are labelled L_α and L_β, and so forth. The electrons in each shell, apart from the K-shell, do not have exactly the same energy. This is because of the different levels, or sub-shells, within each shell. This gives rise to fine structure in the X-ray emissions. The sub-divisions are labelled $K_{\alpha 1}$, $K_{\beta 1}$, $K_{\beta 1}$, and so on; however, this has no practical implications for the topic of this book. Characteristic X-ray energy spectrometry may be used to identify the daughter element, but not the isotope, as with γ-ray emission.

The emission of characteristic X-ray photons is known as *fluorescence*, and the probability of fluorescence, as opposed to the Auger effect, is called the *fluorescence yield*. In general the fluorescence yield increases with the atomic number. We will discuss this in more detail in Section 3.3.1.

The second process by which an unstable daughter atom can revert to its original state is the *Auger effect*. Here the energy released in rearranging the electron does not appear as an X-ray, but is used to free an electron from the atom as a whole. The emitted electron is called an *Auger electron*.

2.1.6 Bremsstrahlung

Bremsstrahlung is a German word meaning 'slowing down' radiation. It is electromagnetic radiation that is produced when fast electrons or β-particles are deflected in the coulombic field of the nucleus. Their energy loss appears as a continuum of photons with energies, in principle, ranging up to that of the particle. Because their energy is largely in the region of that of characteristic X-rays, bremsstrahlung is often incorrectly regarded as one type of X-rays. Other energetic charged particles lose energy in a similar way, but bremsstrahlung is significant only with light particles since these are deflected more easily. Radioisotopes that decay by β-emission produce some bremsstrahlung, particularly when the β-particles interact with elements of high atomic number. This may be in the source itself, or in the surroundings of the source. However, bremsstrahlung succeeding radioisotope decays is of little practical importance compared to that produced in X-ray tubes (Section B.2).

2.1.7 Activity and Half-life

The probability that an atom of a given radioisotope will decay in a certain time is independent of the decay of other atoms around it, the length of time it has existed, the chemical state of the atom and physical conditions like temperature and pressure. It is an entirely random event, and may therefore be treated by statistical methods. The probability of a nucleus decaying with time is a fundamental property of each radioisotope and is called the *decay constant* λ. The prediction among a large number N of nuclei of the same radioisotope is that dN nuclei will decay in a period of time dt:

$$\mathrm{d}N = -\lambda N\,\mathrm{d}t \tag{2.8}$$

By integration this becomes

$$\int_{\mathrm{N}_0}^{\mathrm{N}} \frac{\mathrm{d}N}{N} = -\lambda \int_0^t \mathrm{d}t \;\Rightarrow\; N = N_0\,e^{-\lambda t} \tag{2.9}$$

where N_0 is the number of radioactive atoms present at time $t = 0$ and N is the number present at time t. The *decay rate* or the *activity* A of a radioactive isotope is the number of disintegrations per second and is thus the time derivative of the number of nuclei:

$$A(t) = \left|\frac{\mathrm{d}N(t)}{\mathrm{d}t}\right| = \lambda N(t) \tag{2.10}$$

The activity is a function of time since N is so. The SI unit of activity is the becquerel, such that 1 Bq = 1 disintegration per second. However, the old unit Curie is still frequently used and is related to becquerel as $1 \text{ Ci} = 3.7 \times 10^{10}$ Bq.

It is more convenient to express this exponential decay in the number of radioisotope atoms in terms of the *half-life* $T_{1/2}$, rather than the decay constant. The half-life is the time required for the activity to fall to half of its initial value such that

$$N = \frac{N_0}{2} = N_0\, e^{-\lambda T_{1/2}} \Rightarrow T_{1/2} = \frac{\ln(2)}{\lambda} \tag{2.11}$$

The half-life is consequently a fundamental property of each radioisotope. The activity may then be expressed as

$$A = \lambda N_0\, e^{-\lambda t} = A_0\, e^{-0.693t/T_{1/2}} \tag{2.12}$$

There are a couple of facts worth mentioning regarding the design and application of nuclear measurement systems:

- From the exponential nature of the decay law it follows that after 2 half-lives the activity is reduced to $A_0/4$, and after m half-lives it becomes $A_0/2^m$. This means that radioisotopes with half-lives of a few hours and a fairly 'standard' initial activity will have virtually no activity after a few weeks, or even days. For this reason these are known as *short-lived isotopes*.
- Secondly, it must be strongly emphasised that the activity is the decay or disintegration rate, and not the emission rate of β-particles, γ-photons or any other radiations or particles. In the case of ^{137}Cs for instance, the emission rate of the 661.6-keV γ-photons is 85.1% of the activity. This is evident from the disintegration scheme shown in Figure 2.2.

2.1.8 Radiation Energy

Although the SI unit of energy is joule (J), *electronvolt* (eV) is the common energy unit used for all ionising radiation. One electronvolt is the energy required to move one electron across a potential of one volt, i.e. $1 \text{ eV} = 1.6 \times 10^{-19}$ J. In the context of this book, the energies of interest are in the kiloelectronvolt (keV) region and in a few cases in the million electronvolt (meV) region.

γ-Rays, X-rays, bremsstrahlung and annihilation radiation are high-energy electromagnetic radiation emitted in discrete bundles or quanta most often referred to as *photons*. A photon may be regarded as a massless particle. This is a very useful approach when treating the interaction of ionising electromagnetic radiation with matter (Chapter 3), and the measurement of this (Chapters 4 and 5). There are, however, some X-ray applications where it is more useful to apply wave theory to explain phenomena like diffraction and interference. The energy of each photon is then related to the electromagnetic wave properties as

$$E = h\nu = h\frac{c}{\lambda} \tag{2.13}$$

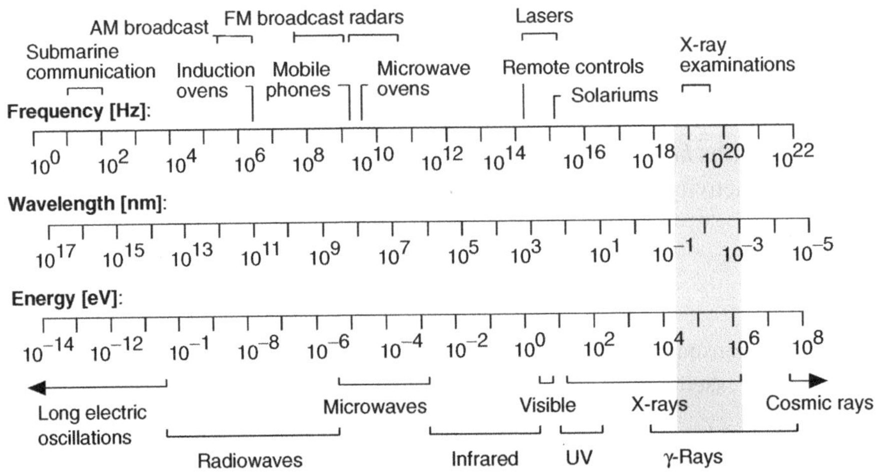

Figure 2.3 The electromagnetic spectrum with some examples of applications. The shaded region indicates the range of interest for nucleonic industrial measurements

where h is Planck's constant and ν, c and λ are the frequency, velocity (in vacuum) and wavelength of the radiation, respectively. Figure 2.3 shows the electromagnetic spectrum and the relationship between radiation energy, wavelength and frequency.

Note that X-rays and γ-rays are named after their source of emission, and not their energy. The possible energies of γ-rays are, however, roughly 1 order of magnitude higher than those of X-rays since the energy levels in the nucleus are larger than those in the atom. Also note that although the energy of a beam of ionising particles and photons is small compared to typical thermodynamic energies, this is targeted energy with high impact on matter. Further, it is possible to quantify with a much higher degree of accuracy.

2.1.9 Summary of Radioisotope Emissions

Table 2.1 gives a summary of the particle and electromagnetic emissions from radioisotope disintegrations. The last column refers to whether a large number of the same radioisotope disintegration results in radiation with a continuum of energies (spectrum), or a single energy line (discrete).

Theoretically, neutrinos from β^+-decay and EC and antineutrinos from β^--decay are crucial in maintaining the universality of the conservation laws of energy and angular momentum. In other words they explain why β-particles are not mono-energetic. The neutrinos and antineutrinos have very small interaction probabilities with matter and are hence undetectable for all practical purposes.

All the other radiation types listed in Table 2.1 may be and are used in a variety of measurement systems in a variety of applications. The main focus of this book is on electromagnetic emissions, which in all cases may be regarded as secondary effects succeeding other processes. Here these processes are nuclear disintegrations; however, particulate and electromagnetic emissions may also be caused by radiation exposure: In thermal neutron reactions, an element captures a neutron and emits a prompt γ-ray photon.

Table 2.1 Summary of the particle and electromagnetic emissions from radioisotope disintegrations

Radiation	Source, process of emission	Charge [*e*]	Mass [*u*]	Energy properties
Particles				
β^- (electron)	Nucleus, β-decay	−1	≈0	Spectrum
β^+ (positron)	Nucleus, β-decay	+1	≈0	Spectrum
α (^{4}He nucleus)	Nucleus, α-decay	+2	4	Discrete
n (neutron)	Nucleus, nuclear reactions, spontaneous fission	0	1	Spectrum
ν (neutrino)	Nucleus, β^+-decay, EC	0	0	Spectrum
$\bar{\nu}$ (antineutrino)	Nucleus, β^--decay	0	0	Spectrum
Electromagnetic				
γ (gamma)	Nucleus, de-excitation succeeding all decay modes and IT	0	0	Discrete
Annihilation	Positron–electron annihilation succeeding β^+-decay	0	0	Discrete
Characteristic X-rays (fluorescence)	Atom, de-excitation succeeding EC and IT	0	0	Discrete
Bremsstrahlung	Deflection of β-particles in the field of the nucleus	0	0	Spectrum

This is known as *neutron activation*. Likewise atoms may be excited to emit characteristic X-rays through interactions with particles and ionising electromagnetic radiation. This is the foundation of X-ray fluorescence spectroscopy. We will taker a closer look at these topics in Section 5.5.3.

γ-Rays and characteristic X-rays are properties of the daughter, but these are accessed from the parent. For this reason, these, particularly γ-rays, are often regarded as properties of the parent, and are often listed in data tables as such. The 661.6-keV γ-emission from ^{137m}Ba is, for instance, most often referred to and known as ^{137}Cs γ-rays, i.e. the parent nuclide.

Nuclide indices are often used to summarise the radiation emission properties of radioisotopes (see Section A.2). These are very useful, for one thing because the average number of photons emitted relative to one disintegration, is tabulated. The γ-ray energy and intensity distributions of the photons are displayed graphically in Figure 2.4.

2.2 RADIOISOTOPE SOURCES

2.2.1 Important Source Properties

Selecting the right radiation source for the application is not only a technical question. It also has to comply with the *ALARA* (As Low As Reasonable Achievable) *principle*. This, which basically involves a risk–benefit analysis to achieve a low radiation dose level, will be further discussed in Sections 6.2 and 8.2. Keeping in mind the ALARA principle, these

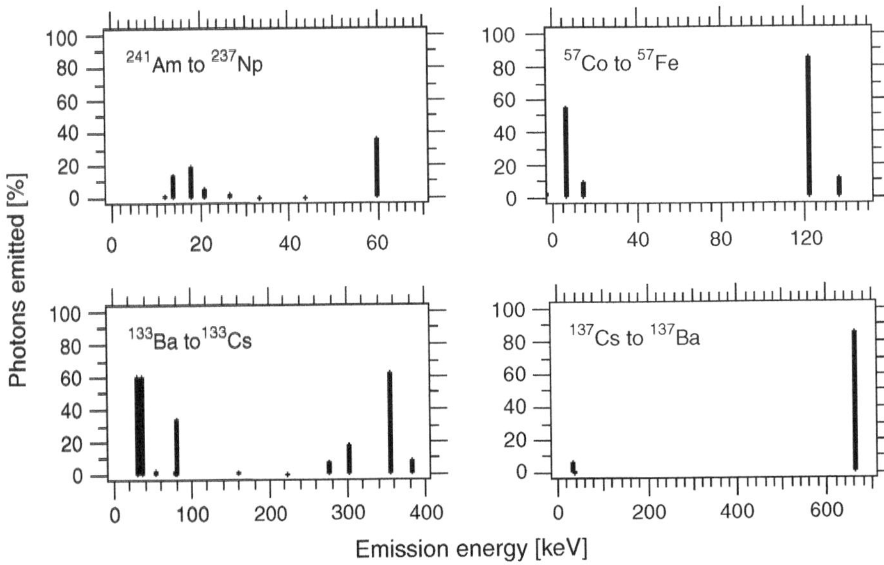

Figure 2.4 Spectral representation of the electromagnetic emissions of four of the γ-ray sources (and their daughters) listed in Section A.2

are the source properties to be considered when selecting a radiation source:

- Category or physical form
- Radiation type
- Energy and spectral purity
- Intensity
- Half-life
- Chemical form and compatibility with process stream (tracers)
- Availability, classification and cost

Some of these are obvious, others not. There are basically three different categories of radioisotope sources in use in industrial measurement systems: *natural sources*, *sealed sources* and *tracers* (unsealed sources). Sealed sources are, with a few exceptions, the most suitable for permanently installed gauges. Tracers are most often used for process diagnostics instrumentation. The major difference in the isotopes applicable for these two categories is their half-life. Isotopes with a long half-life are preferred for permanently installed gauges so as to achieve constant operating conditions throughout the instrument's life, without the need for source replacement. For tracer applications the radioactivity should ideally drop to zero once the measurement has been performed. This reduces the level of residual tracer in the exit stream. Hence, short-lived isotopes are preferred in this case. For tracers the chemical form is also important to ensure that it behaves in the same way as the material under investigation.

γ-Radiation with its relatively high penetration capabilities is the most applicable radiation type for permanently installed gauges. There are, however, applications where β-radiation is used; these will be discussed in Chapters 5 and 7. In the context of radioisotope gauges, neutron sources are mainly used for process diagnostics applications.

The radiation energy is closely related to the penetration capability, whereas the intensity is decisive for the performance or measurement accuracy of the system. This will be discussed in Section 5.3.4. Further, it is very often desirable to have a spectrum uncomplicated by interfering emission lines. With reference to Figure 2.4, it can easily be seen that the *spectral purity* of the emission spectra is high for ^{137}Cs and less for ^{133}Ba.

Finally, the availability of different radioisotopes is important as this in turn influences the cost. This means that although a nuclide meets requirements such as emission energy and spectral purity, it is not necessarily feasible to use it, for instance because of manufacturing costs. *Classification* of radioisotope sources reflects their potential hazards; this will be discussed in Chapter 6.

2.2.2 Natural Sources

Natural sources of radiation are seldom used in industrial measurement systems and are more often encountered as a nuisance providing unwanted background radiation. There are three families of naturally occurring radioactive elements, each consisting of a parent isotope and several daughters. The three parent isotopes are ^{235}U, ^{238}U and ^{232}Th, each of which decays eventually to stable isotopes of lead. In addition to the three families there are two other natural isotopes that commonly occur: ^{40}K and ^{14}C. The isotopes ^{238}U, ^{235}U, ^{232}Th and ^{40}K have half-lives of 4.5×10^9, 7×10^8, 1.41×10^{10} and 1.3×10^9 years respectively, i.e. between about 1 and 10 billion years. This means they were all formed as the stars condensed from the universe. ^{14}C on the other hand has a half-life of only 5760 years and so obviously must be being replaced constantly or it would all have decayed away. ^{14}C is formed through the interaction of cosmic radiation with the nitrogen in the atmosphere by the nuclear reaction $^{14}N(n, p)^{14}C$, which simply means that a neutron replaces a proton (see Section 2.3.2). Figure 2.5 shows a chart of the so-called uranium–radium natural radioactive family.

Many processes and habits of humans accumulate or concentrate natural radioactive material, often known as NORM (naturally occurring radioactive material). This is a problem particularly in the oil industry where concentration of salts in the produced water builds up as radioactive scale in the pipework. In areas of the world where the surface is made up of igneous rock rather than of sedimentary rocks, radon gas gets collected in buildings, particularly in unventilated basements.

The natural radiation of rocks can be useful though, when the need arises to determine remotely where there is an interface between rock strata. Boreholes are logged for natural γ-radiation in order to determine the stratigraphical layers. This method is known as *lithology*. In potash mines where the seams of potassium chloride undulate, its ups and downs can be predicted by monitoring drill holes for the γ-emission from the naturally occurring potassium-40, thus avoiding the wasteful removal of other rocks that are low in potassium content and therefore unwanted.

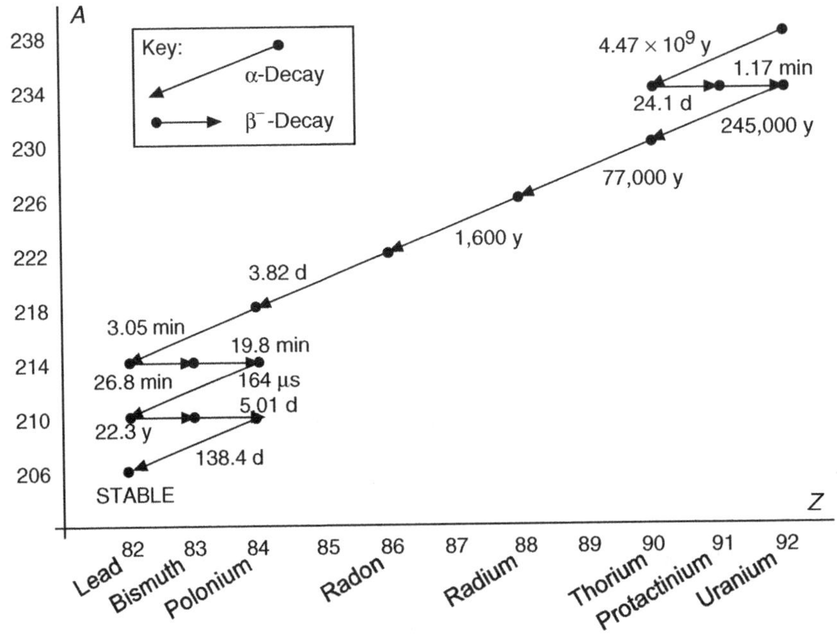

Figure 2.5 The main decay chain of the naturally occurring ^{238}U. Ultimately, through a succession of α- and β^--decays, the radioactive nucleus reaches stability in the form of ^{206}Pb. The ^{222}Rn gas is part of this family and the major contributor to our natural radiation background dose, as we shall see in Section 6.2.5

The isotope ^{14}C is commonly used to date archaeological artefacts. If we accept that the ratio of ^{14}C to ^{12}C (the stable form) in the atmosphere is known at the time the carbon is fixed into vegetable or animal matter, then by knowing the decay rate of ^{14}C, and measuring the ratio of the two isotopes in the artefact, we can, with reasonable accuracy, calculate the age of the artefact.

Cosmic rays are a unique natural source of particles (mainly protons) of (very) high energies that pass through thick layers of matter in the atmosphere and undergo a complicated chain of transformations. The radiation reaching the Earth's surface is thus quite different from that incident from outer space. At sea level it is still dominated by particles although there is some γ-radiation and bremsstrahlung. We will not dwell on this here, but merely note that cosmic radiation is one of four dominant sources to our natural background radiation, as will be discussed in Section 6.2.5.

2.2.3 Tracers

Radioactive sources used as *tracers* are unsealed radioactive material. They can be solid, liquid or gas, depending on the application. The sources are used in medicine as an aid to diagnosis, in research as a marker for the destination of chemicals particularly in drug development and in industry as an indicator of the behaviour of process fluids. The use of radiation in medicine is outside the scope of this book but one of the devices originally developed for medical sources, the *radioisotope generator*, is a useful source of industrial

Table 2.2 Some commonly used tracer nuclides and their key data

Nuclide	Half-life	Radiation emitted	Useful energy [keV]	Tracer's physical form
^{24}Na	15 h	γ-rays	2754 and 1369	Aqueous
^{82}Br	35.3 h	γ-rays	776, 554 and 619	Aqueous/organic
^{85}Kr	10.76 y	γ-rays	151 and 305	Gas
^{79}Kr	36 h	γ-rays	398 and 606	Gas
^{3}H	12 y	β^--particles	186 maximum	Aqueous/organic
^{113m}In	99 min	γ-rays	393	Aqueous/organic
^{137m}Ba	2.5 min	γ-rays	662	Aqueous
^{140}La	2.4 d	γ-rays	329 to 1596	Solid
^{14}C	5760 y	β^--particles	156 maximum	Any

tracers. The radioisotope generator uses a long-lived isotope produced in a reactor or cyclotron such as ^{113}Sn, which is strongly attached chemically to an ion-exchange resin. When ^{113}Sn decays a daughter isotope is produced, which is ionically opposite to tin and can therefore easily be washed from the base resin. This daughter is ^{113m}In, which has a shorter half-life than does the parent tin, and is therefore ideally suited for use as a tracer as it soon decays to background and is therefore safe to place into the environment. This type of tracer source is particularly interesting when the application site is some distance from a reactor or an accelerator. Because the daughter isotope is 'milked' off the base resin containing the parent isotope, the radioisotope generator is also known as a *gamma-cow*. This concept also has a potential use in permanently installed gauges, because of the long half-life of the parent. This will be discussed in Chapters 5 and 7.

Other useful generators exist such as ^{137}Cs/^{137m}Ba (caesium/ barium) where the daughter has a half-life of 2.5 min, a very safe tracer for short experiments, and the long life of the parent (30 years) allows many years of elution of the daughters. Most tracers used in industrial plants must be γ-emitters so that they can be detected outside of the process vessels and pipework. Tracers such as ^{24}Na (11 h half-life) and ^{82}Br (35.3 h half-life) are commonly used to trace liquids for residence time measurements, leakage tests or carry over tests. Tracers commonly used for such measurements in the gas phase are ^{85}Kr, ^{79}Kr and ^{133}Xe, all of which give off useful γ-rays (Table 2.2).

The chemical form of a tracer can of course be changed by reaction with substances. The tracer behaves chemically in exactly the same way as the stable isotope, and therefore compounds that actually follow reactions in a chemical plant or the research laboratory can be formulated in the tracer laboratory. Other basic requirements of a tracer are as follows: it should be easily detectable at low concentrations; detection should be unambiguous; and finally, injection, detection and/or sampling should be performed without disturbing the system.

2.2.4 Sealed Sources

Sealed sources are simply radioactive materials that are encapsulated usually in stainless steel capsules. These capsules are so designed as to allow the radiation to escape while securely containing the radioactive material. Sealed sources emit the ionising radiation only

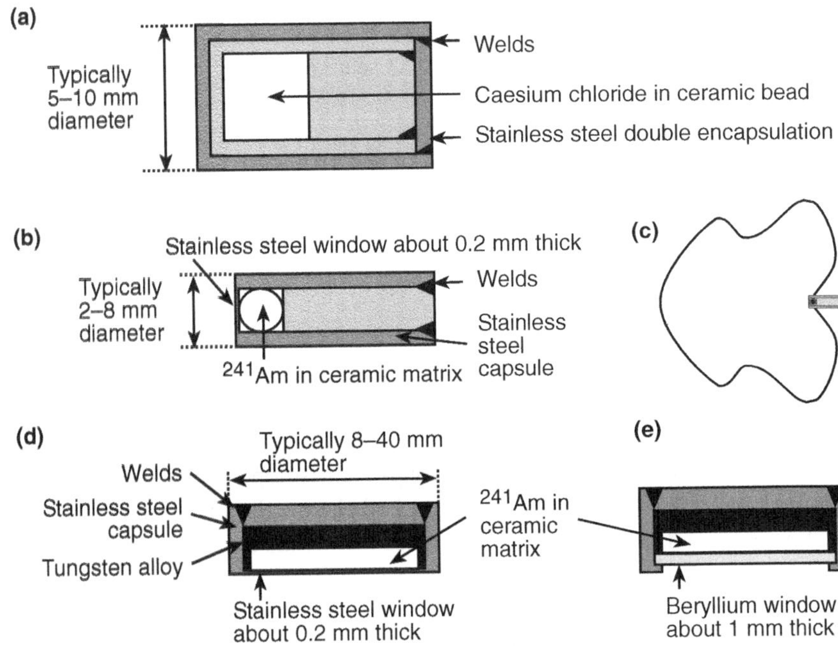

Figure 2.6 Typical encapsulation of γ-ray emitters: (a) high-energy pellet source, (b) low-energy point source, (c) a typical isodose curve (angular emission intensity) for a 60-keV γ-ray emission from a ^{241}Am point source, (d) low-energy and high-activity ^{241}Am disc source (>45 mCi) and (e) a very low energy disc source using Be window to enable transmission of characteristic X-ray emissions

through the capsule wall, with no release of the radioactive material itself. For high-energy γ-emitters such as ^{137}Cs, ^{192}Ir and ^{60}Co, the capsule can be made from reasonably thick stainless steel as the γ-radiation can still exit the capsule without significant absorption (see Figure 2.6a). With low-energy sources the absorption of the capsule is significant; so a thin-end window is required and therefore the sources are directional (see Figure 2.6b). The photon output from the back of the source is very low in this case.

With low-energy sources the self-absorption of the active ceramic is significant, and so to achieve large activities the materials must be arranged in a flat disc rather than a sphere (see Figure 2.6d). When even lower energies are required, such as in sources emitting characteristic X-rays, a low-density (-Z) window is employed. The window is typically of beryllium or aluminium. In the case of ^{241}Am, lower energy photons in the 12–26 keV range are now usable in addition to the 60 keV photons (see Section A.2). The γ-ray source encapsulations shown in Figure 2.6 are the most common, but depending on the isotope there are other configurations available, such as line and annular sources.

β-Sources may also incorporate the active material in a ceramic matrix, such as the point and disc sources shown in Figures 2.6b and 2.6d. However only a very thin window can be used. Its thickness depends on the energy of the source, but it is typically less than 50 μm of stainless steel. Depending on the isotope these sources may also use a block of tungsten alloy to stop backwards radiation leakage. Very often foil sources are used for β^--emitters, such as the one shown in Figure 2.7: The isotope is incorporated in the surface layer of a metal foil, often silver. The foil thickness is some hundred micrometres, whereas

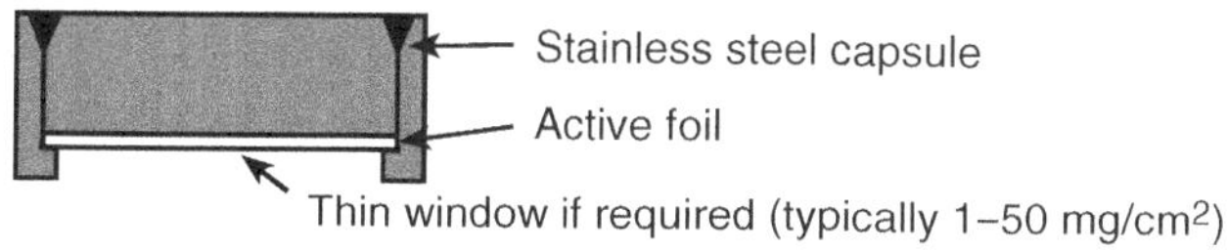

Figure 2.7 Typical design of a beta particle foil source

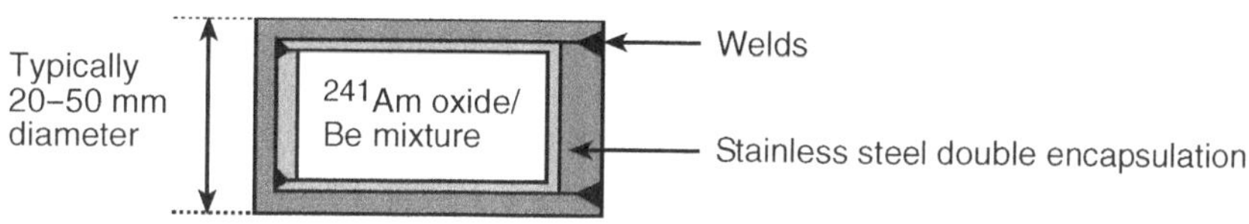

Figure 2.8 Typical design of a ^{241}Am/Be neutron source. The ^{252}Cf source uses the same type of encapsulation, but with smaller dimensions: the diameter is typically less than 10 mm

the face thickness with the isotope is about 50 μm or less. If required, these sources also incorporate a thin window of stainless steel, aluminium or polymeric material.

The ^{85}Kr isotope is sometimes used as β^--particle source, but because krypton is a gas, the encapsulation is a hermetically sealed stainless steel or titanium container with internal pressure less than atmospheric pressure.

The ^{241}Am isotope is also available as *α-particle source*. This is also a disc source (as shown in Figure 2.6.e), but only with a very thin metal window, e.g. 2 μm gold/palladium, because α-particles are easily stopped in any solid-state matter.

There are two categories of radioisotope *neutron sources*. One is the ^{252}Cf spontaneous fission source and the others are sources based on the (α, n) nuclear reaction. The most popular one in the latter category is ^{241}Am, which emits α-particles and 60-keV γ-rays. Americium is mixed with beryllium as target material and the α-particles eject a neutron from the beryllium nucleus through the (α, n) reaction (see Figure 2.8).

Often a lead shield is used around ^{241}Am/Be sources to almost completely absorb the 60-keV γ-rays, which are unwanted in a neutron source. This lead has no effect on the neutron output. Important properties of the ^{241}Am/Be and the ^{252}Cf sources are listed in Table 5.6.

Some sources combine a target within the source capsule with the primary radiation emitter in order to produce secondary radiation from the target. This makes it possible to make isotopic characteristic X-ray emission sources (see Figure 2.9). Here also ^{241}Am is often used as the primary radiation source. Depending on the design, the X-ray emission is induced by both the γ-rays and the α-particles, or only the former, in which case the target material is restricted to elements with K-edge below 60 keV. Typical targets are (with their K_α X-ray energies in parentheses) Cu (8.04 keV), Rb (13.37 keV), Mo (17.44 keV), Ag (22.10 keV), Ba (32.06 keV) and Tb (44.23 keV). There are also sources with interchangeable target materials, for instance by rotating a target disc containing all the elements just listed.

The manufacturers of radioisotope sources specify a recommended working life for the sources on the basis of the assessment of several factors such as toxicity of the nuclide, total activity, source construction and half-life. Typical values are between 5 and 15 years. The most commonly used sources for radioisotope gauges are listed in Table 2.3.

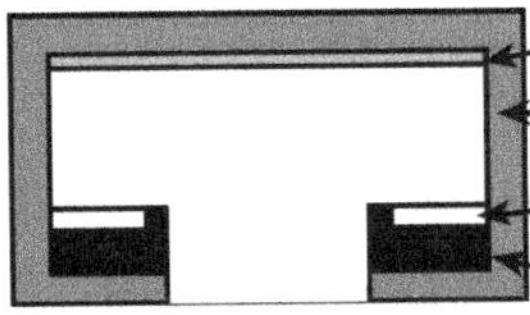

Figure 2.9 Typical design of an α-particle target source emitting characteristic X-rays. The α-particles and γ-rays emitted from the annular source (^{241}Am) induce emission of characteristic X-rays from the target. Tungsten is used to absorb the (unwanted) γ-rays (60 keV) from the ^{241}Am source in the forward direction. Not shown is the Be window used to seal the source. This is put either in front of the source or on top of the annular source. In the latter case the characteristic X-rays produced are only due to excitation of the γ-rays

Table 2.3 Commonly used sealed sources of the different categories and their typical applications in industrial measurement[a]

Isotope	Half-life	Radiation emitted	Useful energy [keV]	Typical application
^{241}Am	433 y	γ- and X-rays	60	Density/multiphase
^{133}Ba	10.5 y	γ-rays	81, 303 and 356	Density/multiphase
^{137}Cs	30.2 y	γ- and X-rays	662	Density/level
^{60}Co	5.27 y	γ-rays	1173 and 1333	Level
^{192}Ir	74 d	γ-rays	316 and 468	Radiography calibration/ reference
^{241}Am-target	433 y	X-rays	8 to 44	Radiography calibration/ reference
^{90}Sr (^{90}Y)	28.6 y	β$^-$-particles	546 and 2274	Thickness
^{85}Kr	10.37 y	β$^-$-particles	672	Thickness
^{147}Pm	2.623 y	β$^-$-particles	225	Thickness
^{241}Am/Be	433 y	Neutrons	~12,000 (maximum)	Level/analysis/ borehole logging
^{252}Cf	2.7 y	Neutrons	~8000 (maximum)	Level/analysis/ borehole logging

[a] See the nuclide index in Section A.2 for more data.

2.3 OTHER RADIATION SOURCES

There are a variety of radiation sources where the emission of ionising particles and electromagnetic radiation is caused by the acceleration, bombardment or exposure to particles and electromagnetic radiation. These are briefly mentioned here for the sake of completeness.

The emitted radiation may be fluorescence and bremsstrahlung as for the *X-ray tube*, or it may be nuclear radiation caused by nuclear reactions. The sequence of events in many nuclear reactions is that an incident particle enters the nucleus with the formation of a 'compound nucleus', which subsequently decays to give a product nucleus and an emitted particle or radiation. There are nuclear reactions caused by high-energy electromagnetic radiation, sometimes called *photonuclear reactions*, but these are much less common. Most often the reactions are caused by thermal neutrons, as is the case with *nuclear*

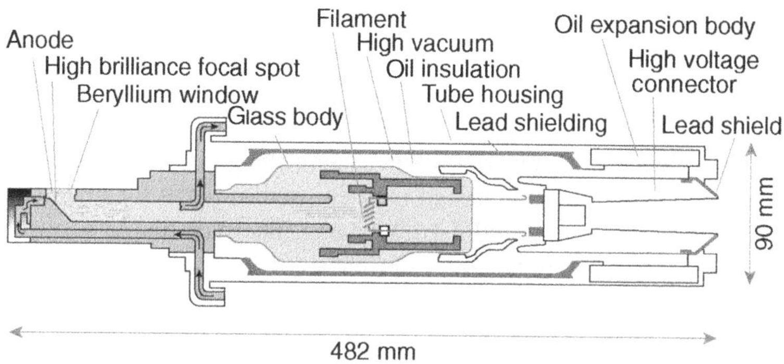

Figure 2.10 Cross-sectional view of a typical side window X-ray tube
Source: Courtesy of PANanlytical

reactors, or fast particles such as protons and α-particles. The latter set of reactions is achieved through the use of *particle accelerators*.

2.3.1 X-ray Tubes

Figure 2.10 shows a schematic representation of a standard X-ray tube with a side window. A tungsten alloy filament is heated by means of a current (the filament current, which typically is a few amperes), causing thermionic emission of electrons and thereby the formation of a region of high electron density around the filament. In a typical tube the filament temperature is about 1800°C. Part of this electron cloud is accelerated by means of a large potential difference towards the anode. Values between 50 and 200 kV are most common for the acceleration voltage, the tube high voltage. The electron beam current, the tube current, is typically in the range of a few tens to several hundreds of milliamperes. The filament is placed inside a focusing cup to direct the electrons onto a small area of the anode, the focal spot, producing characteristic X-rays and bremsstrahlung. The area of the focal spot and the anode angle determine the width of output X-ray beam exiting the window. Some tubes have two filaments of different length, making it possible to select between two different focal spot areas and beam widths.

Although the basic design of an X-ray tube may appear relatively simple and has remained essentially unchanged for a long time, there are certain critical design considerations: The most efficient bremsstrahlung emitters are materials with high atomic number (see Section 3.1.1). Even with these the conversion of electrons to bremsstrahlung is a very inefficient process where only about 1% of the total applied power emerges as useful radiation. The majority of the power emerges as heat, and the temperature at the focal spot may consequently be up to 3000°C. Careful choice of anode material and efficient heat dissipation are thus crucial. Tungsten alloys are common as anode material because of the combination of relatively high atomic number and high melting point (>3400°C). These are, however, relatively poor heat conductors, meaning that other measures have to be taken for heat dissipation. Industrial tubes often use a small tungsten alloy target embedded on top of a copper block, which is cooled by circulating water or oil in a closed loop system with fan cooling. Modern tubes used for medical imaging often use rotating disc anodes driven by an inductive motor with the rotor suspended inside the glass envelope, but with stator windings outside. The heat is then spread over a circular track and larger area rather than a small spot, and transported to the walls of the tube housing by radiation. In these tubes the glass envelope is surrounded by oil contained in a metal housing, which is fan cooled. This design is often used to allow for high tube currents combined with small focal spots and narrow beams. The radiation window also has to be carefully designed because of heating from scattered electrons. Beryllium is often the preferred window material to achieve high transmission, particularly in the low-energy region. This is because of its high mechanical strength combined with low density and atomic number.

The tube current and high voltage control the X-ray beam properties. The tube current, which is controlled by the filament current, determines the radiation intensity of the beam, whereas the high voltage determines its energy spectrum. In tube terminology the energy properties of the beam are often referred to as the *beam quality* since it affects its penetration ability. Correspondingly the intensity is spoken of as the *beam quantity*. All tubes have to be operated within certain heat limit curves that define how long a tube can be run with a certain current and voltage combination. Many tubes also have well-defined warm-up procedures that have to be followed to avoid damage. Most tube failures are in one way or another related to excessive heat.

In addition to the high voltage and current of the tube, the output window and filters affect the properties of the output beam. The high voltage of the tube determines the maximum bremsstrahlung energy, as can be seen in the example in Figure 2.11. Below this energy there is a continuum with increasing intensity, all the way down to the energy of maximum emission intensity. Below this energy the intensity drops off because the low-energy X-rays are attenuated in the tube window. By applying filters at the output window, e.g. a few millimetres of Aluminium, the low-energy end of the spectrum will be more attenuated than the high-energy end. Thus the effect is that the average energy increases and the energy of maximum emission intensity is shifted upwards in the spectrum.

The characteristic X-rays of the anode are superimposed on the bremsstrahlung continuum, as can be seen in Figure 2.11. Tungsten is a very common anode material; however, other materials such as Molybdenum are also used when characteristic lines with other energies are required.

Recent developments have led to tubes with pure characteristic X-ray emission without any bremsstrahlung. This is achieved by exposing a secondary target to the X-ray beam from the primary target, the anode. Production of bremsstrahlung is avoided at the secondary target by keeping it shielded from the electron beam. Figure 2.12 shows the schematic design of such a tube: Electrons emitted from a circular filament are electrostatically accelerated towards a large water-cooled anode block with a conical geometry. The design of the primary anode is such that a maximum power loading of several kiloWatts is possible. In order to increase the efficiency of the generation of X-rays the anode is covered with a layer of a high-Z material. A secondary target is positioned inside the cylindrical primary target, separated from this first target by a beryllium window. This secondary target is located outside the vacuum envelope, and can be exchanged rapidly, thus increasing the flexibility of the design. The cone angle of the secondary target is chosen such that the intensity of the collected radiation emitted by the primary target is

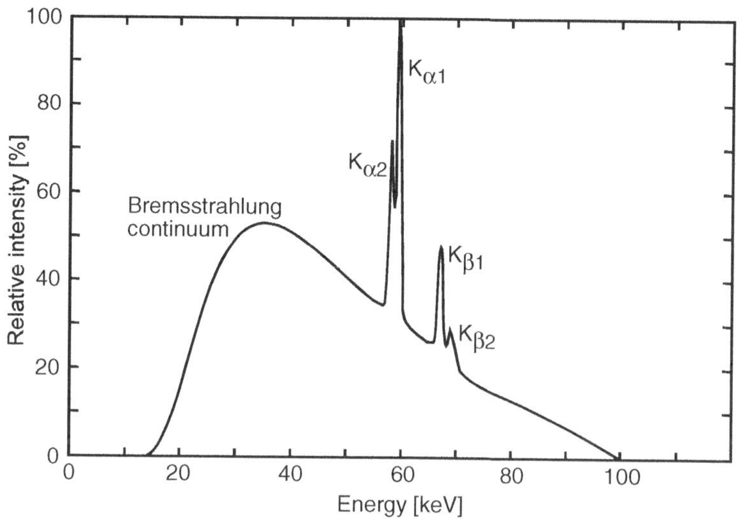

Figure 2.11 Typical X-ray tube output spectrum, with 100-kV tube voltage and Tungsten anode. The output beam is filtered in 2-mm aluminium and 3-mm beryllium

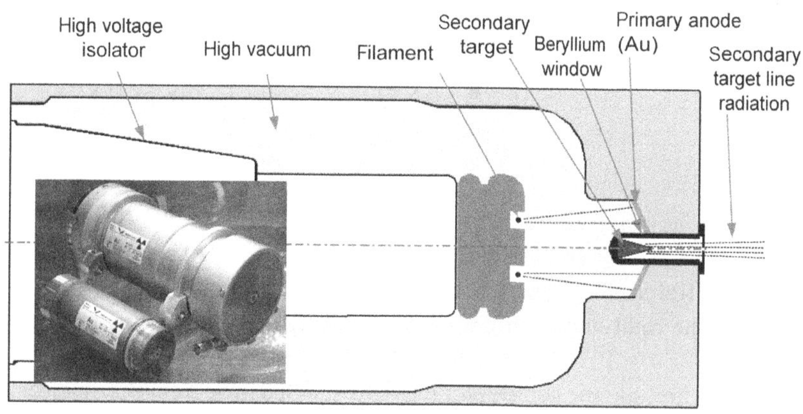

Figure 2.12 Cross-sectional view of the Fluor'X tube, a secondary target tube with low bremsstrahlung output (see Figure 2.13). The photograph in the inset shows two different power versions of the tube *Source*: Courtesy of PANanlytical

maximised. The bremsstrahlung background in the output spectrum of this tube is very low, as can be seen from Figure 2.13.

It is also possible to produce tuneable near-monochromatic X-rays using Bragg scattering of a polychromatic X-rays in a crystal [5].

2.3.2 Nuclear Reactors

In a nuclear reactor the incident particle is often a thermal neutron, in which case the reaction may be written as

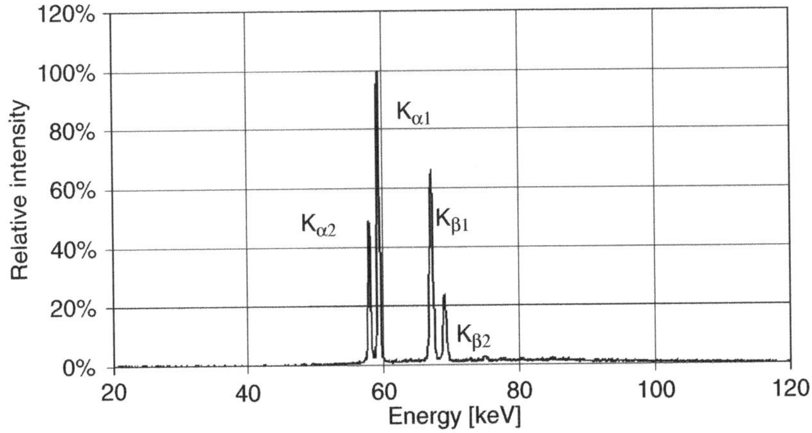

Figure 2.13 Typical emission spectrum of the 160-kV Fluor'X tube with a gold-covered primary anode and tungsten secondary target. These are the K-edge emission lines of tungsten with very low bremsstrahlung background
Source: Courtesy of PANanlytical

$$
{}^{A}_{Z}\mathrm{X} + \mathrm{n} \rightarrow {}^{A+1}_{Z}\mathrm{X}^{*} \rightarrow
\begin{cases}
{}^{A+1}_{Z}\mathrm{X} + \gamma & {}^{A}_{Z}\mathrm{X}\,(\mathrm{n},\gamma)\,{}^{A+1}_{Z}\mathrm{X} \\
{}^{A}_{Z}\mathrm{X} + \mathrm{n} & {}^{A}_{Z}\mathrm{X}\,(\mathrm{n},\mathrm{n})\,{}^{A}_{Z}\mathrm{X} \\
{}^{A}_{Z-1}\mathrm{X} + \mathrm{p} & {}^{A}_{Z}\mathrm{X}\,(\mathrm{n},\mathrm{p})\,{}^{A}_{Z-1}\mathrm{X} \\
{}^{A-3}_{Z-2}\mathrm{X} + \alpha & {}^{A}_{Z}\mathrm{X}\,(\mathrm{n},\alpha)\,{}^{A-3}_{Z-2}\mathrm{X}
\end{cases}
\tag{2.14}
$$

This is known as *neutron activation*. The asterisk indicates that the compound nucleus is left in an unstable state with a characteristic half-life. This nucleus may decay in one or more different ways, and each possible mode of decay has its own probability (cross section). The right-hand column shows the short-hand notations for the respective reactions.

Although there are isotopic neutron sources as discussed in Section B.2, nuclear reactors are a lot more efficient. The reactors are capable of producing a high flux of both fast and thermal neutrons. A typical neutron flux in a reactor is in the order of 10^{12} $cm^{-1}\cdot s^{-1}$ [6], whereas that of an isotopic source is $10^{-6} - 10^{-8}$ times less [7]. Needless to say reactors are complex and expensive. Their only relevance to industrial measurement systems is in the production of isotopic sources. Actually, radioisotopes were initially considered as a by-product of research and development in the nuclear power reactor industry, which has found industrial utilisation [8].

2.3.3 Accelerators

While nuclear reactors can give a flux of only neutrons and γ-rays, accelerating machines can use many other types of bombarding particles. There are several types of accelerating machines, each one named according to the particle accelerated or the acceleration method used. The principle involved here is that a beam of charged particles, usually positively charged, although electrons are also used, is injected into the machine. It is then accelerated in electric and magnetic fields, either in a straight line (linear accelerator), or, more commonly, in a spiral (cyclotron). The high-energy particle thus created is then directed onto a target, transferring sufficient energy to cause a nuclear reaction. Particle accelerators may also be used to produce neutrons; these are so-called neutron generators [9, 10].

Synchrotron radiation is emitted by a beam of fast electrons when they are deflected by magnetic fields, and is so called after the type of accelerator in which it was first observed. This previously unwanted electromagnetic radiation is now produced in accelerators specially constructed for the purpose. These give intense, highly collimated radiation beams both in the conventional X-ray energy region and outside it. Like bremsstrahlung, synchrotron radiation has a continuous distribution of energies.

As for nuclear reactors, the production of isotopic sources is the major relevance of accelerators to industrial measurement systems.

2.4 SEALED RADIOISOTOPE SOURCES VERSUS X-RAY TUBES

In comparison to sealed radioisotope sources, X-ray tube systems are often considered too complex and fragile for operation as part of permanently installed gauges. However, rugged X-ray tubes, with features such as high-speed switching and relatively pure, near-monochromatic emission spectra, are being developed. The main differences between isotopic sources and X-ray tubes may be summarised as follows:

- The output intensity of an X-ray tube is adjustable and typically 10^5 times higher than that which is practically obtainable with an isotopic source [11].
- The very stable emission rate of isotopic sources cannot be achieved with X-ray tubes.
- The emission energies of γ-ray sources cannot be chosen with the same flexibility as X-ray energies because of the limited selection of isotopes. Isotopic X-ray target sources allow for more flexibility, but with the drawback of very limited intensity.
- Standard X-ray tubes have polychromatic output spectra with a continuum, whereas γ-ray sources are monochromatic or emit at discrete energies. The exception to this is the new twin target X-ray tubes with near-monochromatic (pure) emission.
- An X-ray tube system is more complex and fragile and requires high voltage for operation.
- A γ-ray source cannot be switched off (although it can be shut and locked). Some X-ray tubes also allow for high-speed switching or pulsing, which may be advantageous in some applications.

3

Interaction of Ionising Radiation with Matter

The interaction of ionising radiation with matter is the foundation of every nuclear measurement principle. We therefore need to treat this subject in some detail in order to understand and fully utilise these principles. As a starter it is helpful to consider ionising radiation in groups after their interaction properties: heavy charged particles, light charged particles, electromagnetic radiation and neutrons. These are listed in Table 3.1 along with their major interaction mechanisms, which will be discussed in more detail in this chapter. Before doing so, note the third column in the table, listing secondary radiation generated by the various interaction mechanisms: Regardless of the type of radiation, electrons are always generated at some stage or another.

The main focus of this chapter will be on electromagnetic interactions; however, this will also have a basic treatment of charged particle interactions. Finally, the basics of neutron interactions will be covered, partly for the sake of completeness, and partly because neutron measurement methods have potential for use in permanently installed gauges.

3.1 CHARGED PARTICLE INTERACTIONS

3.1.1 Linear Stopping Power

Charged particles such as electrons (β^--particles), protons, α-particles and ions loose their kinetic energy continuously along their track in the absorber, primarily because of interactions with the absorber's atomic electrons. The rate of energy loss with distance, $-\mathrm{d}E/\mathrm{d}x$, is known as *linear stopping power* or *specific energy loss* and was first predicted by Bethe. For our purposes this may be expressed as

$$-\left(\frac{\mathrm{d}E}{\mathrm{d}x}\right)_{\mathrm{c}} = \frac{z^2 NZ}{v^2} f(v, I) \tag{3.1}$$

where N and Z are the number of atoms present per unit volume and the atomic number of the absorber. Further, v is the particle velocity and z is the number of charge units it holds in terms of e (the electron charge); i.e., z is unity for electrons, positrons and protons, and 4 for α-particles etc. The function $f(v, I)$, which is different for light (electrons and

Radioisotope Gauges for Industrial Process Measurements. Geir Anton Johansen and Peter Jackson.
 ISBN 0-471-48999-9

Table 3.1 Summary of interaction mechanisms of the four groups of ionising radiation, and the secondary radiation these produce

Radiation type	Major interaction mechanism	Secondary radiation	Common interaction terminology
Charged Light			
β^--particles (electrons)	Excitation, ionisation	Electrons, bremsstrahlung	Energy loss, absorption, transmission, range
β^+-particles (positrons)	Excitation, ionisation, annihilation	Electrons, annihilation	Energy loss
Heavy			
α-particles and ions	Excitation, ionisation, nuclear reactions	Electrons, particles, γ-rays	Stopping power, energy loss, range
Uncharged			
n (neutrons)	Collisions, nuclear reactions	Particles, γ-rays, electrons	Moderation, crosssection
electromagnetic (photons; γ-, X-rays, annihilation bremsstrahlung)	Scatter, photoelectric effect, pair production	Electrons, fluorescence, annihilation	Attenuation, absorption/stopping efficiency, transmission

positrons) and heavy (protons and ions) charged particles, depends on the mean excitation energy of the absorber, *I*, and the particle velocity. For various reasons, this term is of little importance unless the velocity of the particle approaches the speed of light.

The main factors affecting the stopping power are that it depends on how dense the absorber is (NZ), and that it is strongly dependent on the charge (z) and the velocity (v) of the particle. For a given energy the velocity of a light particle will be much higher than that of a heavy one, since the energy of a particle (with mass m) is given as $(1/2)mv^2$ in the non-relativistic case. Altogether this means that heavy particles (α-particles) loose a lot more energy per unit path length than light ones (electrons) at a given energy, and that heavy absorbers (solids) absorb more energy than light ones (gases).

Bremsstrahlung was introduced in Section 2.1.6 as electromagnetic radiation produced when charged particles are deflected in the coulombic field of the absorber's nucleus. This radiative energy loss may be expressed as

$$-\left(\frac{\mathrm{d}E}{\mathrm{d}x}\right)_\mathrm{r} \propto \frac{ENZ^2}{m^2} \qquad (3.2)$$

where E and m are the energy and mass of the particle, respectively. This loss is, because of the inverse-square dependency on the particle mass, far more important and actually only relevant for light particles, i.e. electrons. Unless the energy is very high, nuclear interactions are rare for heavy charged particles. The square dependency to the atomic number of the absorber explains why high-Z materials are used as anodes in X-ray tubes.

The total stopping power of electrons is the sum of the loss contributions to collisions and bremsstrahlung. The relative importance of these may be approximated as

$$\frac{(\mathrm{d}E/\mathrm{d}x)_\mathrm{r}}{(\mathrm{d}E)/\mathrm{d}x_\mathrm{c}} \cong \frac{EZ}{700} \qquad (3.3)$$

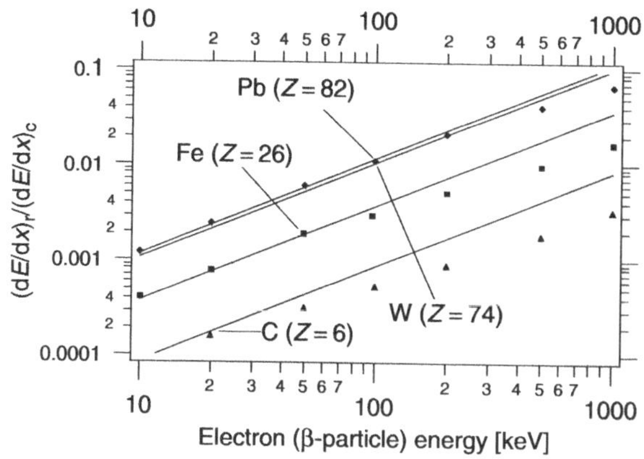

Figure 3.1 Relative importance of the bremsstrahlung loss to the collision loss for a few elements, as predicted by Equation (3.3). The same is shown also for tungsten (W) which is a commonly used anode material in X-ray tubes. For C (graphite, triangles), Fe (squares) and Pb (diamonds), data are taken from Reference [12]

where the electron energy E is in units of MeV (million electronvolts). From the plot of this ratio in Figure 3.1, it is evident that the radiative losses are always small compared to collision losses. The plot also shows that the approximation in Equation (3.3) is most accurate for heavy element absorbers, such as lead and tungsten.

The mean energy loss of positrons (β^+-particles) is slightly higher than that of electrons for energies below about 200 keV, whereas at higher energies we have the opposite case. The difference, which may be a few percent, has no practical significance in the context of this book.

3.1.2 Range

Another important property of particles is their *range* (R) in absorbers. This may be defined as the total path length a particle travels in an absorber until it loses all its energy. The range can thus be determined by integrating Equation (3.1) for $x = 0$ to $x = R$, provided the function $f(v, I)$ is known accurately. This method is known as the *continuously slowing-down approximation* (CSDA). This is used to calculate the range data presented in Figures 3.2 and 3.3. Data are taken from Reference [13]. For electrons (Figure 3.2) the total path length is presented. This may be considered as the maximum range of electrons (see also Figure 3.4). For α-particles (Figure 3.3) the *projected range* is plotted. This is the average depth to which a charged particle will penetrate measured along the initial direction of the particle.

As can be seen from the plots α-particles generally experience high stopping power and have very short range in matter. This makes them attractive for high sensitivity measurements in laboratory instrumentation and smoke detectors, but far less usable for permanently installed industrial gauges. In this context their properties are a lot more important for radiation safety issues: The short range means that if, for instance, α-emitters become internal to the body, e.g. through contamination of air or food, all the particle energy will be deposited in the body. This affects the classification and use of such sources, and will be further discussed in Chapter 6.

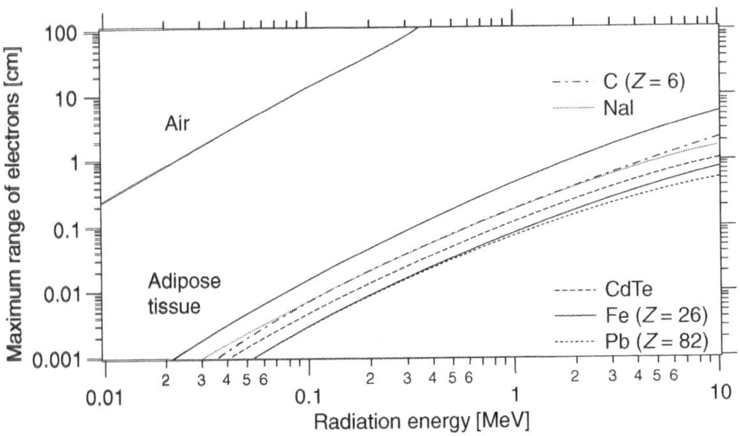

Figure 3.2 CSDA path length of electrons in various materials. The carbon is graphite. Data are taken from Reference [13]

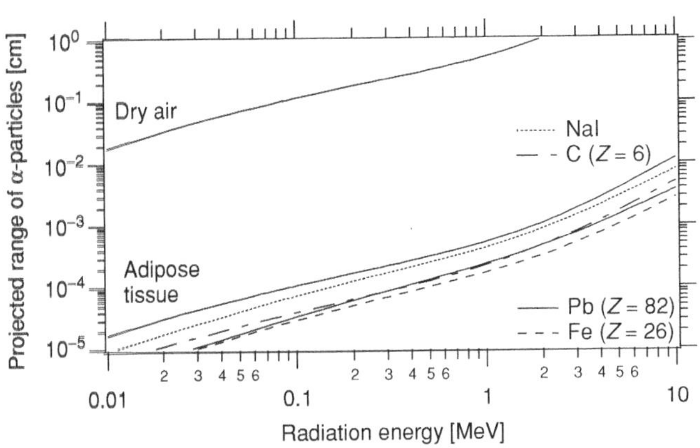

Figure 3.3 Projected range of α-particles in various materials on the basis of the CSDA approximation. The carbon is graphite. Data are taken from Reference [13]

The longer range of electrons, however, makes them applicable for some permanently installed gauges in some industrial applications. This will be treated in Chapters 5 and 7. From the point of radiation safety and contamination risks, β-emitters are considered to be less hazardous than α-emitters when ingested.

In the context of radiation detection it is important to note that all radiation energy (as illustrated in Table 3.1) one way or another is converted to electron energy. All electronic detection principles rely on this energy being converted to electric charge through ionisations in the radiation detector. This means that it is important that the range of these secondary electrons is shorter than the dimensions of the detector so as to avoid so-called *electron leakage* by which part of the signal is lost. Figure 3.2 reveals that this normally is not a problem, particularly for solid-state detectors; the energy of the secondary electrons may, for most practical purposes, be regarded as being deposited at the point of generation.

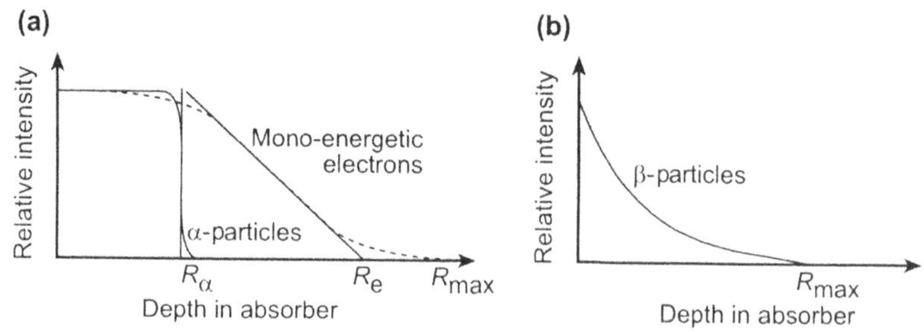

Figure 3.4 (a) Illustration of transmission of α-particles and mono-energetic electrons into an absorber. For α-particles, which always have mono-energetic emission, the range R_α is defined at the depth of half-intensity. For mono-energetic electrons R_e is the extrapolated range whereas R_{max} is the maximum range. The ranges of electrons and α-particles are not in scale relative to each other. (b) Illustration of the transmission of (poly-energetic) β^--particles into an absorber

A useful empirical relationship for estimating relative ranges of a charged particle (ion) in materials with different mass numbers (A) is the *Bragg–Kleeman rule*:

$$\frac{R_1}{R_2} \approx \sqrt{\frac{A_1}{A_2}} \tag{3.4}$$

where the subscripts refer to the different materials. The ranges are in units of length. This is an approximation whose accuracy is best for absorbers with close mass numbers.

3.1.3 Charged Particle Beam Intensity

An electron colliding with other electrons looses a much greater fraction of its energy in a single collision than does a heavy particle. This causes more sudden changes in direction, making its range much less well defined. Its linear distance of penetration will be very different from the length of the path it actually follows through the medium. This is illustrated in Figure 3.4a, where the drop in relative intensity of mono-energetic beams of α-particles and electrons is illustrated.

The situation is different for the transmission of β^--particles, as can be seen from the illustration in Figure 3.4b. This is because β^--particles, unlike α-particles, are not emitted at one single energy, but with a spectrum of energies all the way up till a maximum energy, E_{max} (see Chapter 2). The shape of the transmission curve in Figure 3.4b, i.e. the relative intensity I/I_0, may be approximated as

$$\frac{I}{I_0} = e^{-\mu_\beta x} \tag{3.5}$$

provided x, the depth in the absorber, is less than R_{max} by some margin. Here μ_β is the linear absorption coefficient of the particular β^--particles in the actual absorber (often denoted n). Unlike attenuation of γ-ray photons, which is presented in the next section, this is a pure empirical relationship. It turns out to be very convenient for the use of β-particles in industrial measurements, as will be discussed in Sections 5.5 and 7.4. The

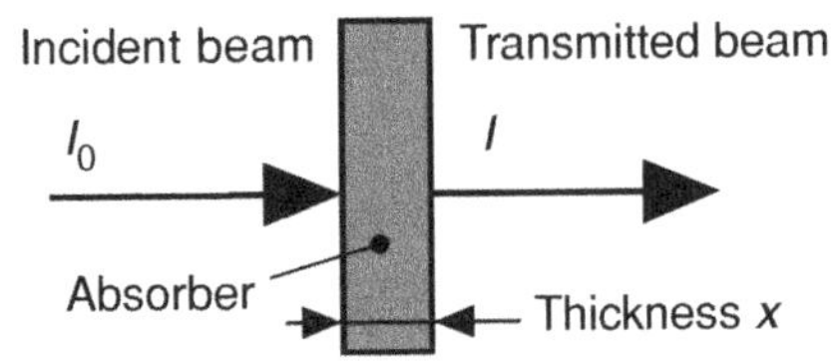

Figure 3.5 Attenuation of a parallel beam of γ-ray photons in an absorber

linear absorption coefficient may be approximated as [14, 15]

$$\mu_\beta(\text{cm}^{-1}) = 22\rho E_{\text{max}}^{-4/3} \tag{3.6}$$

with E_{max} given in MeV and the density ρ of the absorber in g/cm^3.

3.2 ATTENUATION OF IONISING PHOTONS

The absorption mechanisms of X-rays and γ-rays are totally different from those of particles. An energetic photon may travel a long distance in a material without being affected at all, but its history is terminated once it interacts. Energetic photons are said to interact catastrophically. Therefore we need to consider the number of photons removed from *a beam*, penetrating an absorber: The attenuation of a narrow and parallel beam of mono-energetic photons penetrating a thin slab of homogeneous material (as illustrated in Figure 3.5) follows Lambert–Beer's exponential decay law:

$$I = I_0\, e^{-\mu x} \tag{3.7}$$

Here I_0 is the incident or initial intensity, x is the thickness of the absorber, I is the remaining beam intensity and μ is the linear attenuation coefficient (usually with unit cm^{-1}). This expresses the photon interaction probability per unit path length in the absorber. The derivation of Equation (3.7) is included in Appendix B.

3.2.1 The Intensity and the Inverse-Square Law

The *intensity* of the beam is the number of photons per second through a given cross section. In the simplified case of a parallel beam this cross section is equal at any distance from the source. However, a frequently encountered case is the use of a radiation detector positioned at a distance *d* from an isotropic point source. A circular detector aperture with area A_d defines a cone whose cross section increases with *d*. However, provided there is no attenuation and that the source radius is small compared to *d*, the intensity in the solid angle is constant and given by the *inverses-square law* as

$$I_0 = S_0 \frac{A_d}{4\pi d^2} \tag{3.8}$$

Here S_0 is the isotropic emission intensity of the source. Keep in mind that this is not necessarily equal to the activity (A); in the case of ^{137}Cs 661.6-keV γ-photons, $S_0 = 0.851A$ (see Section B.2). The expression in Equation (3.8) is valid for any detector aperture shape and also for disc sources, provided the source radius is small compared to d.

3.3 THE ATTENUATION COEFFICIENT OF IONISING PHOTONS

The linear attenuation coefficient expresses the photon interaction probability per unit path length. It is strongly dependent on the radiation energy and the density and the atomic number of the absorber. It is composed additively of contributions from several independent interaction mechanisms: the *photoelectric effect* (μ_τ), *Compton scattering* (μ_σ), *pair production* (μ_κ) and *Rayleigh scattering* (μ_{σ_R}), which will be explained below. In the literature there is some confusion over the terminology; the terms *coefficient* and *cross section* are often taken to be identical. The cross section, however, gives the interaction probability per target atom and is related to the linear attenuation coefficient as

$$\mu = \mu_\tau + \mu_\sigma + \mu_\kappa + \mu_{\sigma R} = \frac{N_A}{A}\rho(\tau + \sigma + \kappa + \sigma_R) = \frac{N_A}{A}\rho\sigma_{TOT} = N\sigma_{TOT} \quad (3.9)$$

where τ, σ, κ and σ_R are the cross sections of the respective interaction mechanisms, and σ_{TOT} is the sum of these. Their unit is barn, which is equal to 10^{-24} cm^2. Further, N_A is Avogadro's number, N is the number of atoms per unit volume, A is the average atomic mass (or molecular mole weight) of the absorber in units of u, and ρ its density. Because the *linear attenuation coefficient* depends on the density of the absorber, and therefore to some degree on its physical state, the mass attenuation coefficient $\mu_M = \mu\rho$ is preferred in many cases. Equation (3.9) may then be rewritten as

$$\mu_M = \frac{\mu}{\rho} = \frac{1}{\rho}(\mu_\tau + \mu_\sigma + \mu_\kappa + \mu_{\sigma R}) = \frac{N_A}{A}(\tau + \sigma + \kappa + \sigma_R) = \frac{N_A}{A}\sigma_{TOT} \quad (3.10)$$

The unit of μ_M is cm^2/g. There are thus three ways of expressing the interaction probability of energetic photons. As already stated, it is customary to use c.g.s. units for all these quantities. Using the mass attenuation coefficient in Equation (3.7) this can be rewritten as

$$I = I_0\, e^{-\mu x} = I_0\, e^{-\mu_M \rho x} \quad (3.11)$$

where the product ρx is known as the *mass thickness*.

The contribution from one of the attenuation mechanisms, say the photoelectric effect, to the total attenuation, is found by

$$\frac{I}{I_0} = \frac{\mu_\tau}{\mu}(1 - e^{-\mu x}) \quad (3.12)$$

The relationship of the different cross sections to the radiation energy and atomic number of the absorber is shown in Table 3.2.

Table 3.2 Approximate proportionalities of γ-ray cross sections to the radiation energy and the atomic number of the absorber[a]

Interaction mechanism	Cross section	Approximate proportionality to		Comment
		Atomic number Z	Radiation energy E	
Photoelectric effect	τ	Z^4 to Z^5	$E^{-3.5}$ to E^{-1}	
Compton scattering	σ	Z	E to E^{-1}	μ_σ proportional to ρ at low Z ($\neq 1$)
Pair production	κ	Z^2	E to $\ln(E)$	Requires $E >$ 1022 keV
Rayleigh scattering	σ_R	$Z^{2.5}$	$E^{-0.5}$ to E^{-2}	

[a] Also shown is the variation in the energy proportionality at low (≈10 keV) and higher (≈10 MeV) energies.

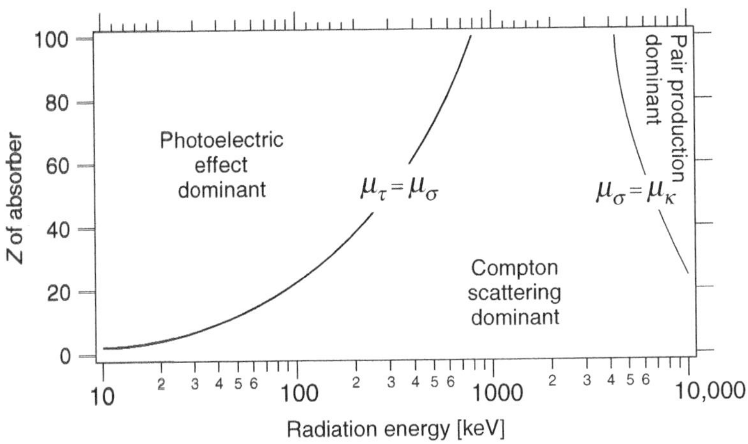

Figure 3.6 Boundary regions of the three major types of γ-ray attenuation mechanisms as a function of radiation energy and atomic number. Data are taken from Reference [12]

The tabulated proportionalities must be considered as approximate values only. They are nevertheless useful when considering the relative influence of the different interaction mechanisms. The E–Z map in Figure 3.6 is also useful in this context. The energy dependence is even better demonstrated in the sample plots in Figure 3.7. The latter also shows Z-dependence since these plots are examples of low-Z (=6), intermediate-Z (=26) and high-Z (=82) materials. Altogether it is evident that the photoelectric effect dominates at low energies and high Z; Compton scattering dominates at intermediate energies and low Z whereas pair production is dominant at high energies and high Z.

To explain the peculiar energy dependence of the attenuation coefficient, it is necessary to study different interaction mechanisms in some detail. In addition to the four mentioned here, there are a few other mechanisms; however, these have no practical importance in the context of this book. At higher energies (>10 MeV), for instance, photonuclear absorption may take place. This means that the photon interacts with the nucleus, causing nuclear reactions. A thorough treatment of this subject is given in Reference [16].

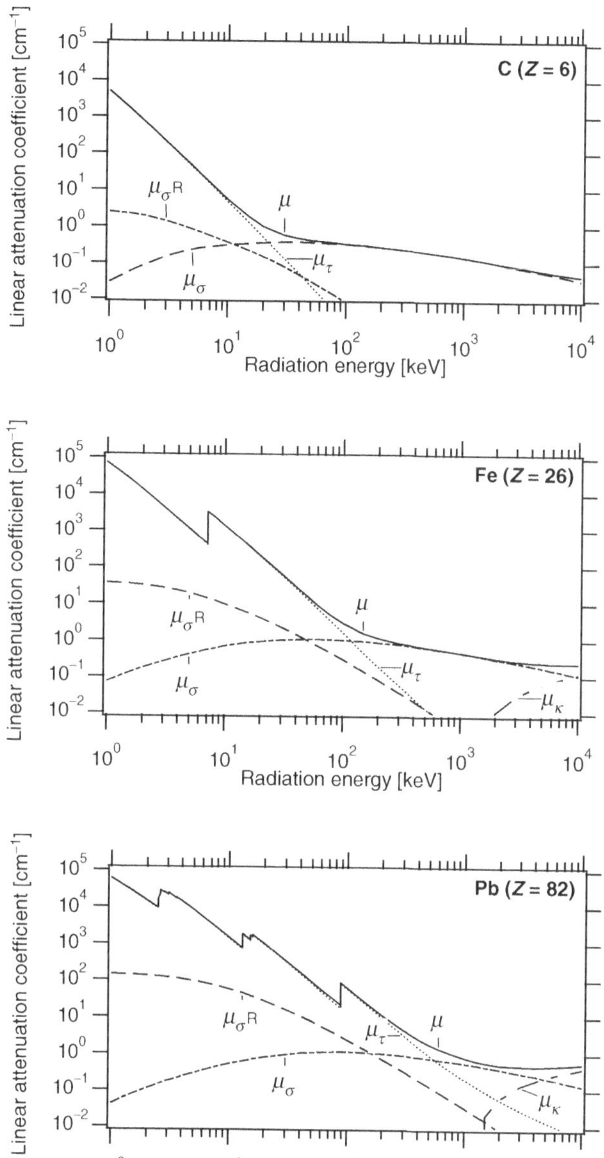

Figure 3.7 The composition and energy dependence of the linear attenuation coefficients of C (graphite, $Z = 6$), Fe ($Z = 26$) and Pb ($Z = 82$). Data are taken from Reference [12]; however, similar data are also available in printed tables [17, 233]

3.3.1 The Photoelectric Effect

The photoelectric effect can occur when a material is exposed to visible light, ultraviolet radiation or more energetic electromagnetic radiation such as γ-rays. As the name implies a photon collides or interacts with an orbital electron, which is ejected with a certain energy transferred from the photon. This is one of the outer electrons in the case of visible light.

The target material is then often a photocathode as in the photomultiplier tube described in Section 4.6.3.

In the case of γ-rays, however, the photons have sufficient energy E_γ to interact with one of the inner atomic electrons. This electron is ejected from the atom with a kinetic energy of

$$E_{\mathrm{kin}} = E_\gamma - E_{\mathrm{b}j} \tag{3.13}$$

where $E_{\mathrm{b}j}$ is the binding energy of the jth shell. The atom is left in an excited state and will within a short time, typically less than 10 ns, return to a stable state through electron rearrangement. This may happen in one of two ways: The energy released in rearranging the electron structure is used to free an electron from the atom as a whole. This is the Auger effect, and the emitted electron with energy $E_{\mathrm{b}j}$ is known as an *Auger electron*. The alternative to the Auger effect is *fluorescence* where a characteristic X-ray is emitted (line emission) when one of the outer electrons in shell i fills the vacancy in shell j. The energy of this X-ray photon will be

$$E_{\mathrm{X}} = E_{\mathrm{b}j} - E_{\mathrm{b}i} \tag{3.14}$$

where $E_{\mathrm{b}i}$ is the binding energy of the ith shell. This may, as explained in Section 2.1.5, be a cascade of electron transitions. However, with inner shell vacancies the major part of the energy is carried away by the first emission. The probability of fluorescence occurring is called the *fluorescence yield* ω_{a}. This increases with the atomic number of the absorber, as can be seen from the plot in Figure 3.8. Also plotted in this figure is the average weighted energy of K-shell X-ray emissions.

The photoelectric contribution to the linear attenuation coefficient is plotted for three different elements in Figure 3.7. The strong dependence on the atomic number and photon energy is clearly demonstrated (see also Figure 3.6). The so-called absorption edges or steps in the plot are due to the fact that once the incident photons have sufficiently high energy to eject a more tightly bound electron, the interaction probability increases in a step. For the K shell of lead, for instance, the binding energy $E_{\mathrm{bK}} = 88$ keV, the fluorescence yield $\omega_{\mathrm{aK}} = 95.5\%$ and the average energy of the fluorescence X-ray photons $E_{\mathrm{K}} = 77$

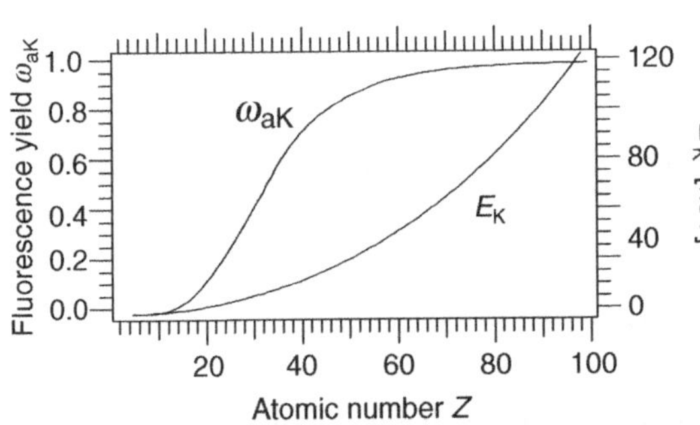

Figure 3.8 K-shell fluorescence yield (ω_{aK}) and average weighted X-ray emission energy (E_{K}) plotted versus atomic number of the absorber. The data, which are taken from Reference [17], are listed in Section A.3

keV [17]. Note that these absorption edges also imply that all elements are relatively transparent to their own fluorescence emissions because the emission energy is always lower than the binding energy and thus on the lower side of the step.

It is important to note that photoelectric interaction implies that the major part of the incident energy is transferred to electrons that, according to Section 3.1.2, have relatively short range in matter. The exception to this is energy carried away by fluorescence X-rays. All the electron energy may, for most practical purposes, be regarded as being deposited at the point of interaction. Photoelectric interaction is for this reason regarded as an absorption process.

Essential properties of the photoelectric effect are as follows:

- This effect is predominant at low energies and high Z.
- All the initial energy is, with the exception of fluorescence X-rays, deposited in the immediate surrounding medium of the interaction.
- Fluorescence is predominant in high-Z absorbers.
- The fluorescence emission is isotropic.
- Elements are relatively transparent to their own fluorescence emissions.

3.3.2 Compton Scattering

Compton scattering is an inelastic or incoherent scattering process where the incident photon interacts with one of the outer and loosely bound atomic electrons. The incident photon is scattered at an angle θ to its original direction. A fraction of its energy is transferred to the so-called *recoil electron* emitted at an angle φ to the incident photon's direction. The energy dependence of the Compton attenuation coefficient is shown in Figure 3.7. It is the predominant interaction mechanism at intermediate energies. However, the energy range of this dominance decreases with increasing atomic number of the absorber (see also Figure 3.6) because of the increasing dominance of the photoelectric interaction mechanism.

The Compton scattering process is shown schematically in Figure 3.9. The energy of the scattered photon can, from conservation of energy and momentum and on the assumption that the electron's binding energy is negligible, be expressed as (see Section B.2.1)

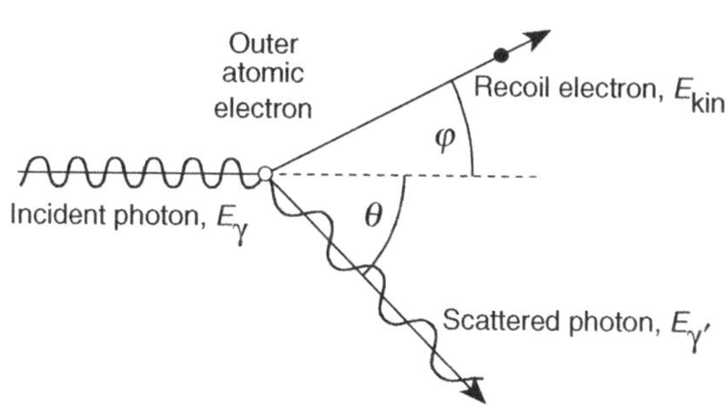

Figure 3.9 The process of Compton scattering

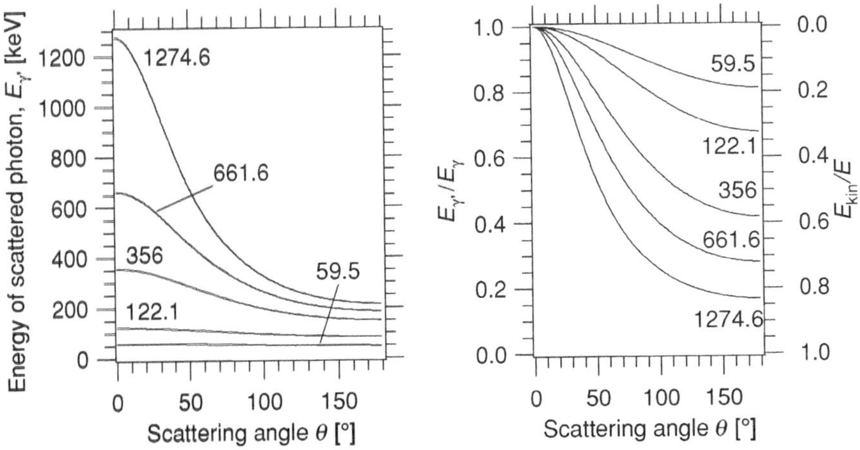

Figure 3.10 Energy of Compton scattered photon ($E_{\gamma'}$) as function of scattering angle for five different incident γ-ray energies. These are the main emission lines of ^{241}Am (59.5 keV), ^{57}Co (122.1 keV), ^{133}Ba (356 keV), ^{137}Cs (661.6 keV) and ^{22}Na (1274.6 keV). The latter is also representative for ^{60}Co (1173.2 and 1332.5 keV). Data are generated using Equations (3.15) and (3.16). On the right-hand side the energies, including that of the recoil electron (E_{kin}), are plotted relative to the incident γ-ray energy

$$E_{\gamma'} = \frac{E_\gamma}{1 + E_\gamma/m_e c^2(1 - \cos\theta)} \tag{3.15}$$

where $m_e c^2 = 511$ keV is the electron rest mass energy. The kinetic energy of the recoil electron can then be derived as

$$E_{kin} = E_\gamma - E_{\gamma'} = \frac{E_\gamma}{1 + m_e c^2/(1 - \cos\theta)} \tag{3.16}$$

The energies of the scattered photon and the recoil electron are plotted as functions of scattering angle and incident γ-ray energy in Figure 3.10. These plots help illustrate some important implications of Compton scattering: Very little energy is transferred to the recoil electron at small scattering angles, say less than about 10°. This means that the energy of the scattered photon is close to that of the incident. Further, for scattering angles above about 150°, there are only marginal changes in the amount of energy transferred. This implies a correspondingly small change in the energy of the scattered photon at these angles. Also note that some energy is always retained by the scattered photon, even at 180° scattering. Finally, one very important property is that the energy transfer to the recoil electron is significantly less at low incident photon energies than at high ones, both in absolute and relative terms.

In some texts on this subject the Compton cross section is split into two parts, exactly to express the sharing of energy between the recoil electron and the scattered photon, and its energy dependence. The energy transfer to the recoil electron is considered as absorption because of the relatively small range of the electron. Like the Auger electron and

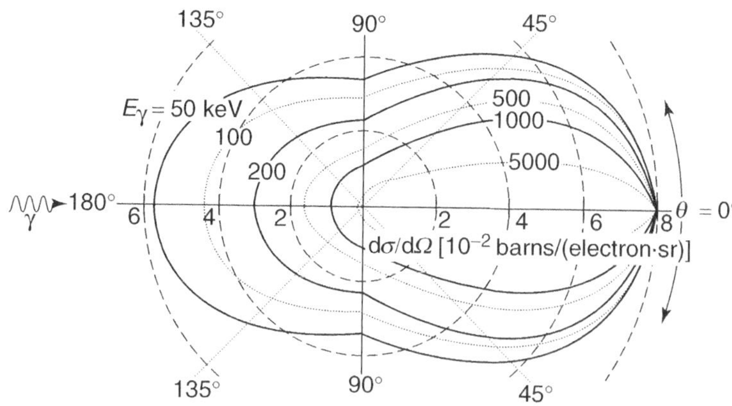

Figure 3.11 The differential cross section per unit solid angle for the number of photons scattered at an angle θ for six incident photon energies ranging from 50 to 5000 keV, as predicted by Equation (B.10) ($Z = 1$). The γ-photons are incident from left. The radius expresses the probability of a photon to be scattered into a unit solid angle at the scattering angle θ

photoelectron in the photoelectric effect, the recoil electron energy is considered deposited in the immediate surrounding medium of the interaction. The scattered photon, however, may escape the absorber and not deposit its energy in it. This is detailed in Section B.2.3.

In Section B.2.2, the differential form of the *Klein–Nishina formula* is presented. This expresses another important feature of Compton scattering; the probability of a Compton photon being scattered into a unit solid angle Ω at scattering angle θ. The Klein–Nishina formula is inaccurate at low values of E_γ because of the assumption that the electron's binding energy is negligible. The angular distribution of Compton scatter exhibits, as can be seen from the plot in Figure 3.11, a strong energy dependence: High-energy photons are mainly scattered in the forward direction, whereas the possibility for backward scatter increases as the energy decreases.

The angular distribution of Compton scattered photons plays an important role in the design of measurement systems. In some cases it is exploited, and in other cases it is regarded as an unwanted effect, such as when designing effective shielding for high-energy γ-rays.

The Compton cross section is a function of the electron density and therefore increases linearly with Z. The linear Compton attenuation coefficient, μ_σ, is then, according to Equation (3.9), proportional to $\rho Z/A$. Now, according to Figure 2.1, the ratio Z/A for low-Z elements, except hydrogen, is close to 1/2. This means that μ_σ is approximately proportional to the density of the absorber and independent of its composition (Z), a property that is exploited in γ-ray densitometry systems.

3.3.3 Rayleigh Scattering

Rayleigh scattering is an elastic scattering process with only a negligible energy transfer. The electron neither ionises nor excites the atom but interacts coherently with all its atoms.

The direction of the incident photon is changed, and it is reported that at least 75% of Rayleigh scattering is confined to angles smaller than a characteristic angle:

$$\theta_c = 2 \tan^{-1}\left(13.286\frac{Z^{1/3}}{E_\gamma}\right) \tag{3.17}$$

where E_γ is the incident γ-ray energy given in keV [16]. As can be seen this angle is largest for low energies and large atomic numbers. For carbon ($Z = 6$), for instance, it is 51.6° at 50 keV and 5.5° at 500-keV γ-ray energies. Rayleigh scattering is often neglected because its attenuation coefficient is less than those of photoelectric absorption and Compton scattering at low and intermediate energies, respectively. This can be seen from the plots in Figure 3.7, where it is also evident that Rayleigh scattering cannot be ignored in accurate models, particularly at low energies and for high-Z absorbers.

3.3.4 Pair Production

Pair production is less important in the context of industrial gauging because, according to Figure 3.7, radiation energies of several MeV are required before it plays any significant role. The reason for this is that pair production, where a γ-ray is converted to an electron–positron pair, is impossible unless the incident γ-ray energy is in excess of 1022 keV, that is, twice the rest mass energy of the electron. The process takes place within the coulombic field of the nucleus, although it may also be within the field of an electron. The energy in excess of the pair production energy of $2m_ec^2$ is shared between the electron and the positron as kinetic energy. The energy balance is then

$$E_\gamma = E^-_{\text{kin}} + E^+_{\text{kin}} + 2m_ec^2 \tag{3.18}$$

The positron is the anti-particle to the electron and thus has a very short life, typically less than 1 ns. Its range is typically in the order of a few millimetres. Once it reaches a low energy it will inevitably come close to an electron and the pair will *annihilate*. In this process two photons, each with energy close to the electron rest mass energy $m_ec^2 = 511$ keV, are emitted at approximately 180° to each other. These photons are called *annihilation radiation*.

3.3.5 Attenuation Versus Absorption

The terms *attenuation* and *absorption* are often used interchangeably for ionising electromagnetic radiation, however, there is an important difference that needs to be made clear. We discussed this in Section 3.3.2, and for Compton scattering it is further elaborated in Section B.2: The term attenuation is associated with the radiation beam or photon quantity and expresses the (relative) number of photons interacting. Absorption, on the other hand, is associated with the energy of the interacting photons. The amount of beam energy absorbed is dependent on the amount of energy carried out of the absorber by fluorescence, scattered and annihilation photons and partially absorbed secondary photons. This depends on a series of statistical processes and cannot be calculated analytically; radiation transport simulation is required as will be discussed later. In the context of radioisotope gauges the correct term is attenuation, for instance for beam intensity measurement.

Table 3.3 Approximate values for the linear attenuation coefficient (μ), the mean free path (λ) and the half-thickness ($x_{1/2}$) of C (graphite, Z = 6), Fe (Z = 26) and Pb (Z = 82) at two γ-ray energies[a]

	At 60 keV (^{241}Am)			At 662 keV (^{137}Cs)		
	C (Z = 6)	Fe (Z = 26)	Pb (Z = 82)	C (Z = 6)	Fe (Z = 26)	Pb (Z = 82)
μ [cm^{-1}]	0.4	10	60	0.2	0.6	1.3
λ [cm]	2.5	0.1	0.02	6	1.7	0.8
$x_{1/2}$ [cm]	1.8	0.07	0.01	4	1.2	0.6

[a] Data are taken from Reference [12].

3.3.6 Mean Free Path and Half-thickness

The attenuation properties of γ-rays in matter are sometimes quoted in terms of *mean free path* or *half-thickness* instead of the attenuation coefficient. The mean free path λ is defined as the average distance a photon travels in an absorber before it undergoes an interaction:

$$\lambda = \frac{\int_0^\infty x e^{-\mu x}\, dx}{\int_0^\infty e^{-\mu x}\, dx} = \frac{1}{\mu} \tag{3.19}$$

The half-thickness $x_{1/2}$ is defined as the average thickness in an absorber required to attenuate the beam to the half of its initial intensity, i.e. $I = I_0/2$. Inserting this in Equation (3.7) yields

$$x_{1/2} = \frac{\ln(1/2)}{\mu} = \frac{0.693}{\mu} \tag{3.20}$$

The mean free path and the half-thickness are given as reciprocals of the linear coefficient and thus have the dimension of length (cm), or mass thickness (g/cm^2), if the mass attenuation coefficient is used. Table 3.3 gives approximate values of μ, λ and $x_{1/2}$. All these are statistical and average values. True values of individual photons, which may be shorter or longer than these average values, are given by statistical distributions.

3.4 ATTENUATION COEFFICIENTS OF COMPOUNDS AND MIXTURES

3.4.1 The Attenuation Coefficient of Homogeneous Mixtures

The total mass attenuation coefficient of a homogeneous mixture of n elements can be found as

$$\mu_{\mathrm{M\ mix}} = \left(\frac{\mu}{\rho}\right)_{\mathrm{mix}} = \sum_{i=1}^{n} w_i \left(\frac{\mu}{\rho}\right)_i = w_1 \left(\frac{\mu}{\rho}\right)_1 + w_2 \left(\frac{\mu}{\rho}\right)_2 + \cdots + w_n \left(\frac{\mu}{\rho}\right)_n \tag{3.21}$$

when the *weight fractions* w_i and the mass attenuation coefficients $(\mu/\rho)_i$ of the different components in the mixture are known. The *volume fraction* of each component is equal to the product of the component's weight fraction and density: $\alpha_i = w_i \rho_i$. Equation (3.21) can thus be expressed in terms of the linear attenuation coefficient as

$$\mu_{\text{mix}} = \sum_{i=1}^{n} \alpha_i \mu_i = \alpha_1 \mu_1 + \alpha_2 \mu_2 + \cdots + \alpha_n \mu_n \tag{3.22}$$

3.4.2 The Linear Attenuation Coefficients of Chemical Compounds

The linear attenuation coefficient of a chemical compound can be determined from the chemical structure and density of the compound, and the mass numbers (atomic weights) A_i, the densities ρ_i and the linear attenuation coefficients μ_i of the n elements:

$$\mu_{\text{compound}} = \rho_{\text{compound}} \frac{\sum_{i=1}^{n} A_i x_i (\mu/\rho)_i}{\sum_{i=1}^{n} A_i x_i} \tag{3.23}$$

Here x_i is the number of atoms of the ith element in the compound molecule. As can be seen from Equation (3.10), there is no need to know the values of ρ_i and μ_i if the mass attenuation coefficients $(\mu/\rho)_i$ of the elements are known. Equation (3.23) is an approximation because it is assumed that each element has the same attenuation coefficient in the compound as if it were alone.

3.4.3 Attenuation in Inhomogeneous Materials

So far we have discussed attenuation in homogeneous materials. And also, if the material constitutes several components, we have assumed these are homogeneously mixed. In cases where this is not true we need to consider the attenuation along the path length l through the absorber:

$$I = I_0 \, e^{-\int_0^x \mu(l) \, dl} \tag{3.24}$$

This is then Equation (3.7) in a more general form. Two different cases help facilitate the consequence of this for a gauge based on attenuation (transmission) measurements: For an absorber constituting several layers of different materials and with a parallel beam incident perpendicular to these layers, the result of the integration in Equation (3.24) is similar to that given in Equation (3.22). Hence this case does not pose any problem for a radioisotope gauge based on measurement of μ_{mix}. The problem arises when the beam is incident parallel to the layers because the total attenuation then is the sum of the attenuations in the different layers, and not a product of these as in the former case. The expression of μ_{mix} in Equation (3.22) is thus not valid, and nor will it be for any intermediate cases between these two. We will study the consequences of this for transmission gauges later.

3.5 BROAD BEAM ATTENUATION

The narrow beam set-up in Figure 3.5 is referred to as *good geometry* and is seldom achieved in a realistic measurement system: a portion of the photons interacting outside the beam defined by the source/detector geometry is scattered towards the detector aperture

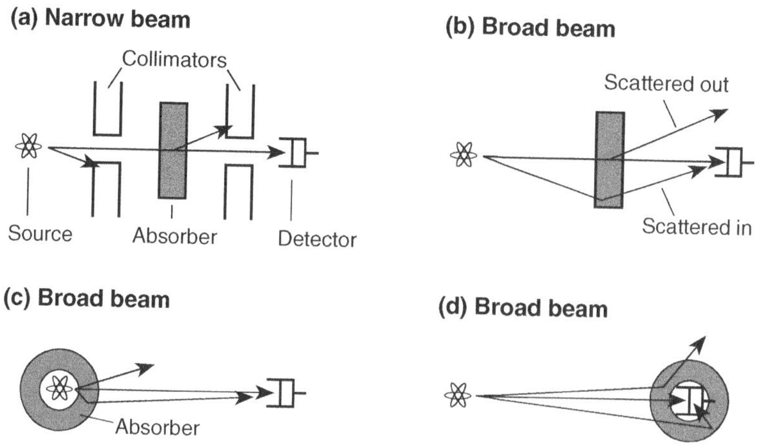

Figure 3.12 (a) Narrow beam or good geometry as assumed in Figure 3.5. Typical examples of broad beam attenuation or poor geometry is illustrated in (b), (c) and (d), but is often combinations of these

and contributes to the measured intensity. Whenever a significant fraction of the scattered or secondary photons can reach the detector, the arrangement is called *broad beam* or *poor/bad geometry* (see Figure 3.12).

3.5.1 The Build-Up Factor

The scatter contribution is known as *build-up*, and may be accounted for by introducing the *build-up factor* $B(\mu, x)$ into Equation (3.7):

$$I = B(\mu, x) I_0 \, e^{-\mu x} \tag{3.25}$$

Generally, $B(\mu, x)$ depends on the linear attenuation coefficient and its composition, and the thickness and geometry of the material in which the scatter is generated. This may be the shielding of the source and the detector, the measurement object itself, or even the source and detector encapsulation. The source of the build-up is normally Compton scattering. However, Rayleigh scattering is sometimes an equally important contributor to forward directed scatter, particularly for high-Z materials at low energies (see Figure 3.7).

In the case of good geometry $B(\mu, x)$ is unity. Otherwise it cannot be calculated analytically; it has to be determined from experiments, simulations or models thereof. Semi-empirical models of build-up are also implemented in software, allowing easy estimation of its magnitude in various situations [12]. Monte Carlo simulation is a powerful tool in accurate calculation of build-up and is being increasingly used. We will discuss this in Section 8.5. The build-up factor in lead for a point isotropic γ-ray source of energy E_γ is plotted in Figure 3.13. Data on build-up in water, aluminium and concrete are given in Reference [19].

3.5.2 Build-Up Discrimination

All Compton scattered events have, as discussed in Section 3.3.2, lower energy than do the incident events. This makes it possible to do some degree of scatter discrimination by

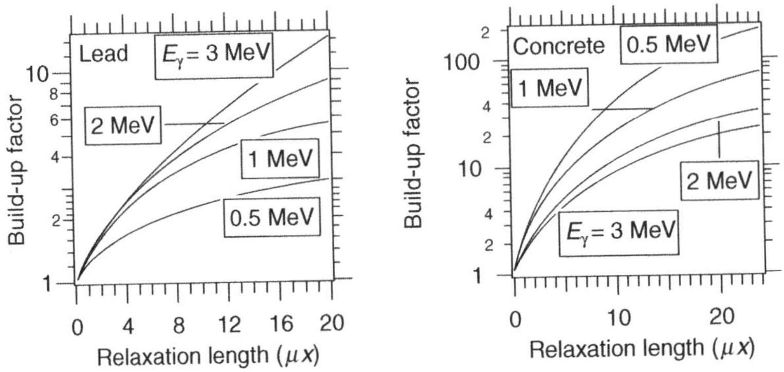

Figure 3.13 The build-up factor for lead and concrete as function of radiation energy and the μx-product, also known as the *relaxation length* [18]

using energy-sensitive detectors. This is, however, tricky at low energies where the energy difference between direct and scattered photons is low (see Figure 3.10); nor will Rayleigh scattered events be accounted for since these have no energy loss. This will be discussed in more detail in Chapter 5.

3.5.3 The 'Effective' Attenuation Coefficient

Another way of coping with build-up is to define an *effective linear attenuation coefficient* μ_{eff} such that

$$I = BI_0\, e^{-\mu x} = I_0\, e^{-\mu_{\text{eff}} x} \tag{3.26}$$

This implies that

$$\mu_{\text{eff}} = \mu \ln(B) \tag{3.27}$$

The effective coefficient is thus fairly complex as it also depends on the thickness of the absorber. But it turns out that this is a sufficiently accurate scheme in many measurement situations with a fixed set-up and where the measurement range boundary values of the attenuation coefficient are found through calibration.

3.6 NEUTRON INTERACTIONS

Neutrons carry no charge, nor are they electromagnetic radiation. They interact solely with the nucleus of the absorber, basically through collisions, i.e. scattering, or nuclear reactions. Scattering processes do not change the identity of the target whereas nuclear reactions do by adding an extra neutron to the target nuclei, which in turn often initiates secondary reactions. There are two types of scattering: *elastic* and *inelastic scattering*. With elastic scattering the sum of kinetic energies of the neutron and the target nucleus particles remains constant, whereas with inelastic scattering some energy is spent on excitation of the target nucleus. This will quickly de-excite by the emission of one or several characteristic γ-ray

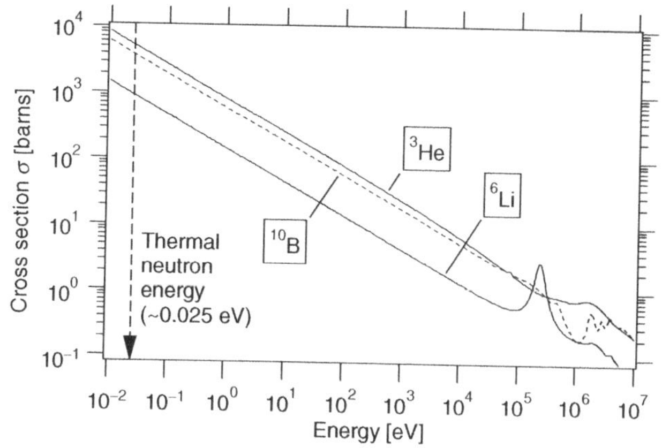

Figure 3.14 Atomic neutron absorption cross sections for ^{10}B, ^{3}He and ^{6}Li [20]

photons, so-called *prompt* γ*-rays*. Inelastic collisions are possible if the fast neutron energy is sufficiently high, but is far less probable than elastic scattering for light nuclei.

Elastic scattering occurs at all energies and is the mechanism by which energetic, so-called *fast neutrons* are slowed down until they approach thermal energy of about 0.025 eV. The succession of collisions is often referred to as *moderation* of the neutrons. This process is most efficient if the nucleus of the absorber is of equal mass to that of the neutron, because more energy can then be transferred per collision. The fractional energy transfer from the neutron to the target nucleus of mass M (atomic weight) averaged over all scattering angles is [15]

$$f = \frac{2M}{(M+1)^2} \tag{3.28}$$

Consequently hydrogen ($M = 1$) and hydrogen-rich materials, such as water, paraffin wax and polyethylene, are all efficient neutron *moderators*.

Moderated thermal or *slow neutrons*, also called *thermal neutrons*, generally are absorbed or *captured* in nuclei into which they diffuse. Depending on the type of absorber, this happens through a large set of neutron-induced reactions (see Section 2.3.2). The most probable in most materials is the (n, γ) reaction in which a γ-ray photon is emitted. In order to detect neutrons and utilise them in measurement systems, however, energetic electrons must be generated one way or another. This is possible with other reactions such as (n, α) and (n, p) because α-particles and protons ionise the detector material. Reactions utilised for neutron detection are $^{10}B(n, \alpha)^{7}Li$, $^{3}He(n, p)^{3}H$ and $^{6}Li(n, \alpha)^{3}H$. The total atomic absorption cross section of these elements is plotted in Figure 3.14. Another capture reaction sometimes used in detector systems is $^{157}Gd(n, \gamma)^{158}Gd$. Its thermal neutron capture cross section is among the highest found in any material – 49,000 barns [21]. This so-called *internal conversion* process is often (39%) followed by the emission of a *conversion electron* rather than a γ-ray emission. This may be considered as the nuclear equivalent to the atomic emission of Auger electrons. The utilisation of these reactions for neutron detection will be further discussed in Section 4.10.

The attenuation of a narrow beam of neutrons has strong analogy to the attenuation of γ-rays, as described previously. The beam intensity follows Lambert–Beer's exponential decay law:

$$I = I_0 e^{-\Sigma x} \tag{3.29}$$

and the attenuation coefficient Σ is called the *macroscopic cross section*. This is, like the photon attenuation coefficient, equal to the product of the atomic density and the atomic cross section, i.e. $\Sigma = N\sigma$. It can also be broken into components, yielding the contributions of the different interaction mechanisms, i.e. $\Sigma = \Sigma_{\text{scatter}} + \Sigma_{\text{absorption}}$. The former accounts for elastic and inelastic scatter, and the latter accounts for nuclear reactions and fission. Finally, the mean free path $l = \Sigma^{-1}$ is also used to characterise the attenuation properties of absorbers. A neutron beam is, however, very much a poor geometry case. Neutrons are moderated through a series of scatter interactions until they become thermal and start to diffuse randomly within the absorber. The practical use of neutron transmission is thus a lot more complex than that of photons. On the other hand, there are other ways by which neutrons can be utilised in measurement systems (see Chapters 5 and 7 for more about this).

3.7 EFFECTIVE ATOMIC NUMBER

In some cases it is very useful to know the 'effective atomic number' of mixtures or chemical compounds, for instance for evaluation of attenuation or stopping power properties. For attenuation of γ-rays this is particularly convenient because of the strong dependence of the photoelectric effect on the atomic number. In this context the effective atomic number has been defined as [22]

$$Z_{\text{eff}} = \left(a_1 Z_1^m + a_2 Z_2^m + \cdots + a_n Z_n^m\right)^{1/m} = \left(\sum_{i=1}^{n} a_i Z_i^m\right)^{1/m}$$
$$a_i = \frac{p_i(Z_i/A_i)}{\sum_{i=1}^{n} p_i(Z_i/A_i)} \qquad p_i = \frac{n_i A_i}{\sum_{i=1}^{n} n_i A_i} \tag{3.30}$$

where n_i is the number of atoms with atomic mass and number A_i and Z_i, so that p_i is the weight fraction of each element. The exact value of m depends on the γ-ray energy and is, according to Table 3.2, between 4 and 5 for the photoelectric effect. It is highest at low energies. Likewise, an effective atomic number can also be calculated with respect to attenuation by pair production using $m \approx 2$. Finally, it may also be calculated for the collision stopping power of charged particles, which, according to Equation (3.1), is proportional to Z. In this case ($m \approx 1$) a simpler expression may be used for effective atomic number [23]:

$$Z_{\text{eff}} = \frac{\alpha_1 Z_1^2 + \alpha_2 Z_2^2 + \cdots + \alpha_n Z_n^2}{\alpha_1 Z_1 + \alpha_2 Z_2 + \cdots + \alpha_2 Z_2} = \frac{\sum_{i=1}^{n} \alpha_i Z_i^2}{\sum_{i=1}^{n} \alpha_i Z_i} \tag{3.31}$$

where α_i is the atomic fraction of each element, i.e. the number of atoms of an element divided by the total number of all atoms present.

3.8 SECONDARY ELECTRONS

We have seen in this chapter that interactions of ionising photons, charged particles or neutrons very often result in a chain production of radiation as listed in Table 3.1. The number of stages in this chain depends on the radiation type and energy, but is generally unpredictable because of the random nature of radiation transport in matter. Nevertheless, we know that at some stage energetic electrons are produced. These are known as *secondary electrons*, but are sometimes referred to as *δ-rays*. These play an important role in all radiation detector principles because the detector performance relies on efficient conversion of the secondary electron energy into electric charge, which can be sensed by the read-out electronics.

4

Radiation Detectors

The sensing element in nuclear measurement systems is commonly referred to as the radiation detector, and not the sensor as otherwise referred in measurement science. This does not comply to the 'International vocabulary of basic and general terms in metrology' [24], but we will use 'detector' because this is the established term. The radiation detector may in its simplest form be considered as a unit that converts radiation energy into an electronic signal. There are also detectors based on blackening of plastic films (e.g. in dental X-ray radiography) and phenomena like thermoluminescence or cloud generation, but these will not be dealt with here.

4.1 PRINCIPLE OF OPERATION

Detection of radiation is closely related to the absorption of radiation dealt with in the previous chapter. Any of the electromagnetic interaction mechanisms except for Rayleigh scattering generate secondary electrons such as photoelectrons, Auger electrons, Compton recoil electrons and positrons in the case of pair production. These ionise and excite atoms along their track in the absorber, in this case the detector. Their range is typically a few millimetres in gaseous absorbers and a few micrometres in solid-state absorbers, see Figure 3.2. Thus, from a practical point of view, their total energy is regarded as deposited on the spot where the interaction took place. There are different detector principles; however, what they all have in common is that it is the energy of the secondary electrons that is detected. This means that the total energy of an interacting γ-photon, an *event*, can be detected only in cases where all its energy is transferred to electrons. The energy of scattered photons, fluorescence photons and so forth, is lost unless these undergo further interactions in the detector. If they do, no radiation detector is capable of separating charge from such secondary interactions from the initial one; it is all recognised as one event. In summary a radiation detector may be considered to have a threefold function, namely

1. *stop γ-photons* in their active volume
2. *convert the energy* of each of these photons to energy of secondary electrons and finally
3. *collect or sense the charge* generated by these electrons.

Radioisotope Gauges for Industrial Process Measurements. Geir Anton Johansen and Peter Jackson.
 ISBN 0-471-48999-9

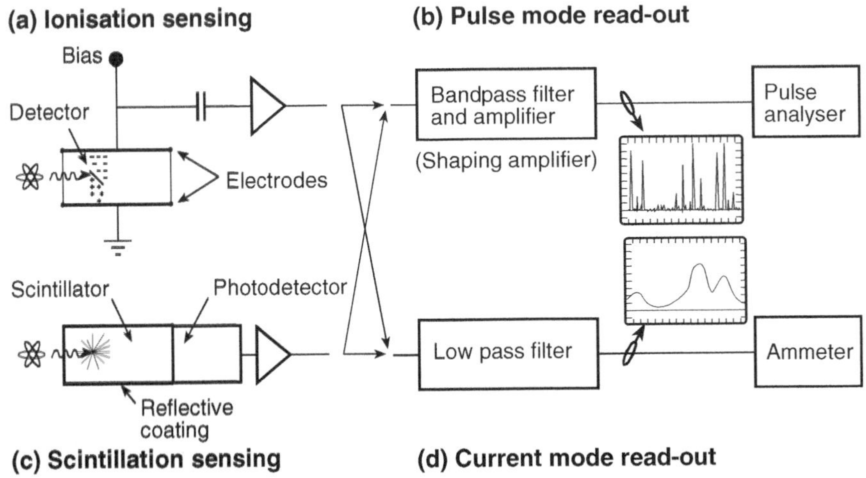

Figure 4.1 The two categories of radiation detectors: those sensing the ionisation of the detector material directly (a) and those sensing the scintillation light generated by the ionisation (c). Both may be connected to either a pulse mode (b) or a current mode (d) read-out system. In pulse mode, the pulse width is typically between 0.1 and 10 μs; however, for each particular system it is constant and independent of the pulse height

Regardless of detector type, the first two functions are mainly determined by the absorbing material's attenuation coefficient and its composition. This is because there are normally practical restrictions as to how thick a detector can be. The efficiency of the third point, however, depends very much upon the type of detector.

One category of detectors collects and senses the charge directly through an electric field across the absorbing material. These are the *ionisation sensing* detectors. In this category we find gaseous detectors and semiconductor detectors, see Figure 4.1a for principle of operation. A voltage referred to as *bias* sets up the field, which causes the electrons and ions (or holes) to be separated and swept towards their respective electrodes where they are collected. The other category of detectors uses a scintillator or scintillation crystal as absorbing material. These are the *scintillation sensing* detectors (illustrated in Figure 4.1c). The scintillator generates rapid flashes of light when ionised and excited absorber atoms de-excite. This scintillation light is in turn directed towards a photodetector where it is detected. There are a variety of photodetectors available, but their common mission is to convert light to an electric charge signal.

This means that both detector categories produce a charge output with some relationship to the energy initially deposited in the detector. There are two distinct alternatives for further processing of this charge signal: *current mode* and *pulse mode read-out*. In current mode the detector is connected to an ammeter or equivalent circuitry that has a long time constant compared to the response time of the detector. It thus measures the average energy deposition in the detector (see Figure 4.1d). In pulse mode, the total charge resulting from each event is processed separately: The detector charge signal is often integrated, then amplified and filtered so that a series of events produce a pulse train at the output of the detector electronics, as illustrated in Figure 4.1b. In most detector systems the amplitude of the output signal is proportional to the energy deposited in the detector.

Pulse mode processing thus gives information on both timing and energy of the individual events. Further information like intensity and energy spectrum of the incident beam is then readily available. The pulse analyser may be anything from a system counting pulses above a certain threshold to a *multichannel analyser*, which most commonly is used as a *pulse height analyser* (PHA) to sort the pulses according to their magnitude. We will take a closer look at read-out electronics in Section 5.1.

The difference in the output signals, as they typically would appear on an oscilloscope with identical time axis, is also illustrated in Figures 4.1b and 4.1d. The information content associated with pulse mode is higher than that of current mode; however, the cost is more sophisticated read-out electronics. Pulse mode read-out is most commonly used with radioisotope sources because of the relatively limited radiation intensities produced by these. On the contrary, current mode read-out is the only option for X-ray tubes and high-intensity beams: No radiation detector is capable of distinguishing signals from successive events if these are too close in time, as they would be with high-intensity beams. Nevertheless, the current mode output signal is proportional to the average energy deposition in the detector, which is further proportional to the radiation intensity, and this often is the only information required in many situations.

There is one additional way of categorising radiation detectors: those with internal gain and those without. This is particularly important for pulse mode read-out systems, which by far is the most common for permanently installed gauges. As will be shown later the initial charge liberated by a γ-photon interacting in a detector is relatively small. This charge signal thus needs to be amplified in the read-out system. Some detectors (and photodetectors) have build-in charge amplification so that their output signal has a relatively high signal-to-noise ratio (SNR). Detectors without gain are dependent on a high performance preamplifier to achieve the required signal amplitude and SNR. These detectors thus need to be considered as a system consisting of the detector and the preamplifier.

4.2 DETECTOR RESPONSE AND SPECTRUM INTERPRETATION

Before discussing different types of radiation detectors, it would be very useful to study how they respond to γ-ray exposure when operated in pulse mode. This basically means a consideration of their threefold function: to stop the radiation, convert the radiation energy to secondary electron energy and finally to collect or sense the charge generated by the secondary electrons.

4.2.1 Window Transmission and Stopping Efficiency

Every radiation detector has an active volume where every radiation interaction contributes to the output signal. Interactions outside, such as in the walls, do not contribute unless secondary radiation from these reaches the active volume. The detector wall or encapsulation facing the radiation beam is often referred to as the *entrance window* or simply the window. Some detectors also have additional windows to shield or protect the detector against external interference from light, pressure, etc. Inevitably, some of the radiation beam energy is lost in this window. In a charged particle beam every particle looses some

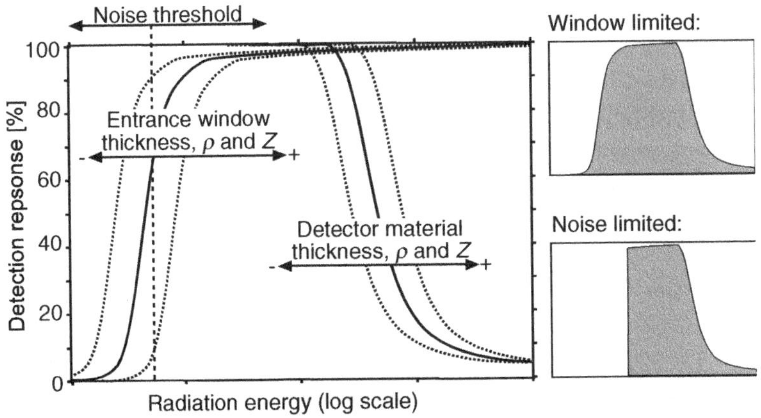

Figure 4.2 Spectral γ-ray response of a radiation detector, relative to the incident beam intensity. The high-end drop-off is determined by the stopping efficiency of the active volume. The low-end response is limited either by attenuation in the entrance window or the noise threshold in the detector, depending on which is the highest. This is illustrated in the right-hand side of the figure; the noise threshold is less than the window drop-off (see top) and vice versa (see bottom)

of its energy, whereas a photon beam will be attenuated causing the intensity to be reduced. Using Equation (3.7), the transmitted intensity fraction of a narrow beam will be

$$\frac{I}{I_0} = e^{-\mu x} \tag{4.1}$$

where x is now the window thickness. A typical plot of this as a function of energy is shown in Figure 4.2.

The magnitude of the window attenuation depends on the window's thickness, density and atomic number. If these increase, so does the attenuation, as indicated in Figure 4.2.

In some cases it is not the window attenuation that limits the low energy detection limit, but the noise level. All detector systems exhibit some degree of noise as will be explained later. It is not unusual to find that the signal produced by low-energy photons is buried in the system noise, that is, the signal-to-noise level is less than unity. In these cases the low-end response is noise limited rather than window limited as demonstrated in Figure 4.2.

Every charged particle entering the active volume of a detector produces a pulse at the detector output. This is not the case for ionising photons that may penetrate thick absorbers without any interaction at all. For a γ-ray beam it is, therefore, convenient to introduce the *radiation stopping efficiency*, also known as the *detection efficiency*, which expresses the radiation detector's ability to attenuate the beam. It is simply defined as the ratio of the number of photons interacting in the detector, and thus producing a pulse at its output, to the number of photons incident to the detector. The detection efficiency of a narrow beam may thus be expressed as

$$\frac{I_0 - I}{I_0} = 1 - \frac{I}{I_0} = 1 - e^{-\mu x} \tag{4.2}$$

It is, as may be expected, dependent on the attenuation coefficient and through that on the radiation energy and the density and atomic number of the detector material. Further, to the first approximation it depends on the thickness of the detector in the direction of the incident beam. But as will be shown in the succeeding sections, it is also affected by the detector volume. Consequently, the high-end energy response of the detector is limited by the stopping efficiency as illustrated in Figure 4.2. The detection efficiency concept also applies to neutron beams, which like photons are neutral particles with a certain interaction cross section, see Equation (3.29).

For γ-ray beams it is also common to talk about the *total efficiency* and the *peak efficiency*. The former means that all events producing a pulse on the detector output are counted, regardless of whether their full energies are deposited in the detector or not. For the peak efficiency, only events depositing their full energy in the detector are counted. The meaning of this will be clarified in Section 4.2.2.

4.2.2 The Noiseless Detection Spectrum

The next step in the detection sequence is to convert the radiation energy of the interacting photon or particle to secondary electron energy. The whole process, from the first interaction to the energy deposition of all secondary electrons, happens within a very short time, typically a few picoseconds. The detector thus recognises this as one event, i.e. it all contributes to the same output signal.

Let us first study how a beam of mono-energetic γ-photons interacting in a single-element detector gives rise to a characteristic spectrum of detected energy, as illustrated in Figure 4.3. This spectrum may be regarded as a histogram sorting a large number of output signals from a pulse mode system according to their amplitude (which is proportional to energy deposition). Effects from statistical fluctuations in energy deposition and charge collection, electronic noise, etc. are ignored. It is also assumed that the kinetic energy of all secondary electrons is fully absorbed in the detector.

Ideally one would like this spectrum to solely contain one single line corresponding to the energy of the incident γ-photons, the so-called *full energy peak*. The detector system would in other words output the full energy of each and every detected event. In reality, however, the escape of secondary photons, that is scattered photons, fluorescence photons, etc. gives rise to a detection spectrum as shown in Figure 4.3. Its composition is evident from the possible energy depositions resulting from each interacting mechanism:

In photoelectric interactions succeeded by fluorescence there is a possibility that the characteristic X-ray escapes the detector. This gives rise to the *X-ray escape peak*. Its energy equals the kinetic energy of the photo-electron as given in Equation (3.14). The height of this peak depends on the fluorescence yield which, as can be seen in Figure 3.8, increases with the atomic number. It is also influenced by the degree of re-absorption.

The situation is more complex with Compton scattering: The energy of the scattered photon, which may escape the detector, depends on the scattering angle. This is evident from Equation (3.15) and Figure 3.10. The maximum energy transfer to the recoil electron, $E_{kin,m}$, takes place at head-on collisions where the photon is scattered 180° (backwards). This gives rise to the *Compton edge* in the spectrum. Its energy, which is independent of

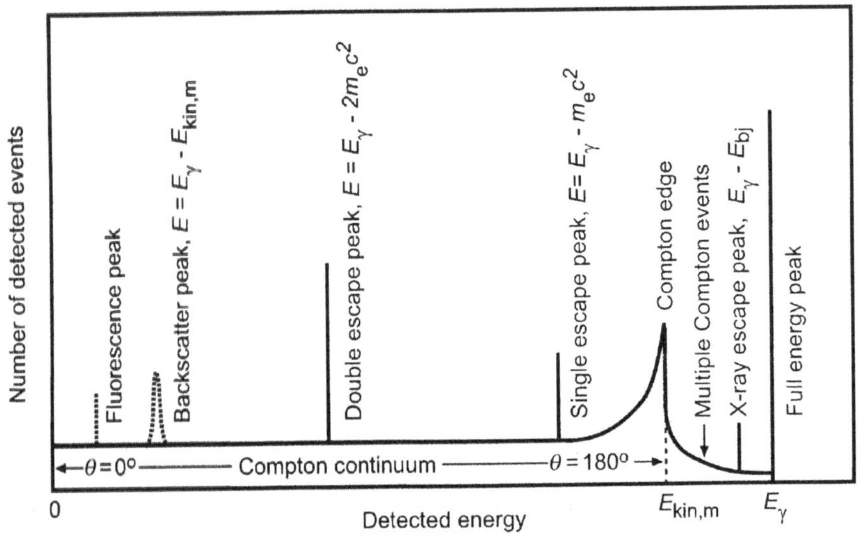

Figure 4.3 Noiseless detector energy spectrum produced by monochromatic photons of energy $E_\gamma(>2m_ec^2)$. The single and double escape peaks may also have associated X-ray escape peaks and Compton edges not shown here. The dashed peaks are due to secondary radiation from interactions in the detector surroundings, such as the housing

the absorber properties, is found by inserting $\theta = 180°$ in Equation (3.16):

$$E_{\mathrm{kin,m}} = \frac{E_\gamma}{1 + \frac{m_ec^2}{(1-\cos(180°))}} = \frac{E_\gamma}{1 + (1/2m_ec^2)} \tag{4.3}$$

Below the Compton edge there will be a continuous energy spectrum, the *Compton continuum*, from recoil electrons with lower energies, i.e., from interactions where the incident photon are scattered at smaller angles ($0° < \theta < 180°$). The peculiar shape of the Compton continuum and edge is explained by the angular distribution of scatter [Equation (B.10)] and its energy distribution [Equation (3.15)]. Multiple Compton interactions from one and the same initial event result in energy depositions between the Compton edge and the full energy peak. This explains the top-end tail of the Compton edge in Figure 4.3.

If the scattered photon undergoes further interactions in the detector there is a possibility that the total energy of the initial event is detected, thus making a contribution to the full energy peak in the spectrum.

Rayleigh scatter does not contribute to the detection spectrum because of the negligible energy transfer involved in this process. The effect of interaction by pair production is explained by Equation (3.18). Annihilation radiation escaping the detector gives rise to two peaks: the *single-* and *double-escape peaks* emerging when one and both annihilation photon escapes, respectively.

The effect of multiple Compton interactions is shown in Figure 4.3; however, there are several other multiple interaction possibilities not shown in the figure, including combinations of interaction mechanisms. The annihilation escape peaks, for instance, also have associated X-ray escape peaks and Compton edges. The degree of multiple interactions is closely related to the attenuation properties of the detector, the interaction position in

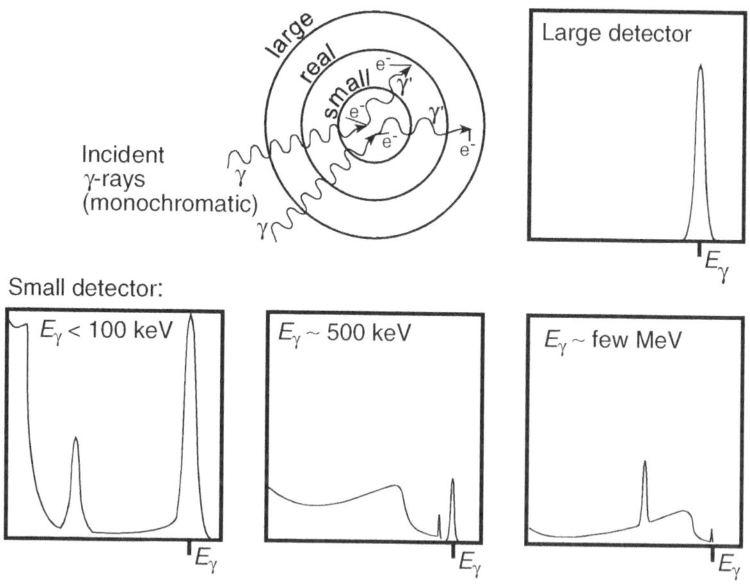

Figure 4.4 Illustration of the three detector models as circles with increasing diameters. All of the incident monochromatic γ-rays interact in the centre. All secondary photons escape the small detector; some of them interact in the intermediate (real) detector, whereas everything is detected in the large detector. The detection spectrum of the latter thus contains only the full energy peak (top). In the small detector extreme the appearance of the detection spectrum very much depends on the energy of the incident γ-photons (bottom). These spectra are only illustrative

the detector volume and the detector size. A photon, which undergoes multiple Compton interactions, will gradually loose energy and increase the possibility of photoelectric absorption to take place. Full energy deposition is most likely at low energies where the mean free path is smallest. In most cases photoelectric absorption is the predominant contributor to the full energy peak. For this reason it is often called the *photopeak.*

4.2.3 Detector Models

The fact that the exact appearance of a detection spectrum depends very much on the absorption properties of the detector and the γ-ray energy makes it difficult to interpret it. A common method to illuminate these rather complex issues is to consider the detector extremes:

The *small detector* by definition is so small that all secondary photons escape the detector. The other extreme is the *large detector*, which by definition is large enough to absorb all radiations generated by the initial interacting γ-photon. This is illustrated in Figure 4.4 by using circular representation of the detectors. These are exposed to monochromatic γ-photons that interact in the centre of the detectors.

The large detector's response is simple since the full energy of all events is detected. A typical real detection spectrum of a large detector is shown in Figure 4.4. Also shown are typical spectra of a small detector at three different energies. All energy of secondary photons is lost in this case. At the highest energy, where pair production is possible,

Compton interaction is still the dominant mechanism. There is virtually no photoelectric absorption, which in the small detector case is the only contributor to the full energy peak. There are some pair production interactions. However, both annihilation photons escape the detector and give rise to just one peak: the double escape peak. At intermediate energies, where pair production is no longer possible, the contribution of photoelectric absorption starts to increase. In addition to the full energy peak, this also gives rise to the X-ray escape peak. The relative height difference of these peaks is determined by the fluorescence yield (see Figure 3.8). In low-Z detectors there is virtually no fluorescence whereas the opposite is true for high-Z detectors. The relative separation between the Compton edge and the full energy peak increases with decreasing energy. This is because the relative energy transfer to scattered photons increases (see Figure 3.10).

The relative importance of the different interaction mechanisms is also dependent on the atomic number of the detector. This can be readily seen from Figures 3.6 and 3.7. Low-Z detectors may virtually have no full energy peak at all, whereas this often dominates the detection spectrum of high-Z detectors. The latter thus brings us closest to the ideal situation with one peak in the spectrum. On the other hand, high-Z detectors have a drawback in that the emission of characteristic X-rays increases with Z. The ratio of the area below the full energy peak to that of the full spectrum is called the *photofraction*.

A real detector is an intermediate size detector somewhere in between the two extremes. In most real cases some of the secondary photons are absorbed and some are not. In spite of the complexity added by this, the small detector approach is very often a useful first step in interpreting a real detection spectrum. Actually the spectrum in Figure 4.3 is representative of a real detector if spectral distortions are taken into account, see next section.

4.2.4 The Real Detection Spectrum

In a real spectrum, *noise* and *statistical fluctuations* in energy deposition and charge (or light) collection in the detector make single energy lines appear as distributions. Noise has been added to the peaks shown in Figure 4.4. This also affects the Compton edge, which is not as sharp as indicated in Figure 4.3. This is also smeared out by another effect; the binding energy of the recoil electron cannot always be neglected as it is in Equation (4.3). In some cases there is partial energy loss because some secondary electrons escape the active detector volume before they are slowed down. This *electron leakage* is most likely either at high energies where the electrons have longer range or with small volume detectors and low-density (Z) detectors or combinations of these. Finally, for energetic secondary electrons some energy is lost to bremsstrahlung, see Figure 3.1. The influence of these effects are thus most pronounced at the high end of the spectrum.

So far we have considered only single-element detectors and monochromatic radiation. With *multiple-element detectors*, such as compound semiconductor detectors, each chemical element has its associated X-ray escape peak. Needless to say, *polychromatic* incident radiation makes the spectrum a lot more complex. Further, even though a radioisotope source has monochromatic emission, the radiation entering the detector may constitute several energies because of the following:

- Many γ-ray sources are β^--emitters. These are encapsulated to absorb all the energy of the β^--particles and this will generate some bremsstrahlung, particularly in the case of high energy β^--particles.
- Fluorescence generated in the surrounding material illuminated by the beam is also a common interference in detection spectra (see Figure 4.3). The most prominent fluorescence sources are high-Z materials because of their higher photoelectric absorption and fluorescence yield. In this connection these are also most problematic because their emission energy is high enough to interfere with the measurement. Lead is a common fluorescence source because it is often used as collimator material for the source and the detector. Fluorescence may be suppressed by using *graded shielding* (this will be explained in Section 5.4.2).
- Scatter from surrounding material also often interferes with the measurement. Scatter originating from behind the detector gives rise to the *backscatter peak* (illustrated in Figure 4.3). Actually, scatter is likely to occur over a broad energy range in the detection spectrum, but the backscatter peak is so distinct because there are only marginal changes in the energy of the scattered photon for scattering angles above about 150° (see Figure 3.10). All photons scattered right behind the detector in a relatively large solid angle thus have about the same energy. The detector housing is a common scatter source. The effect of scatter may be reduced by careful design of the system: One should prevent unnecessary illumination of material by proper collimation, use high-Z materials rather than low-Z ones in housing, etc. so as to increase the probability of full absorption, and finally use efficient detector collimation. But then again, this may increase fluorescence background.
- *Background radiation* from naturally occurring isotopes, cosmic radiation, etc. may also be a problem, particularly in low-level applications. This means that the measured intensity is very low and may be comparable with the background intensity. This is seldom the case for permanently installed gauges; however, the treatment is normally to measure the background contribution and correct for it.
- *Electronic distortion* needs mentioning in this connection because in some situations it gives rise to severe artefacts in the detection spectrum. It has a variety of origins, but is often indirectly related to the radiation beam properties. All pulse mode systems will for instance start to malfunction if the intensity increases beyond the limit it is designed for. Electronic distortion will be treated in Section 5.1.3.

Regardless of all these effects, in most cases where the incident photons have one or a few energies, it is possible to recognise the main characteristics of the detection spectra as outlined in Figure 4.3. A good understanding of the interaction mechanisms, particularly the photoelectric effect and Compton scattering, is crucial for interpreting the response function or detection spectra of radiation detectors. A typical spectrum acquired with a NaI(Tl) scintillation detector exposed to 661.6 keV γ-rays and 32 keV characteristic X-rays is shown in Figure 4.5.

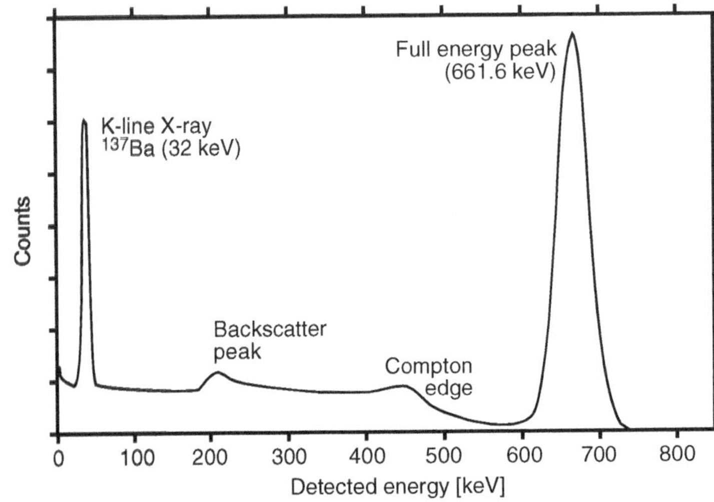

Figure 4.5 Experimental pulse height spectrum of ^{137}Cs 662 keV γ-rays collected with a relatively large volume (Ø = 75 mm × 75 mm) NaI(Tl) scintillation detector

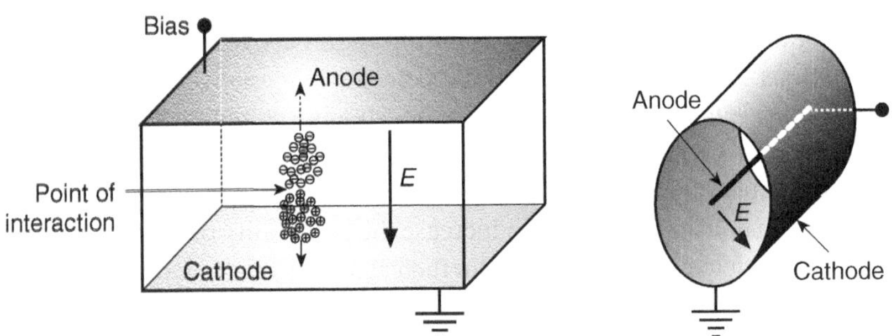

Figure 4.6 The two most common geometries for ionisation sensing detectors: the planar or parallel plate (left) and the coaxial or cylindrical (right). The electric field strength is uniform in the former whereas in the latter it increases towards the anode

4.2.5 Signal Generation in Ionisation Sensing Detectors

An ionisation sensing detector is basically an absorbing medium in between two electrodes: one anode and one cathode, as illustrated in Figure 4.6. Secondary electrons from ionising radiation interactions create *charge carrier pairs* along their track when they are slowed down in the absorber. These pairs are swept towards their respective electrodes by the electric field set up by the bias. This motion adds a transient field to the applied stationary electric field. This transient field, often referred to as the *weighting field*, is sensed by the capacitive coupling of the read-out electronics. The induced signal thus depends on the detector geometry and the efficiencies by which charge carriers are created and transported to their respective electrodes. The two most commonly used ionising detector geometries are shown in Figure 4.6.

There are two types of ionisation sensing detectors: *gaseous detectors* with *electron–ion pairs* as charge carriers and *semiconductor detectors* with *electron–hole pairs* as charge carriers. Although these detector types are very different in many ways and need separate treatment (see Sections 4.4 and 4.5), they also have some basic properties in common. These are discussed in this section.

The first is the average energy, w, required to create one charge carrier pair in the absorber. As a rule of thumb this is about 30 eV for gaseous and 3 eV for semiconductor detectors. This is more than is required to ionise one of the absorber molecules. Some of the incident energy is 'lost' to heat and by other mechanisms such as excitation of electrons to a higher state.

The total number of charge carriers generated by an interacting photon is approximately equal to the energy deposited in the detector divided by w. For the output signal properties, however, the fluctuations in this number are equally important as the number itself (see Section 5.3.6).

The next issue is the *charge collection efficiency*, which basically expresses the proportion of liberated charge carriers collected at the electrodes. This is very much a question of how long the *charge collection time* is compared to the average *carrier lifetime*. The free carriers may disappear or become neutralised for several reasons: For gases it may happen through recombination when a free electron and an ion collide. Further, in so-called electronegative gases there is a probability that a free electron may become attached to a neutral molecule forming a (slow) negative ion. In the case of semiconductor detectors, free carriers may disappear through trapping as well as recombination. Trapping means that the electrons or holes are temporarily caught by impurities in the semiconductor crystal lattice. In summary, the average charge lifetime is a statistical quantity, which ideally should be as large as possible.

The charge collection time, τ_C, is the time it takes for the charge carriers to migrate from their point of generation to their respective electrodes. It should ideally be as short as possible to prevent loss of carriers. Its maximum value is the cathode–anode separation divided by the drift velocity, v, which may be expressed as

$$v = \frac{\mu E}{p} \text{ for gases} \quad \text{and} \quad v = \mu E \text{ for semiconductors} \tag{4.4}$$

where μ is the *mobility* of the charge carriers, E the strength of the electric field and p the pressure of the gas (which controls the number of molecules per unit volume). The maximum charge collection time is thus as given in Table 4.1.

The only inherent material property affecting the charge collection time is therefore the mobility of the charge carriers. In gases the mobility of electrons are, due to their much lower mass, typically three orders of magnitude greater than that of ions. Typical charge collection times are in the microseconds and milliseconds region for electrons and ions, respectively. As for the effect of the electrical field there is a saturation effect in the case of electrons in some gases: Equation (4.4) is valid only up to a certain value of the E/p ratio where v starts to flatten, and in many cases even decrease if E/p is further increased. The exact value and behaviour depends on the gas.

There is a similar saturation effect for both types of charge carriers in the case of semiconductor detectors: The velocities increase with the electric field up till a certain value where it starts to become independent of the field. In a semiconductor material, however, the mobility of electrons and holes is in the same order of magnitude.

Table 4.1 Maximum charge collection times for planar and coaxial gaseous and semiconductor detectors

		Charge collection time, τ_C	
Geometry	Electric field, E	Gaseous detector	Semiconductor detector
Planar	$\frac{V}{d}$	$\frac{d}{v} = \frac{dp}{\mu E} = \frac{d^2 p}{\mu V}$	$\frac{d}{v} = \frac{d}{\mu E} = \frac{d^2}{\mu V}$
Coaxial	$\frac{V}{r\ \ln(r_C/r_A)}$	$\frac{r_C}{v} = \frac{r_C p}{\mu E} = \frac{r_C^2 \ln(r_C/r_A) p}{\mu V}$	$\frac{r_C}{v} = \frac{r_C}{\mu E} = \frac{r_C^2 \ln(r_C/r_A)}{\mu V}$

Note. d is the cathode–anode separation for the planar detector whereas r_C is the inner radius of the cathode (cylinder) and r_A is the anode radius for coaxial detectors. Further V is the high voltage or bias setting up the electric field

But, in contrast to gases, it may be very different from one material to another. Because the mobility generally is greater in semiconductors than in gases, the charge collection time is less and in some cases typically in the order of 10 ns.

For semiconductor detector materials it is common to specify the product of the carrier lifetime and the mobility, $\mu_e\tau_e$ for electrons and $\mu_h\tau_h$ for holes, since these are the two inherent material properties affecting the charge collection efficiency. Typical values for common materials are listed in Section 4.5.

The net motion of charge carriers is caused by the combination of electric field drift and *diffusion*. The latter is a random thermal motion of the carriers away from regions of high carrier density. The effect of diffusion is some spread in the collection time and arrival position (at the electrode) for carriers originating from the same point. Diffusion may normally be neglected for small volume detectors and in some cases its effect is also negligible compared to the spread in the carrier generation position. On the other hand, diffusion may have a significant effect in large volume detectors or when precision timing or position measurements are required. The latter applies to position sensitive detectors.

In summary, the charge collection time and efficiency are very important because they define several of the major properties of a detector: Its speed of response and through that the maximum detection frequency (count-rate, n), the precision by which the detector can measure the energy, timing and position of the interacting events (energy, temporal and position resolution), and finally the size and volume of the detector, and through that also its detection efficiency.

4.2.6 Signal Generation in Scintillation Sensing Detectors

In scintillation sensing detectors the energy of the ionising radiation is converted into energy carried by a certain number of photons in the visible or ultraviolet region. The term charge collection thus does not apply to the scintillation process. Here it is more convenient to study how efficiently the energy of the secondary electrons is converted to scintillation photon energy, and how efficiently this energy is transported to the light detector. The second part which needs consideration is the light detection process and its efficiency. Here the scintillation photons are converted into charge carriers which in most light detectors are fed through an amplification stage before they are collected. Let us first study the scintillation process and its general properties and then move on and do the same with the light detector. The most relevant scintillators and light detectors in the context of industrial measurement are presented in Section 4.6.

As mentioned in the previous section, part of the energy of the secondary electrons is spent in excitation of the absorber molecules. In contrast to ionisation sensing detectors where this energy is considered as a loss, this is the energy that is utilised in scintillation detectors. The *scintillation efficiency*, Q_C, is defined as the efficiency with which secondary electron energy is converted to scintillation photon energy. It is also referred to as the scintillation *light output* or *light yield*, and is then often quoted in terms of number of scintillation photons per MeV incident radiation energy. For typical scintillation materials Q_C is between a few percent up to about 12% at the best. Ideally it should be higher, but unfortunately there are alternative de-excitation mechanisms to photon emission; energy is also lost to vibrations or heat. Another important feature of scintillation materials is their transparency to the scintillation photons. These travel at the speed of light from their point of origin through the material, and for the principle to work they need to reach the light detector with minimal loss. Hence, the materials of choice are either *scintillation crystals*, so-called *inorganic scintillators*, or transparent *plastic scintillators*, so-called *organic scintillators*. There are also liquid and gaseous scintillators; however, these are less relevant for use in industrial gauges and will not be discussed here.

Unfortunately all the excited molecules do not de-excite instantaneously. The development of the scintillation flash is shown in Figure 4.7; there is a rapid increase to a maximum emission and then an exponential like decay with time constant, τ_D, called the *decay constant*. This varies from a few nanoseconds for the fastest scintillators to a few microseconds for the slowest. The decay constant is important because it is the fundamental limitation to the detection rate or count-rate in pulse mode systems. In some

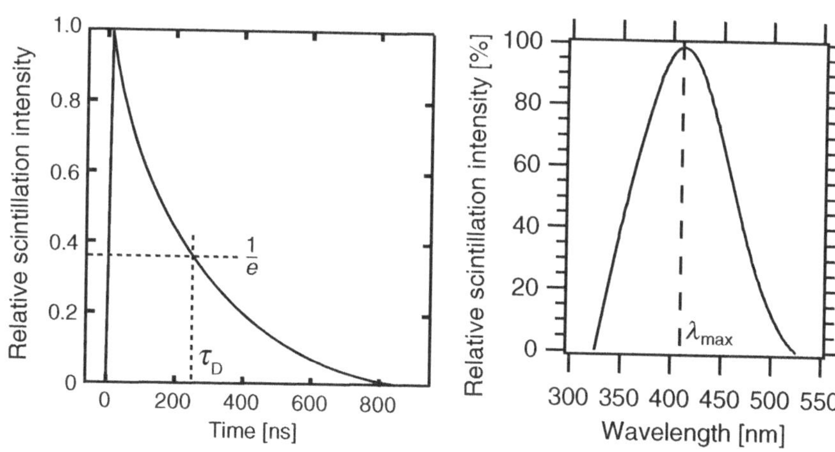

Figure 4.7 Illustrations of typical scintillation signal decay (left) and scintillation emission spectrum (right). Some scintillators have more than one decay constant

scintillators there is also the so-called *afterglow* from transitions of long-lived excitation states. Afterglow may last for several hundred milliseconds and constitutes a significant fraction of the total light output. This may be a problem since it effectively increases the background. In correct terminology the direct emission is known as *luminescence* and afterglow as *phosphorescence*.

The *scintillation emission spectrum* is another important property that is specific to every scintillator. A typical emission spectrum is shown in Figure 4.7. It is characterised by a wavelength of maximum emission, λ_{max}, but its shape need not be symmetric. The *detection spectrum* or *spectral response* of the light detector has to match the emission spectrum to avoid signal loss. We will discuss this in more detail in Section 4.6 where the most relevant light detectors are also presented. The purpose of the light detector is to convert the scintillation light energy to an electrical pulse. This conversion is characterised by the *quantum efficiency*, Q_E, which is defined as the number of electrons produced in the conversion process per incident scintillation photon. Based on the above we can now calculate the average energy deposition in the scintillator, w, required to generate one electron in the light detector:

$$w = \frac{hc/\lambda_{max}}{Q_C Q_E (1 - L)} \tag{4.5}$$

where the expression in the numerator is the average energy of the scintillation photons. This is around 3 eV for a typical scintillator where Q_C and Q_E typically are in the excess of 10% when used with a typical light detector. Some fraction, L, of the scintillation signal is lost for various reasons. Altogether, as a rule of thumb, w is about 300 eV for scintillation detectors compared to about 30 and 3 eV for gaseous and semiconductor detectors, respectively. We shall see in Section 4.6 that this number is very approximate for scintillation detectors because it depends so much on the exact configuration. It is nevertheless quoted because of its importance to the peak broadening in the detection spectrum. This affects the detector's ability to resolve close incident radiation energies (see Section 5.3.6).

The loss fraction L needs some further comments because some fraction of the scintillation light signal will be lost for several reasons: All scintillators exhibit some degree of self-absorption. Its influence increases with the volume of the scintillator, but it is a small problem compared to the incomplete charge collection in many ionisation sensing detectors.* Further, there is some loss due to imperfect reflections in the scintillator walls even though these have special coatings to optimise reflection. Then comes loss due to unwanted reflections in the interface between the scintillator and the light detector. This is minimised by matching the refractive indices and using optical compounds. Finally, there is some loss due to spectral mismatch between the scintillator and the light detector. In many cases the latter is severe and the major loss contributor. For standard size crystals (except long bars) the losses due to imperfect light collection are negligible because of very efficient reflectors. We will come back to these issues when dealing with different types of scintillators and light detectors in Section 4.6.

*Actually, efficient signal transport is the strength of scintillation detectors: High stopping efficiency is achieved exactly by the combination of large volume and solid-state absorber.

4.3 PURPOSES AND PROPERTIES OF DETECTOR SYSTEMS

We have seen how radiation detectors operate. Before proceeding with a more detailed presentation of the different detector types and their characteristics, it is useful to consider which properties are important when selecting the detector for the application. These are the type of properties that appear in data sheets of detector systems, but which are based on the more fundamental properties presented in the preceding sections. A pulse mode operated detector system may be used to measure one or several of these fundamental quantities:

1. The energy of each interacting photon (e.g. for spectrometry).
2. The time at which each photon interacts (e.g. for coincidence).
3. The interaction position of each photon (e.g. for imaging).
4. The radiation beam intensity (e.g. for transmission measurement).

The question is then how good is a detector system in fulfilling its purposes, and how can this be quantified. For the first three points this is done by specifying the detector system's ability to resolve energy, time and position of the interacting events. Bear in mind that in most cases we have to consider the whole detector system, that is the detector and its associated electronics.

4.3.1 Energy, Temporal and Spatial Resolution

The *energy resolution* of a radiation detector system expresses its ability to resolve radiation energy. Consider a case where a large number of mono-energetic (E_0) γ-photons interact and deposit their the full energy in a detector. Ideally the detection spectrum should then contain one single line; the full energy peak as discussed in Section 4.2.2. But in Section 4.2.4 we say that noise causes this peak to be smeared out as illustrated in Figure 4.8.

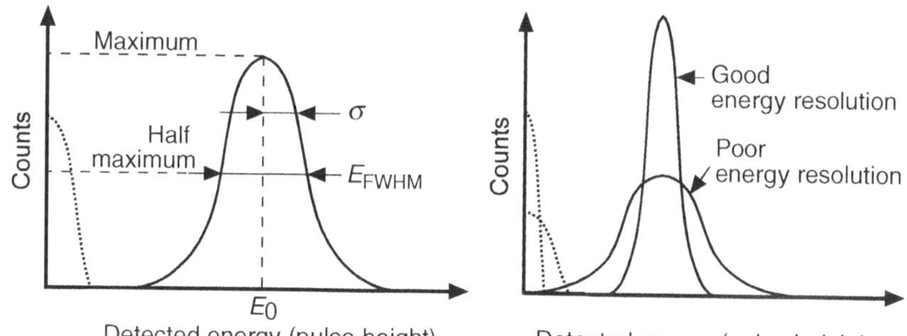

Figure 4.8 Illustration of line width (E_{FWHM}), standard deviation (σ) and energy resolution for a gaussian pulse height distribution. The dotted curves indicate how the always present electronic noise, which is one of the contributors to the line width, produces separate pulses forming the noise threshold (this will be elaborated in Section 5.4.6)

This broadening of the peak is expressed by the *line width*, E_{FWHM}, at the height where the number of counts is half of that in the centroid of the peak. This is also referred to just as the FWHM (Full Width at Half Maximum).* The energy resolution, R, is now defined as

$$R = \frac{E_{\text{FWHM}}}{E_0} 100\% \tag{4.6}$$

on the assumption there is a linear relationship between pulse height and detected energy. Ideally the value of R should be as small as possible. We then say we have good energy resolution, meaning that the system has good ability to resolve two peaks of different but close energies. With poor resolution the peak content is distributed over a wider range in the spectrum as is shown in Figure 4.8. The latter also implies fewer counts in the peak centroid since the peak integral is the same. Note that good energy resolution sometimes is referred to as high-energy resolution. This means high ability to resolve energies and not that R has a large value.

With a sufficient number of counts, the differential pulse height distribution is often gaussian. The standard deviation, σ, is for this reason sometimes quoted instead of the line width. The relationship is (see Appendix B.6)

$$E_{\text{FWHM}} = 2.35\sigma \tag{4.7}$$

The *temporal resolution* expresses how accurately a radiation detector system is able to determine the time of interaction. It is readily appreciated that, for instance, large variations in the charge collection time make it difficult to obtain good time resolution. Although the time resolution is more difficult to determine experimentally than the energy resolution, it is possible [25]. It may be presented as in Figure 4.8, but with time along the abscissa rather than energy. Most often it is quoted in terms of FWHM.

Spatial resolution applies only to position sensitive detector systems. This expresses the detector system's ability to determine the interaction position in the detector or in an array or matrix of detectors. Again the illustration in Figure 4.8 may be used, but now with distance along the abscissa. Its magnitude is also often quoted in terms of FWHM.

4.3.2 Important Properties

The list provided below contains important properties and features we typically look for when selecting the correct radiation detector system for the application:

- Radiation stopping efficiency, particularly peak efficiency.
- Entrance window transmission.
- Energy resolution.
- Linearity between output pulse amplitude and detected energy.

*In applications such as spectrometry the line width may also be specified at other heights such as FWTM, which is at tenth maximum.

- Temporal resolution.
- Speed of response or count-rate capability.
- Spatial resolution.
- Geometry (available sizes and shapes).
- Complexity (operation, mechanical, electronic).
- Sensitivity to ambient disturbances such as electromagnetic noise, electric or magnetic fields, light, chemicals, pressure, temperature and vibrations.
- Reliability (MTBF – mean time between failure).
- Power consumption.
- Cost.

Most of the listed properties are somehow related to each other meaning there always are trade offs between them forcing compromises to be made.

4.4 GASEOUS DETECTORS

Detectors using gas as the absorbing medium are among the oldest radiation detectors. The first electroscope presented in Chapter 1 uses air as the absorbing gas whereas in modern gaseous detectors special gas mixtures are used to optimise performance. A gaseous detector is basically a gas filled metal case or chamber with two electrodes: one anode and one cathode, as illustrated in Figure 4.4. Secondary electrons generated by ionising radiation create electron–ion pairs when they are slowed down in the gas. These pairs are swept to their respective electrodes by the electric field set up between the electrodes and the charge is sensed by the read-out electronics.*

4.4.1 Detector Types

As a rule of thumb it takes about 30 eV to create one electron–ion pair in gaseous detectors. This means that if the full energy of a 300 keV γ-ray photon is deposited in the detector, roughly 10^4 electron–ion pairs are created. This is equivalent to about 1.6×10^{-15} C, which is a very small charge to be sensed by the read-out electronics in pulse mode operation. The magnitude of the output signal of a pulse mode operated gaseous detector is, however, not always dependent only on the initial liberated charge: If the electric field is sufficiently high, *charge multiplication* may take place. The liberated electrons are, because of their low mass, easily accelerated by the applied field and may acquire enough kinetic energy to ionise the gas molecules. Charge multiplication, which often is called *gas multiplication*, takes place when each of these electrons on average produces two or more new electron–ion pairs. The electrons from these secondary ionisations will also be accelerated and an *avalanche* is formed.

*There are also scintillating gaseous detectors; however, these are not considered here.

Gaseous detectors are categorised by their charge multiplication properties. This is illustrated in Figure 4.9 where the output pulse amplitude is plotted as a function of the applied bias, which most commonly is referred to as the *high voltage* for gaseous detectors. Using this plot six different regions are defined:

1. *The recombination region*: Here the electric field is insufficient to prevent a fraction of the liberated electron–ion pairs from *recombining* before they drift apart and reach their respective electrodes.
2. *The ion saturation region*: The electric field is now sufficiently high to separate the electron–ion pairs so that virtually all of them are collected at their electrodes. The *ionisation chamber* operates in this region, which for this reason is also called the ionisation chamber region.
3. *The proportional region*: Here the applied voltage sets up a field sufficiently strong for charge multiplication to occur. One very important property of this region is that the output pulse amplitude is proportional to the initial charge. The *proportional counter* operates in this region.
4. *The limited proportionality region*: The field is now so strong that the output pulse amplitude to an increasing degree becomes independent of the amount of initial charge deposited.
5. *The Geiger–Müller region*: Because of the high field, the pulse amplitude is now constant and independent of the amount of charge initially deposited in the detector. Any initial charge results in a complete discharge of the detector. The *Geiger–Müller tube* operates in this region.
6. *The continuous discharge region*: Here the field is so high that the continuous discharge happens without any trigger from radiation.

As a rule of thumb, the field threshold for gas multiplication to occur is in the order of 10^6 V/m at atmospheric pressure. Needless to say this is difficult to obtain when using a detector geometry with two parallel plate electrodes, such as indicated in Figure 4.1a. The solution is to use a wire as the anode. Cylindrical detector geometry with the cylinder wall as the cathode and a wire along the cylinder axis is often used as the anode for this reason. The electric field at radius r from the anode is then given, as shown in Table 4.1. Here the electrons experience an increasing field as they approach the anode, and the charge multiplication condition is easily obtained in its immediate vicinity. Another common detector geometry uses several parallel anode wires to allow for larger detector volume. This will be discussed in more detail in the next sections. For the sake of completeness, the charge multiplication properties also depend on the type of gas and its pressure, see Table 4.1.

4.4.2 Wall Interactions

Gaseous detectors have relatively poor radiation stopping or absorption properties because of their low density. Depending on the initial energy the secondary electron energy may not be fully absorbed in the gas. This can be seen from the plot in Figure 3.2 where the maximum range of, for example, 100 keV electrons is predicted to be about 10 cm in air at atmospheric pressure. Even though some detectors operate at higher pressure, electron leakage is likely. For γ-ray detection this will be more pronounced at higher energies where secondary electrons with higher energy are produced. The detection spectrum is consequently distorted because some of the initially deposited energy is lost in the detector walls without contribution to the output signal. This effect is most serious in the high end of the detection spectra.

On the other hand the opposite is often the case: Secondary electrons from γ-ray photons, which interact in the detector walls close to the inner surface, may reach the gas and give rise to an output signal. Only a fraction of the γ-ray energy is detected and yet again spectrum distortion results.

The extent of these effects depends not only on the detector size, fill gas and pressure, but also on the detector design: A thin radiation entrance window may, for instance, be used to reduce the influence of the latter effect. Altogether the consequences are that in cases where energy information is required, the application of gaseous detectors is restricted to relatively low γ-ray energies. This is also partly because the γ-ray stopping efficiency is higher at these energies. On the other hand, if only the intensity of the γ-ray beam is to be measured, wall interactions are advantageous because it increases the number of detected events and thus the overall stopping efficiency. We shall see in Section 4.4.5 that this is the case for Geiger–Müller detectors.

4.4.3 The Ionisation Chamber

The *ionisation chamber*, also known as *the ion chamber*, is in principle the simplest of all detectors. Planar electrodes setting up a uniform electric field is the most common geometry; however, coaxial geometry is also used. The design of the chamber depends on the type of radiation and application it is built for. Low-absorption entrance windows are for instance used for detecting charged particles, such as α-particles. The chamber is most often operated at atmospheric pressure and a variety of fill gases may be used. Air is very common, but in some configurations electronegative gases must be avoided. Dense gases such as argon may be used to increase the sensitivity. The ionisation chamber relies on a sufficient voltage between the electrodes to operate on the first plateau outside the recombination region, as shown in Figure 4.9. All charge carriers generated by the ionising radiation are then collected; a further increase in the voltage does not increase the number of collected carriers. This also explains the name ion saturation region, see Figure 4.9. On the other hand a further increase in the voltage within the plateau will ensure faster charge collection and rapid separation of the generated electron–ion pairs. The latter is important to reduce the extent of recombination, particularly for highly ionising radiation, such as α-particles, where the charge density along the track becomes very high.

Pulse mode operated ionisation chambers are rarely used for anything else than detection of heavy charged particles, such as α-particles. The charge generated by low-energy γ-rays is insufficient to be detected, while higher energy γ-rays rarely interact in the detector because of its low stopping efficiency. Likewise, β-particles with a relative long range deposit only a fraction of their energy in the chamber and thus produce a comparatively

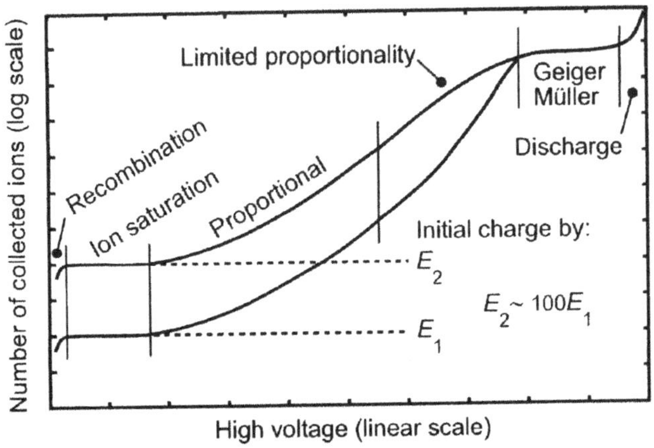

Figure 4.9 The different operation regions of gaseous detectors. The two curves are the output amplitude characteristics produced by two different initial energy depositions. In this example $E_1 \approx 100E_2$ (note the logarithmic Y-axis scale). The high voltage is not quantified because the charge multiplication properties also depend on the detector geometry

weak signal. For heavy charged particle spectroscopy, the ionisation chamber has been replaced by semiconductor detectors in many applications; however, it still has some competitive advantages like unrestricted size, long-term stability and a relatively simple design.

Recent developments within high pressure Xenon ionisation chamber technology has resulted in detectors suitable for higher energy γ-ray detection. These are chambers operating at 40 atm with energy resolution and stopping efficiency close to that of compound semiconductor detectors [26, 27].

The longer the charge collection time of gas ions, typically a few milliseconds, means that the time constant of the read-out electronics has to be correspondingly long if the full signal is to be utilised. The disadvantages of this are that the detector can only be used with low detection rates (count-rates) and that long time constants make the detector system more susceptible to low-frequency noise, such as noise from mechanical vibrations, so-called *microphonics*. To overcome this electronics with shorter time constant is normally used. This introduces a new problem: the signal amplitude and shape now becomes dependent on the interaction position in the detector. A full signal is achieved for interactions close to the cathode where all ions contribute because of the short collection time, whereas the opposite is true for interactions close to the anode. This is solved by introducing the *Frisch grid* as illustrated in Figure 4.10. The chamber is collimated so the radiation enters the volume between the cathode and the grid. The grid is held at a fixed bias somewhat below the anode bias, causing the anode signal to depend only on charge moving in the volume between the grid and the anode. The only charges moving here are electrons that have drifted through the interaction volume and through the Frisch grid. The result is a detector sensitive to negative charge carriers only. This means that it is important to avoid electronegative fill gases where the electrons may become attached to the gas molecules and make these negatively charged.

The ionisation chamber is most commonly used in current mode, also known as DC mode. The average current generated by the ionising radiation is then measured and the

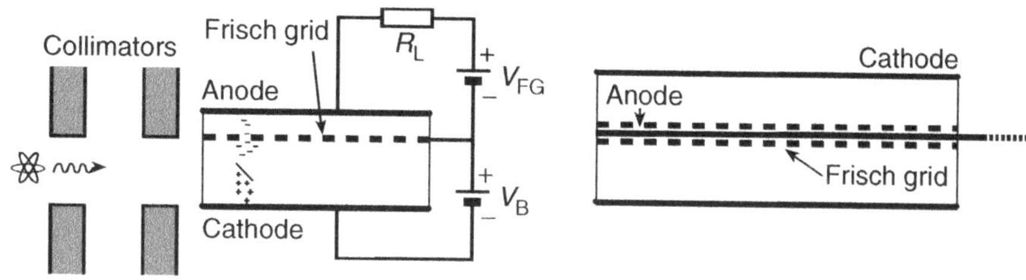

Figure 4.10 Outline of a planar geometry ionisation chamber with Frisch grid. Only Charge moving only between the grid and the anode contributes to the signal over the load resistance R_L. The coaxial version of the Frisch grid detector is shown to the right

long collection time of ions no longer matter. The magnitude of this current depends on the type of radiation, its energy and intensity, but it is very often in the pA region. This requires sensitive read-out electronics and careful design with high electrode insulation to avoid leakage currents. In the context of this book the current mode ionisation chamber is mainly of interest for its use in radiation survey instruments for radiation monitoring purposes. The absorbed dose resulting, for instance, from γ-ray exposure may be derived from measurement with a specific fill gas such as air. This is the foundation of the dose rate meter, which are presented in Section 6.3.2. Another nice feature of the ionisation chamber is that it may be used to measure radioactive gases by incorporating them as a constituent of the fill gas.

4.4.4 The Proportional Counter

To enter the proportional region shown in Figure 4.9 and achieve charge multiplication without using very high voltage, coaxial geometry is used for most proportional counters. By using anode wires with small diameter, typically around 50 μm, the electric field becomes sufficient high close to the anode for charge multiplication. And because this is possible only in the close vicinity of the anode, the gain is virtually independent of the interaction position in the detector; secondary electrons generated anywhere in the detector, even close to the cathode, all drift towards the anode without charge multiplication as in an ionisation chamber. Multiplication starts when the field exceeds the critical value very close to the anode, typically a few anode wire diameters out. The volume in which charge multiplication is possible is thus very small and virtually negligible compared to the total volume of a typical detector. This is an important feature since it ensures that electrons generated by all interactions, independent of position, experience identical multiplication. This can be obtained with other geometries as well, for instance planar geometry where the anode constitutes a set of parallel wires. By using wires also for the cathode and multiple layers of anodes and cathodes, large volume counters can be realised without degradation of the speed of response.

Typical proportional counters operate with high voltages typically between 1000 and 2500 V. The gas multiplication factor depends on the geometry, type of fill gas and its pressure, but in most configurations we are talking about values between 10^2 and 10^5. This makes it possible to detect and measure low-energy γ-rays and X-rays with fairly good energy resolution. These counters often have low attenuation entrance windows

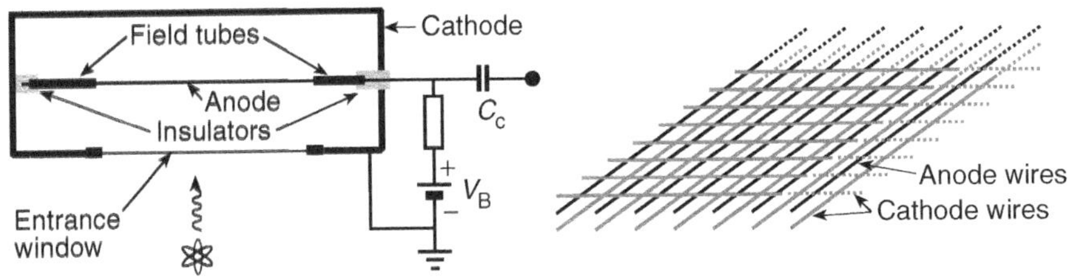

Figure 4.11 Cross-sectional view of a typical coaxial proportional counter with side window. Planar geometry may be realised using multiple wires in parallel (as shown to the right). This is the so-called *multi-wire proportional chamber*, which often also has multiple anode–cathode layers to increase the detector volume. An orthogonal arrangement is then used for the cathode layers to achieve position sensitivity

of materials such as beryllium and aluminium. Both ends of the anode wire are suspended in insulators and vacuum tight feedthroughs. The electric field is distorted in both ends of the tube near the end walls and the insulators. To avoid inhomogeneous electron multiplication so-called field tubes are inserted around the anode wire in these regions, see Figure 4.11. The outer diameter of these tubes is sufficiently large to keep the electric field on their surface to be below the threshold for multiplication. This arrangement also supports the thin anode wire making the system less susceptible to microphonics, that is, spurious pulses produced by the relative movement between the anode and the cathode.

Again it is important not to use electronegative fill gases to avoid electron attachment to neutral molecules and formation of slowly moving negative ions. So removal of any trace of electronegative molecules, such as oxygen, is very important for the detector performance. Air is consequently not an alternative fill gas. Neon and particularly argon are popular fill gases, but krypton and xenon are also used to increase the photoelectric stopping efficiency of γ-rays. The atomic numbers of these gases are 10, 18, 36 and 54, respectively. High-gas pressure is used to increase the stopping efficiency, but this is not straight forward as it also affects the multiplication factor. To achieve multiplication factors in excess of 100 it is necessary to add about 5–10% of so-called *quench gas* to kill gas ionisation by UV and visible photons from de-excitation gas molecules. Those photons reaching the cathode surface and interacting there may generate photoelectrons. These in turn may give rise to delayed avalanches when they enter the multiplication region, and even worse, this process may repeat itself as positive feedback. The excitation of the gas molecules happens through collisions in the multiplication process. The quench gas, methane or ethanol, has complex molecules with high cross section for photon absorption. And even more important, because of their complex structure these molecules do not de-excite by photon emission, but the energy is spent in decomposing of the molecule. Moreover, these molecules also quench excited noble gas molecules directly by collision. In summary, the effect of secondary photon excitation is virtually removed by this treatment. A final remark regarding fill gases is the possibility of operating proportional counters with a continuous flow of gas through them. However, for field applications sealed counters are most convenient.

The so-called micro-strip or micro-pattern gas chamber represents the latest development within proportional counter technology. Here the wires are replaced by

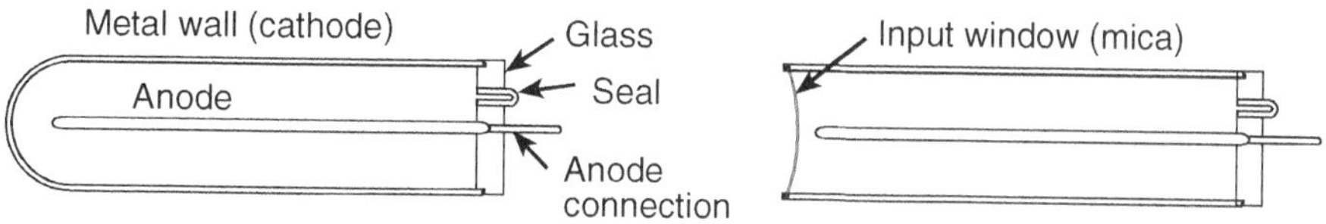

Figure 4.12 Geiger–Müller tubes (GMT) designed for detection of γ-rays (left) and low energy X-rays (right). The latter has a thin mica end window to allow the radiation to enter the gas with a minimum of attenuation. Short GMTs with entrance windows are also used for detection of α- and β-particles. The metal wall, often chrome iron, normally has a strap attached to it, which is used for cathode connection. The anode wire diameter is usually about 1 mm

microstructures with different geometries produced by photolithographic methods on insulating structures [28, 29].

4.4.5 The Geiger–Müller Tube

The Geiger–Müller tube (GMT) was introduced in 1928 and has been and will be a popular detector for industrial measurement systems. It operates in the Geiger–Müller region where the output pulse amplitude is independent of the initial charge deposited in the detector (Figure 4.9). Unlike most other radiation detectors, the GMT thus has no energy sensitivity, and can basically only be used for pulse counting and measurement of beam intensity (see Figure 4.12).

The major difference between the GMT and the proportional counter is the much higher field strength close to the anode. This means that a larger number of electrons are involved in each avalanche. Secondly, a larger fraction of the avalanche electrons excites the gas molecules in a GMT. Upon de-excitation these emit UV photons that undergo photoelectric interactions elsewhere in the detector and give rise to new avalanches there. This develops to the full length of the tube and leaves a high concentration of ions close to the anode. This positive space charge reduces the field strength close to the anode below the threshold for gas multiplication and the signal generation process is terminated. Typically 10^9–10^{10} ion pairs are involved in a Geiger discharge resulting in a large output pulse amplitude, typically in the order of volts. This often completely eliminates the need for further signal amplification. The substantial discharge is, however, also a disadvantage since it takes more time to charge or restore the tube for new events (see Figure 4.13).

The dead time is by definition the time it takes from initialisation to termination of the signal. The recovery time is, as can be seen, much longer; it is determined by the read-out circuitry and how fast the tube can be recharged. The resolution time depends on the trigger level of the read-out electronics. It is basically the minimum time interval between two separate events that enables both to be counted.

The gas pressure in conventional GMTs is typically between 50 and 150 mbar. The reason for the low gas pressure is that for the electrons to rapidly attain sufficient energy between collisions to contribute to excitation of the gas molecules. This starts the chain reaction of UV photon emission, which in turn leads to the Geiger discharge. The use of low-gas pressure enables use of relatively low electric field values, see Equation (4.4). The low gas pressure is not a drawback for γ-ray detection because this almost solely is based

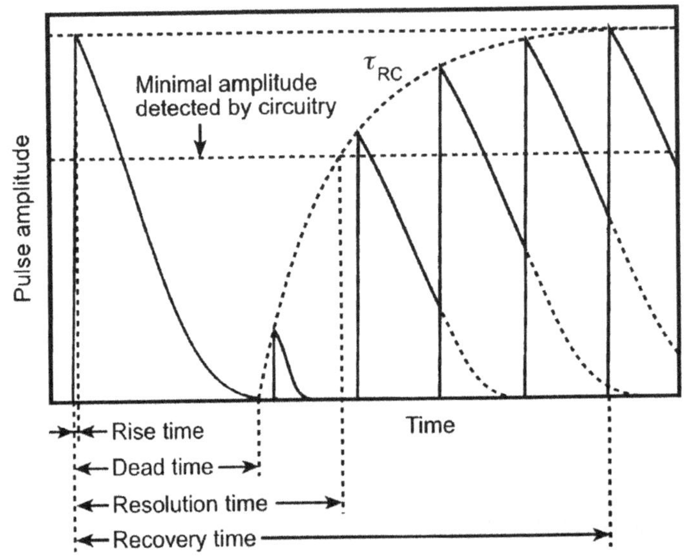

Figure 4.13 Typical pulse development in a GMT. The exact shape of the pulse and values of the recovery and resolution times depend on the read-out circuitry. Also shown is the development of pulses initiated at different times before the tube is fully recharged

on wall interactions, as explained in Section 4.4.2. The exception is GMTs for detection of low energy X-rays, which as a consequence use higher gas pressure, typically between 800 and 900 mbar. These tubes thus need higher voltage for proper operation.

For the same reason as for the proportional counters, noble gases are normally used as GMT fill gas. Helium and argon are often used for conventional tubes, whereas higher Z gases such as krypton and xenon are often used for low energy X-ray detection. A small amount of quench gas is added to the fill gas in GMTs as well, but with a different purpose than for the proportional counters. In a GMT a large number of positive ions drift towards the cathode where they are neutralised. In this process there is a fairly high possibility that at least one electron is released into the gas initiating a new Geiger discharge. To avoid this 5–10% quench gas with lower ionisation potential is added. This promotes positive charge transfer when fill gas ions on their way to the cathode collide with quench gas molecules. These then start to drift, but upon impact with the cathode there is very little electron emission probability. This is because the quench gas has a more complex molecular structure where molecule disassociation is in preference to electron emission. Chrome iron is the most commonly used wall material because this is non-reactive to halogen quenching gases.

Modern tubes usually use halogens such as chlorine and bromine as quench gas. Unlike older types of quench gases, the halogen molecules often spontaneously recombine after some time. There are, however, other mechanisms also by which the halogen molecules are spent hindering them to take part in the quenching. This limits the *operational lifetime* of the tube. The lifetime is expressed as *life expectancy*, which normally is quoted in numbers of Geiger discharges or counts. Typical numbers quoted are in the order of 10^{10} counts, but experience often tells these numbers are conservative. Note that operation at high temperatures and careless soldering may reduce the lifetime and even

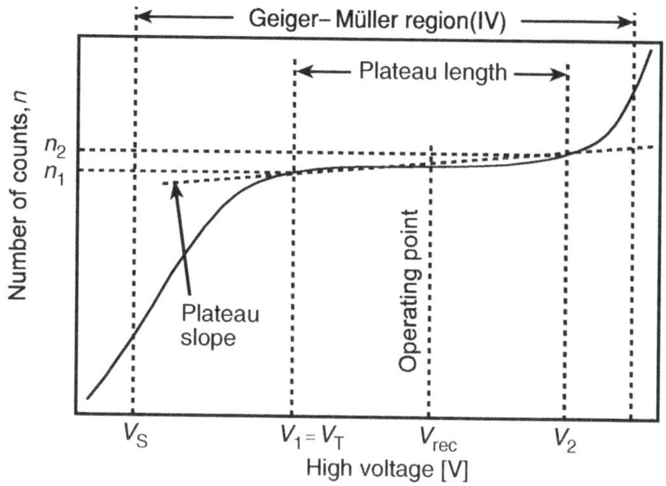

Figure 4.14 Simplified characteristic curve of the GMT showing the number of counts at constant intensity irradiation as a function of the applied high voltage. The *start voltage* V_S is the lowest voltage applied to a GMT at which pulses can be detected by a system with certain characteristics. Further, V_T (V_1) is the *threshold voltage* of the plateau over which the number of counts is relatively constant and independent of the voltage. The plateau ends at V_2 above which the number of counts increases rapidly as one moves into the continuous discharge region

destroy the tube. This may also happen if the anode connection pin is bent or the mica window is touched (see Figure 4.12).

The operating conditions for a GMT and its read-out circuitry are best explained by the simplified plot in Figure 4.14. This is in many ways similar to that in Figure 4.9; however, here it is the number of counts resulting from a beam of constant intensity irradiation that is plotted against the applied voltage. The pulse amplitude is independent of the initial charge deposition, but increasing the field strength causes the avalanche volume of the tube to increase. The consequence is a small but sufficient increase in the pulse amplitude to trigger more counts in the read-out circuitry. The recommended operating voltage of the tube is normally at the centre of the plateau. The plateau length decreases as the quench gas of the tube is spent. The plateau slope should ideally be zero and is defined as

$$\text{Plateau slope} = \frac{n_2 - n_1}{\frac{1}{2}(n_2 - n_1)} \frac{100}{V_2 - V_1} [\%/V] \tag{4.8}$$

There are two options for read-out circuitry: anode and cathode signal detection as illustrated in Figure 4.15. Anode detection is used whenever it is desirable to keep the tube cathode at ground potential. A coupling capacitor is then used to block the high voltage so that only the negative signal pulse is fed to the counter circuitry. This is not necessary for cathode signal detection where the output baseline is at ground potential and the pulse is positive (Figure 4.13). The magnitude of this pulse is approximately given as

$$V_0 = \frac{R_2(V_{rec} - V_S)}{R_1 + R_2} \tag{4.9}$$

This is typically in the order of volts and there is thus seldom need for external

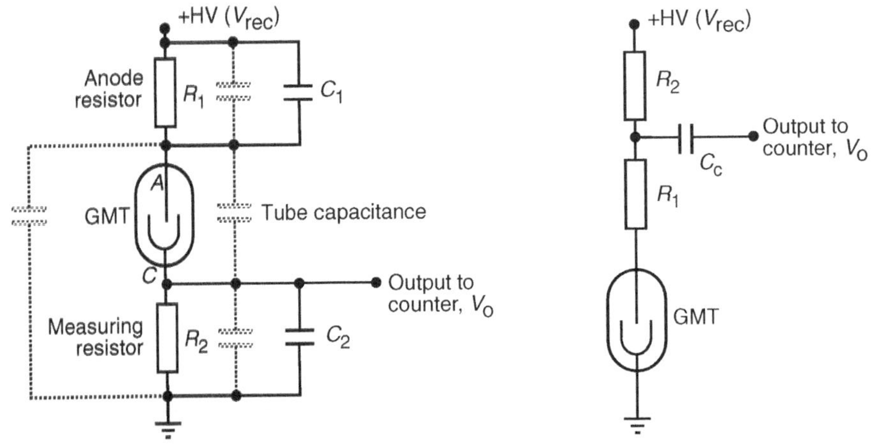

Figure 4.15 Equivalent read-out circuitry for the GMT; anode (left) and cathode (right) signal detection. Stray capacitance and the tube self-capacitance are shown with dotted lines to the left only

amplification. In cases where the signal is fed though a coaxial cable to a counter some distance away from the GMT, a buffer amplifier is often used to avoid pulse shape distortion from the cable capacitance (C_2), which is in parallel to the measuring resistor.

When a Geiger discharge takes place the current flowing through the GMT causes a voltage drop over the anode resistor (R_1), sufficient to bring the GMT voltage below the starting voltage (V_S). This resistance is, therefore, crucial for proper operation of the GMT and recommended minimum values are consequently quoted in the data sheet of every GMT. The recovery time also depends on R_1 and, in addition, the anode-ground capacitance because these determine the recharge time constant for the tube. To obtain a shortest possible recovery time, it is thus important to keep all stray capacitance as low as possible. One important method to achieve this is to connect the anode resistor directly to the anode connector of the GMT. Recommended values for the different components are often given in the GMT data sheet and application notes. The values of R_1 and the anode-ground capacitance also affect the GMT characteristics as shown in Figure 4.14: large values of the latter increase the plateau slope and reduce its length. The plateau length is also made shorter by lowering the value of R_1.

GMT tubes have a fairly flat response in the Compton energy region; the stopping efficiency is typically between 1 and 2%. In the low-energy region where photoelectric absorption becomes dominant, the efficiency increases (Figure 4.29). Because the GMT output signal carries no information about the radiation energy, it is often desirable to have tubes with a flat response over the full-energy range. This may be achieved by so-called *energy compensation*, where a filter is placed around the tube. This filter is made of a high-Z materials, such as tantalum, tungsten, platinum, gold or lead, in which the photoelectric absorption increases substantially more than Compton scattering. The low-energy efficiency peak is then reduced and a nearly flat response is obtained. This is desirable, for instance, when the GMTs are used in survey meters, see Section 6.3.2.

The relatively low stopping efficiency for γ-rays and the limited count-rate capability (~10 kc/s) are considered to be the main drawbacks of GMTs. The stopping efficiency is higher in the photoelectric dominant region (see Figure 4.29), otherwise it may be increased by using a battery of several small diameter tubes rather than one with a large diameter. This is because of the higher effective active wall volume, see Section 4.4.2. Placing this battery inside a metal block where photons are scattered, may further increase the stopping efficiency because some photons are scattered into one of the GMTs.

The count-rate capability of the GMT is significantly improved by using a so-called active quenching circuit. This essentially reduces the dead and recovery times after events by sensing avalanches very early in their development and quickly lowering the high voltage below the start voltage. The events are then detected whereas the recovery time typically is reduced one order of magnitude. This concept was introduced many years ago. However, its success has been dependent on efficient high voltage control. This is now enabled by relatively recently developed fast high voltage transistors. The lifetime of the tube is also extended substantially by avoiding complete discharge.

4.5 SEMICONDUCTOR DETECTORS

Silicon and germanium have been by far the most widely used semiconductor detector materials. However, recent developments in compound semiconductors have resulted in increasing interest in these for a variety of applications. State-of-the-art germanium detectors have several advantages making them superior for spectrometry and analysis applications [30]. The drawback, which basically rules them out as detectors in permanently installed industrial gauges, is the need of cryogenic cooling. This means they must be operated at very low temperatures to reduce noise. Although substantial progress has been made on electrical cooling systems, germanium detectors require more cooling capacity. This means liquid nitrogen flasks must be used, and these are large and need refilling on a regular basis. We shall, therefore, focus on silicon detectors and the most promising compound semiconductor detectors.

4.5.1 Electrical Classification of Solids

Semiconductor materials have resistivities intermediate between the resistivity of *conductors* and *insulators*. In a free atom the electrons are disposed in precisely determined energy levels. Combining a collection of atoms together into a solid structure, a crystal lattice, broadens those energy levels into *energy bands*, each of which contains a fixed number of electrons. Between these bands are energy regions that are forbidden to electrons. The uppermost occupied energy band is known as the *valence band*, and the electrons here are responsible for chemical reactions. Two conditions must be fulfilled to make current flow through the material: Electrons must be able to move out of their current energy state, and an electric field must be imposed on the material causing the electrons to drift.

In insulators and semiconductors the valence band is full and the next available energy states are in a higher band called the *conduction band* separated from the valence band by a *forbidden region*. For an electron to contribute to current it must gain sufficient energy to jump from the valence band across the *band gap*, E_g, into the conduction band.

- In an insulator the band gap is in the order of 10 eV. This means that electrons cannot jump across the band gap causing the resistivity to be very high, typically 10^{14}–10^{15} Ω cm.
- In a conductor, that is a metal, the valence band is not full. The conduction band is continuous with the valence band, meaning there is no band gap. The electrons can thus move freely and the resistivity is very low, typically in the order of 10^{-6} Ω cm.
- In a semiconductor the band gap is in the order of 1 eV, similar to energies achievable by thermal excitation of electrons. Under normal conditions there will consequently always be some electrons in the conduction band. The resistivity of semiconductor detector materials is in the range of 10^{9}–10^{11} Ω cm.

The thermal excitation probability per unit time of an electron has very strong temperature dependence:

$$p(T) \propto T^{3/2} e^{-E_g 1kT} \tag{4.10}$$

Here k is Boltzmann's constant and T is the temperature. This is a very important relationship in the context of semiconductor radiation detectors. Thermal excitation of electrons, and the current this causes, is noise. Our signal is the current caused by electron excitation by secondary electrons from ionising radiation interactions.

4.5.2 Impurities and Doping of Semiconductors

Whenever an electron jumps from the valence band into the conduction band it leaves behind a vacancy, a *hole*, in the otherwise full band. An electron within the valence band may fill this hole so that a new hole is created. Holes are thus effectively able to 'migrate' through the material and therefore play the role as positive charge carriers in semiconductors. Holes may also be filled by electrons from the conduction band – this is recombination. In an absolutely pure semiconductor there are equal numbers of electrons and holes. The material is then referred to as *intrinsic*.

The conductivity or resistivity of semiconductors can be changed by adding small amounts of impurities in their crystal lattice. Adding atoms with five valence electrons, such as phosphorus, to a semiconductor material where the atoms have four valence electrons leaves behind unpaired or free electrons. The excess of free negative charge carriers makes this an *n-type* semiconductor material. The bound impurity atoms are positively charged since they all have lost one electron. Similarly, adding impurity atoms with three valence electrons, such as boron, creates an excess of free holes – positive charge carriers. We now have a *p-type* material.

The impurities may be introduced to the lattice in controlled amounts when the crystal is grown. It is also possible to introduce the impurities into a crystal after it is grown. This is done by using either *diffusion* or *ion-implantation* methods. Exposing a crystal to a vapour of impurity atoms causes impurity atoms to be evaporated onto the crystal surface. These will then start to diffuse into the lattice. The impurity concentration and its profile depends on vapour concentration, temperature and time. In ion-implantation the crystal is placed in vacuum and bombarded with impurity ions that are accelerated to a certain energy. The impurity concentration and its profile are now very dependent on the ion beam intensity, time and ion energy. This bombardment causes crystal defects, which are fixed by *annealing* – a procedure where the crystal temperature is increased, often stepwise in controlled time intervals. Diffusion and ion-implantation are normally used to create shallow layers with different doping in crystals grown as n- or p-type. These layers often have higher impurity concentration and therefore referred to as p^+ and n^+ regions. Manufacturers of semiconductor detectors have special methods whereby they can make very shallow layers (less than 100 nm) with high-doping concentration and an abrupt profile. We shall see the importance of this in the next section.

4.5.3 The pn Junction

The foundation of a diode detector is an n-type and a p-type material in close contact. Since these have high charge concentration of opposite type, electrons will diffuse from the n-side to the p-side and vice versa for holes. The result of these diffusions is the formation of a *pn-junction* with a potential difference across it, the *contact* or *diffusion voltage*, and with electrons and holes in equilibrium.

If now an external voltage is added across the junction with the positive pole to the n-side, the potential across the junction increases and so does the junction width, x_w. In doing so we have added a *reverse bias* to the diode junction. Changing the polarity means we have *forward bias*. This reduces the junction width. If the forward bias is increased beyond the point where it cancels out the diffusion voltage, current will flow though the diode. This is the way the diode component in electric circuits works. It is illustrated by the *IV* characteristics plotted in Figure 4.16. A radiation detector, however, is operated with reverse bias as indicated by the dashed line in the *IV* characteristics.

The major challenge in making low noise semiconductor detectors is to keep the *leakage current*, I_l, at a minimum. This is also called the *dark current* simply because it is the current flowing through a reverse biased detector, which is in complete darkness and not exposed to any radiation. It increases with the reverse bias as indicated in the *IV* characteristics in Figure 4.16. It is due to thermal generation and recombination of electron–hole pairs somewhere in the active volume of the detector. The leakage current has one surface component and one bulk component. Unlike the other electrical properties of a semiconductor detector it is difficult to accurately predict the magnitude of the leakage current and

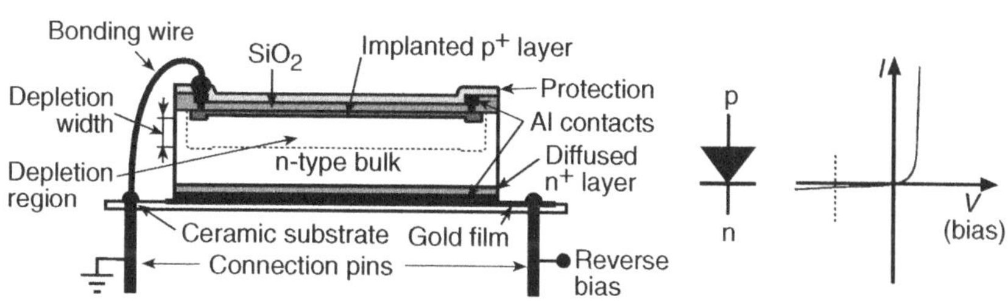

Figure 4.16 Cross section of a planar oxide-passivated PIN silicon diode detector (left) and the *IV* characteristics of a diode (right). This detector is operated with a negative bias as indicated by the dashed line in the characteristics. To minimise stray capacitance to the detector housing the p^+ layer is most often grounded, and the n^+ layer is connected to a positive so-called reverse bias

its dependence on the reverse bias. This is because every crystal lattice to some degree has so-called intermediate centres. These disrupt the perfect periodicity of the crystal and thereby introduce energy levels in the forbidden gap. This favours thermal generation and recombination of electron–hole pairs. These intermediate centres are imperfections in the crystal as a result of impurities, crystal defects or surface states. The surface component of the leakage current also depends on how well the surface is protected or passivated.

4.5.4 The PIN Silicon Detector

State-of-the-art silicon detectors are fabricated using the planar process with oxide passivation and ion implantation. A schematic cross section of such a detector is shown in Figure 4.16. The geometries of the device are easily defined by using photolithographic techniques to produce masks for the implantation area, the Al contact ring and so forth. The radiation enters the topside of the detector where there is a very shallow p^+ layer surrounded by an Al contact ring. The silicon chip is often attached to a gold plated ceramic substrate with a conducting resin.

The n^+ layer facing the rear Al contact serves as a *getter* in the annealing process and captures impurities and contamination diffusing into it. This reduces the bulk component of the leakage current. These are all important features in reducing the leakage current and thereby the noise level in the detector. The efficiency of the getter process is currently limiting the detector thickness to about 1 mm. The SiO_2 passivation of the detector surface helps control the surface component of the leakage current. The surface contact ring may also be surrounded by a guard electrode at the same potential to control the surface current.

This type of radiation detector is normally operated fully depleted. This means that the magnitude of the reverse bias is sufficient to extend the depletion width across the full thickness of the diode. The depletion region is effectively *intrinsic* with equal numbers of electrons, and thus the PIN name. There are three reasons for using full depletion: It gives maximum active volume and thereby the best stopping efficiency for the detector. Secondly, the electrical capacitance and series resistance of the detector is less when it is fully depleted. In Section 5.1.3 we shall see that this is important in order to achieve a high SNR. Thirdly, the charge collection times decrease with increasing reverse bias and the electric field becomes more uniform over the area. The drawback of increasing the reverse bias is that it increases the leakage current. For this reason every detector has an optimal value of the reverse bias where the noise is at minimum. We will discuss this in more detail in Section 5.1.3. The junction capacitance and leakage current for a typical PIN detector is shown in Figure 4.17.

The PIN detector may be used to detect several types of radiation and the front end or entrance window of the detector is tailored to suit. The device shown in Figure 4.16 has a thin protection layer on top of the oxide layer. This configuration is useful for low energy γ- and X-ray detection. Various transparent epoxy resins or opaque polyamides may be used for surface protection layer. For low energy charged particles there is no protection layer and the oxide is removed on top of the active area. This is necessary to minimise the entrance energy loss in what effectively is a dead layer. That is, it is only the energy loss in the active volume that contributes to the signal. The device may also be used for the detection of high-energy charged particles, which generate a high-charge density along

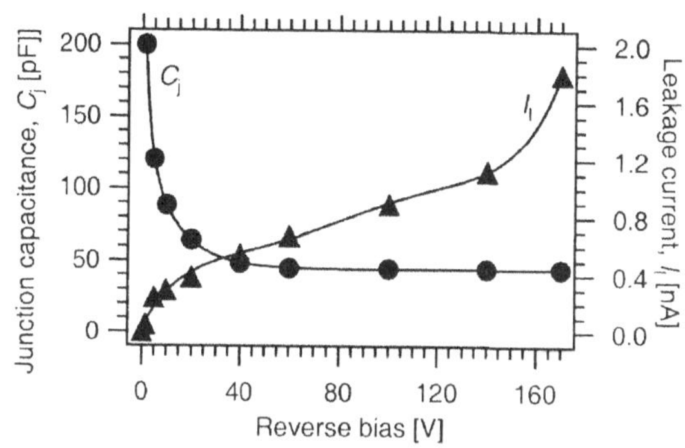

Figure 4.17 Measured (markers) leakage current (right axis) and junction capacitance (left axis) as functions of reverse bias for a planar oxide passivated and ion implanted silicon detector. This is a 280 μm thick AME AE9441 diode with $10 \times 10\,\text{mm}^2$ active area and 4 kΩ cm bulk resistivity. The depletion capacitance is readily calculated from a parallel electrode condenser model, provided the dielectric constant is known. The leakage current, however, is more unpredictable and varies from one detector to the other

the track. The p^+ layer in these devices normally has very low resistivity, but may now require a thin Al or Au layer on top of it to provide sufficient conductivity to ensure uniform response over the detector surface.

PIN detectors are also excellent photodiodes; actually some of them were first designed for this purpose. The oxide layer then also functions as an anti-reflection layer where the thickness may be optimised for particular wavelengths.

4.5.5 Compound Semiconductor Detectors

The youngest in the semiconductor family are monolithic compound materials such as HgI_2, CdTe and CdZnTe. Some configurations of these have potential for use in industrial gauges, particularly CdZnTe has received a lot of attention the past years. A compound semiconductor may be configured as a diode with a rectifying junction, but most often the MSM (metal–semiconductor–metal) configuration is used. This has 'ohmic' metal contact electrodes on either surfaces of the semiconductor block and may thus be regarded as a solid-state ionisation chamber, as shown in Figure 4.6. Gold or platinum contacts are normally applied by evaporation, sputtering or chemical deposition. Planar geometry is by far the most common; however, coaxial or semi-coaxial geometries are also used. The inner electrode is then on the surface of a centre hole through the cylinder axis.

These detector materials have several advantages which make them favourable compared to silicon: Their higher density and atomic numbers give a significant increase in the detection efficiency, particularly in the photoelectric region. Further, their higher resistivities and band gaps result in lower leakage currents and thereby lower noise.

The major problem with compound semiconductor materials is the poor mobility-lifetime product of the charge carriers, particularly that of holes. The result is incomplete charge collection which causes distortion in the detection spectrum: Monochromatic radiation peaks will not be Gaussian shaped as with silicon detectors, but asymmetric with

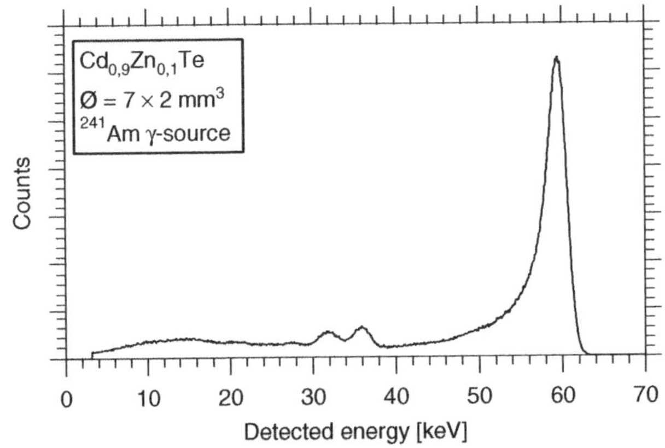

Figure 4.18 Room temperature (22°C) pulse height spectrum of ^{241}Am 59.5 keV γ-radiation acquired with a circular CdZnTe detector with beryllium entrance window [31]. The effect of incomplete charge collection is clearly seen on the asymmetric full energy peak (FWHM = 2.7 keV). The two small peaks at about 31 and 35 keV are the K line escape peaks of Te and Cd, respectively. The low energy emissions of ^{241}Am are lost in the relatively thick source encapsulation

a tailing towards low energies (see Figure 4.18). The distortion is more pronounced for thick detectors where the electrode separation is longer. It may be reduced by increasing the bias and field strength so that the charge collection time decreases. There is, however, a trade-off because higher bias increases the leakage current and noise level in MSM devices. One reason for incomplete charge collection is charge trapping due to impurities and lattice imperfections. The occurrence or density of these may exhibit large variations, also within the same crystal ingot. There may thus be considerable variations from one detector to another. Some manufacturers therefore test and sort the detectors as counter, discriminator and spectrometry grade after their charge collection and noise properties. *Ballistic deficit* is another source of spectrum distortion which in effect also is incomplete charge collection. This will be dealt with in Section 5.1.3, but it is basically signal loss because the charge collection time is long compared to the time constant of the band-pass filter of the read-out electronics.

Methods have been developed to reduce the effect of poor hole collection. The simplest is to illuminate the detector through the cathode where the holes are collected. These will then have the shortest drift distance since the major fraction of the detector attenuation takes place on the entrance side. This is particularly true for low energy radiation where the mean free path often is very small compared to the detector thickness (see Table 3.3). This method is used in the example in Figure 4.18, but the charge trapping effect is still present.

The other approach is to reject the hole contribution to the signal induction and use only that of the electrons [32–38]. Several methods have been developed whereby this can be achieved, for example the coplanar-grid technique. Here the anode is designed as a pair of interleaved electrode grids (strips) whereas the cathode is a normal full area electrode. This is a three terminal device where the interleaved electrode grids have a small voltage difference. Charge carriers drifting throughout most of the detector volume induce equal signals on these grids until the carriers (electrons) reach their vicinity. The electron signal is then induced only on the grid with the highest potential. By processing the signals from the two grids separately and subtracting one from the other, the hole component of the induced signal is virtually removed [34]. These methods may be considered as the semiconductor detector version of the Frisch grid introduced in Section 4.4.3 and shown in Figure 4.10. It is also possible to use a conventional detector with two electrodes and reject the charge induction by holes through signal processing. This requires simultaneous measurements of pulse height and pulse rise time. The energy resolution may be significantly improved with all these methods; however, it is on the cost of more complex fabrication and/or read-out electronics. It is thus a question of balancing better energy resolution against increased cost and complexity.

Several manufacturers have made diode type CdTe detectors where the rectifying junction limits the leakage current when operated with reverse bias. This enables operation with higher bias and higher field for detectors with limited thickness, normally less than 500 μm. The consequence is shorter transit times and better collection efficiency for holes. Chlorine doped CdTe with indium as anode material is most frequently used for this purpose [32, 39]. Because of the electron affinity of this p-type CdTe and the low metal work function of indium, the CdTe/indium interface forms a *Schottky barrier.* This potential barrier, which in effect is a 'pn diode junction', was used in the first Ge and Si detectors, the so-called *surface barrier detectors*. In addition to very good energy resolution with symmetric peaks the Schottky CdTe diode detectors also have excellent timing properties with FWHM time resolution less than nanosecond region for thin detectors. Their drawback is a polarization effect which is a build up of space charge over time. This interferes with carrier collection and effectively also reduces the depletion width and thus active detector volume. To increase the effective thickness of these detectors a configuration using two CdTe detectors back-to-back with one common read-out channel has been proposed and successfully tested [40].

4.5.6 Characteristics of Semiconductor Detectors

Table 4.2 summarises the most important basic properties of the most relevant semiconductor materials. Germanium is shown for comparison; its low band-gap disables Ge detectors from being used at or near room temperature. Si detectors are excellent for particle detection, but for energies exceeding a few tens of keV they have poor γ-ray detection efficiency. This is because the thickness of planar oxide passivated diodes is limited to a few mm, and because of the low density and atomic number of silicon. Compound semiconductor materials are the only option for γ-ray detection above, say 50 keV. Which material and which configuration to choose, however, is very much a question of the application [32, 33, 41–43]. For measurement of radiation beam intensity for instance, high count-rate capability is often more important than energy resolution, whereas the opposite is true for energy measurement applications. We have not commented on HgI_2 so far, but it should be fair to say that this is a detector primarily suited for spectrometry applications and less applicable for permanently installed industrial gauges. Altogether the conclusion is that there still is a considerable development going on in semiconductor detector technology, making it worthwhile checking literature and manufacturers before making any decisions.

Table 4.2 Intrinsic physical properties of semiconductor detector materials at room temperature

Property	Ge	Si	CdTe	$Cd_{0.8}Zn_{0.2}Te$	HgI_2
Effective atomic number[a], Z_{eff}	32	14	50	50	69
Density, ρ [g/cm^3]	5.33	2.33	5.85	5.9	6.4
Band gap [eV]	0.67	1.12	1.4	1.6	2.1
Carrier creation energy, w [eV]	2.9	3.6	4.4	4.7	4.3
$\mu_e \tau_e$ electrons [cm^2/V]	0.8	0.4	$\sim 10^{-3}$	$\sim 10^{-3}$	1×10^{-4}
$\mu_h \tau_h$ holes [cm^2/V]	0.8	0.2	10^{-4}–10^{-5}	10^{-4}–10^{-5}	1×10^{-6}
Resistivity [Ω cm]	$\sim$50	$\sim 10^4$	$\sim 10^9$	$\sim 10^{11}$	$\sim 10^{13}$
Relative dielectric constant, ε_r	16	12	11	11	9
Fano factor, F[b]		0.085		0.08	

Note. The values given for mobility-life time products and resistivity are approximate and vary with the fabrication properties.
[a] Calculated for evaluation of photoelectric attenuation using $m = 4.5$ in Equation (3.30).
[b] The Fano factor will be explained in Section 5.3.6.

In spite of their excellent properties there is probably a twofold reason why compound semiconductor detectors have not found widespread use in industrial gauges: Firstly, because of their piezoelectric properties these detectors are sensitive to microphonic noise such as vibrations in the frequency range of the signal. This may be suppressed and in many cases completely eliminated by proper detector packaging involving embedding the detector in a shock absorbing material such as silicon rubber or even better foam. Secondly, the noise in semiconductor materials is very sensitive to the temperature. As a rule of thumb the noise in Si detectors doubles with every seven degrees temperature increase.

The noise properties of semiconductor detectors cannot be discussed without considering the preamplifier simultaneously, as will be seen in Section 5.1.4. For now it is sufficient to note that the total noise is proportional to the square root of the leakage current, and to the capacitance at the preamplifier input. The latter means that in addition to keeping stray capacitance at a minimum it is also favourable with respect to low noise to use detectors with low capacitance.

4.6 SCINTILLATION DETECTORS

There are two types of solid state scintillators; inorganic and organic. *Inorganic scintillators* are crystals made of alkali halides, such as NaI and CsI, or oxides. These have scintillation properties by virtue of their crystalline structure which creates energy bands between which electrons can jump. Some crystals need activators to enable scintillation emission in the visible range of the spectrum. Thallium is used as activator in the best known and most frequently used scintillation crystal, NaI(Tl). *Organic scintillators* on the contrary are plastics composed of aromatic hydrocarbons. These are non-fluid solutions consisting of fluorescent organic compounds dissolved in a solidified polymer matrix. Unlike inorganic scintillators, organic scintillators scintillate on a molecular level so that each scintillator molecule can act as a scintillation centre. A comprehensive coverage of scintillation physics is found in reference [44], but is also treated more in more detail in references [25, 45, 46]. This section focuses on γ-ray scintillation properties of various scintillators. Most of this also applies to charged particle detection, but key properties such as scintillation efficiency may be slightly different here.

4.6.1 Plastic Scintillators

Inorganic scintillators are usually made of high-Z elements with fairly high density as can be seen in Table 4.3. Some of these are applicable for γ-ray detection up to energies of several MeV. In contrast, plastic scintillators are made of elements (carbon and hydrogen) with low Z and low density. This makes them more suitable for detection of particles than γ-rays. γ-ray spectrometry is virtually impossible because the low Z number means there are very few events with full energy absorption. This improves at low energies, but there the relatively low quantum efficiency makes the SNR critical. Nevertheless, plastic scintillators are used for γ-ray detection, particularly for intensity measurement where the plastic may be loaded with lead to improve the detection efficiency. Their advantages are relatively low cost of plastic scintillators and ruggedness, and in some cases also the very short decay time, see the example in Table 4.3. The former makes them popular for applications requiring large volume detectors. Plastic scintillators are also available in the shape of fibres.

4.6.2 Common Scintillation Crystals and Their Properties

Table 4.3 lists the most important properties of scintillator crystals applicable for permanently installed gauges, and some others. The attenuation properties of crystals are determined by their atomic number and density. But the volume in which the crystal may be manufactured in one block is also important.

There are several properties that impact the signal magnitude (and SNR) and through that the energy resolution of a scintillator detector system, see Equation (4.5). The first is the scintillation light output of the crystal. This may be specified in terms of quantum efficiency or scintillation photons per MeV γ-ray energy as explained in Section 4.2.6. But it is perhaps more common to use the output relative to that of a NaI(Tl) crystal with a *photomultiplier tube* (PMT) with bialkali photocathode as scintillation light detector (see Section 4.6.3). In this case it is important to specify the type of light detector in use since the signal magnitude of the total detector system depends very much on the spectral matching of this and the scintillation crystal. This is reflected in the wavelength of maximum emission, λ_{max}, in Table 4.3; however, the spectral matching is better seen from spectral plots like in Figure 4.19. The relevance of spectral matching is clearly demonstrated by considering CsI(Tl), which according to Table 4.3 is the brightest scintillator ($Q_C = 11.7\%$), but which only have a relative yield of 45% to NaI(Tl) ($Q_C = 11.4\%$) when used with a bialkali photocathode. This is, as can be seen from the plots in Figure 4.19 because of spectral mismatch. The refractive index is also important in this context: The better matching of the refraction indices of the crystal and the light detector, the less loss of scintillation light in their interface. Moreover, the afterglow should be low. In the case of current mode afterglow in effect increases the background level. For pulse

Table 4.3 Important physical properties of scintillation materials at room temperature [25, 45, 47–52, 229], see Section 4.2.6 for definitions. With the exception of the plastic scintillator (included in the rightmost column for comparison) all others are inorganic scintillators, i.e. scintillation crystals

Material	NaI(Tl)	CsI(Tl)	CsI(Na)	BGO[a]	GSO[b]	LSO[c]	CWO[d]	YAP[e]	BaF_2	Plastic[f]
Density, ρ [g/cm^3]	3.67	4.51	4.51	7.13	6.71	7.40	7.90	5.55	4.89	1.03
Effective Z, Z_{eff}[g]	51	54	54	75	59	66	64	33	52	7
Light output (Y), Q_C [%]	11.4	11.7	11.5	2.1	2.5	7.4	3.7	6.4	4.0	3
photons/MeV E_γ	40000	52000	38000	8200	9000	25000	15000	18000	10000	10500
relative[h] [%]	100	45	85	15–20	20	75	40	35–40	16	25–30
Typical energy resolution[i] [%]	7.0	8.5	8.0	11.0	9.0	11.0	8–9	7	9–10[j]	
Decay constant, τ_D [ns]	230	1000	630	300	30–60	40	5000	28	630[k]	3.3
Wavelength, λ_{max} [nm]	415	550	420	480	440	420	475	350	310	434
Refractive index[l]	1.85	1.80	1.84	2.15	1.85	1.82	2.3	1.94	1.50	1.58
Cleavage plane?	No	No	No	No	Yes	No	Yes	No	Yes	– (No)
Hygroscopic?	Yes	Slightly	Yes	No	No	No	No	No	Slightly	No
After glow [%]	0.3–5	0.5–5	0.5–5	0.005	0.005	High	0.1	0.005		
after time [ms]	6	6	6	3	6		3	6		

[a] $Bi_4Ge_3O_{12}$.
[b] Gd_2SiO_5(Ce).
[c] Lu_2SiO_5(Ce).
[d] $CdWO_4$.
[e] $YAlO_3$(Ce).
[f] NE 110/BC-412/EJ-208.
[g] Calculated for the evaluation of photoelectric attenuation using $m = 4.5$ in Equation 3.30.
[h] Relative to 'standard configuration' with NaI(Tl) and bialkali photocathode.
[i] For good crystals at 661.6 keV γ-ray energy (^{137}Cs) and with bialkali photocathode PMT read-out.
[j] When using long shaping (peaking) time so that ballistic error is avoided (see Section 5.1.3).
[k] BaF_2 also has fast decay UV component ($\tau_D = 0.6$ ns) that is absorbed unless quartz window is used.
[l] The refractive index of the entrance window of most common light detectors is between 1.4 and 1.6.

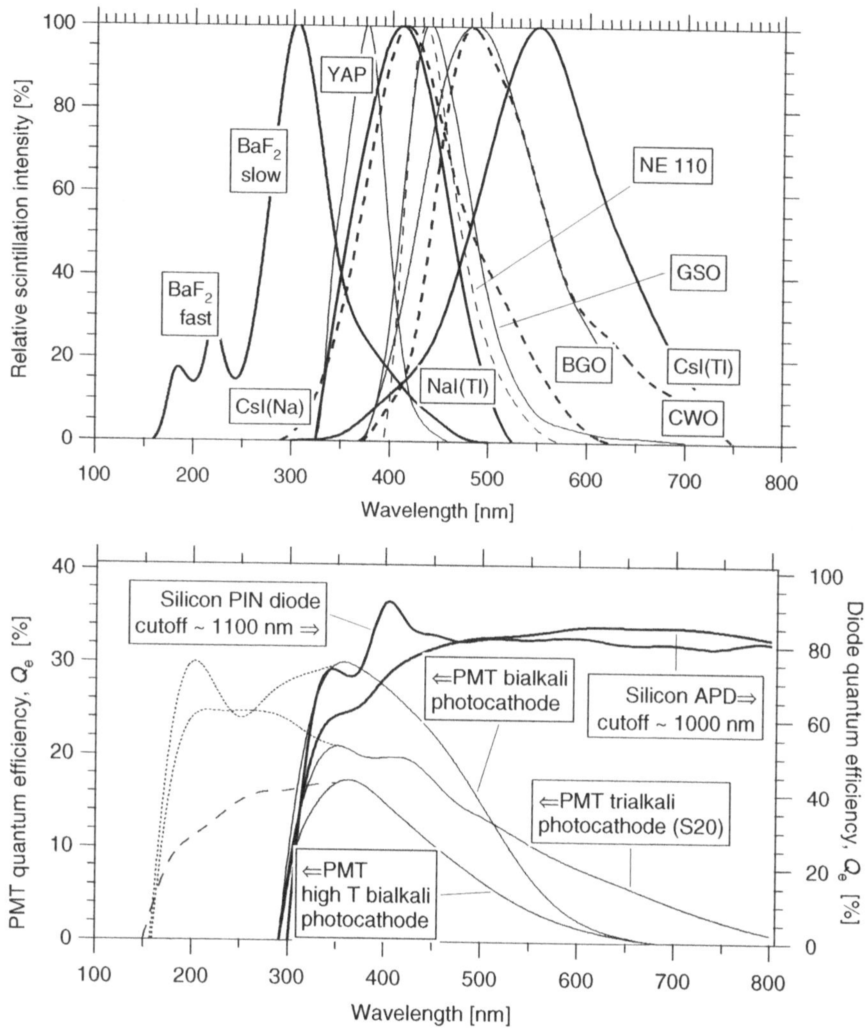

Figure 4.19 (Top) Scintillation emission spectra of the scintillators in Table 4.3 plotted relative to their individual maximum emission intensity [48, 53, 54]. (Bottom) Spectral response curves of photocathodes used in PMTs with borosilicate windows (solid line), and the DC response of an UV-enhanced Si photodiode and that of a APD (avalanche photodiode). The PMT photocathode responses may be extended in the UV region by using quartz (i.e. fused silica) window (dotted line) or sapphire window (dashed line) [53, 55, 56]

mode operation it appears as noise degrading the energy resolution of the detector. After for instance strong UV or light irradiation this effect is severe in some crystals. The scintillation efficiency for most scintillators is strongly dependent on the temperature as will be shown in Section 4.6.7. It is also dependent on the γ-ray energy because the light production by secondary electrons often is different at low and high energies [57]. This of course affects the linearity of the detectors.

The signal magnitude may also be affected by the decay constant of the crystal in cases where it is long compared to the time constant of the read-out electronics. We then have ballistic deficit, which means we loose part of the signal. The main importance of the decay constant is its implications on the timing properties of the crystal: It is the fundamental limitation to the count-rate capability of the detector, see Section 5.1.3. Further, it also tells much about the precision by which the arrival time of events can be measured.

Some scintillation crystals are *hygroscopic* and need sealed assemblies. Moisture leaks produce hydration on the crystal surface and degrade the energy resolution. Hydrate usually appears as a discoloration of the crystal. The sealed assembly requirement is not a disadvantage for most practical purposes, but there are situations where it complicates the design of the detector system. It is for instance more complex to stack separate crystals tightly in an array or matrix. Furthermore, many crystals are brittle with so-called weak cleavage planes in the crystal structure. This means they are more susceptible to mechanical damage and consequently less suited for operation in some harsh environments.

Since the discovery of the scintillation properties of NaI(Tl) in 1948 this crystal still remains dominant in many application areas despite almost five decades of subsequent research devoted to other scintillation materials. This is because it is a low cost crystal which is relatively easy to grow in large volume ingots. The cost of scintillation crystals is normally attributed to the growth process rather than the cost of raw material. Furthermore, compared to other crystals it has a high light output and the best energy resolution when used with a PMT read-out. It is a general purpose crystal which is also suitable for use at high temperatures. CsI(Tl) is a high-Z rugged material with good spectral matching for photodiode read-out. Its major disadvantage is the long decay constant. It is a highly stable material less susceptible to thermal shock than NaI(Tl). It is therefore ideal for use in hostile environments where high counting rate is not of paramount importance. It is for instance used in geophysical applications (borehole logging), bunker level devices, thickness gauging, mining applications and ore sorting. CsI(Na) shares most properties with CsI(Tl), except that its emission spectrum matches to bialkali PMT read-out, and the decay time is somewhat shorter. However, the material becomes more deliquescent than NaI(Tl) and CsI(Tl).

BGO is a high-Z, high density and very rugged crystal with very low afterglow. Its major disadvantage is that the light output (see Figure 4.26) exhibits a very strong temperature dependence. For this reason it is mainly used in controlled environment applications such as PET (see Section 5.5.4). GSO is also a high-Z, high density and fast crystal, but has a drawback in that it cleaves very easily and is thus rather susceptible to mechanical shock. It is very radiation hard material and has found its main application within physics research. LSO is a relatively new crystal that shows a unique combination of high density and Z number, a fast decay time and relatively high light output. The crystal is weakly radioactive giving rise to an inherent background of about 300 c/s per cm^3 of crystal. Even though its practical utilization so far has been hindered by difficulties of high temperature growth of reasonably large size crystals with a uniform light output, it is expected to play an important role in many applications [47]. CWO has relatively high density and Z number and stable light output with temperature. Due to the long decay constant it is mainly used in current mode rather than pulse mode systems, for instance in computerised tomography. It is often used with photodiode read-out.

YAP is a high density crystal with short decay time. The latter has made it attractive for high count-rate application. BaF_2 has relatively high density and Z number. Even though the highest light output is associated with the long decay constant, the crystal is mainly

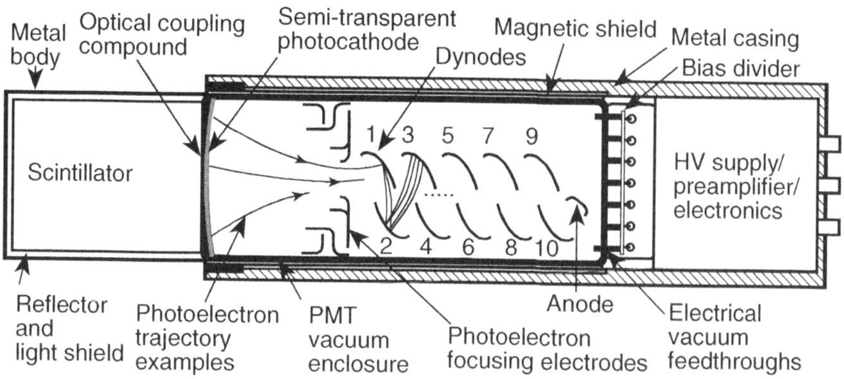

Figure 4.20 Schematic representation of a complete scintillation detector comprising a scintillator, a head-on photomultiplier tube, a bias divider for the dynode voltages and an electronics unit with preamplifier and possibly a high-voltage supply

used in fast timing applications because of the second sub-nanosecond time constant. It then requires quartz window PMTs for read-out.

4.6.3 The Photomultiplier Tube

In the world of electronics the radio tube has, with a few exceptions, been replaced by the transistor many years ago. This is not true in the world of radiation detectors even if semiconductor detectors have gained popularity. Here vacuum technology still has a high standing, particularly in detection of low light levels such as those produced by scintillators; and keep in mind that light barely sensed by the human eye is considered to be a high light level in this context! The photomultiplier tube (PMT) is the most frequently employed light detector. This is a photosensitive device consisting of a photoemissive cathode followed by focusing electrodes, an electron multiplier and an electron collector (anode) in a vacuum tube. A complete scintillation detector with scintillator, head-on PMT and associated electronics is shown in Figure 4.20. There are also so-called side-on PMTs with the entrance window on the side. These are less used for scintillation applications and will not be further discussed here.

When scintillation light impinges the photocathode photoelectrons are emitted into vacuum by the photoelectric effect as briefly explained in Section 3.3.1. The photocathode is optimised for this purpose: It is a very shallow layer of a material with low work function promoting the escape of electrons. With respect to light absorption the cathode should be thick. However, this would make it impossible for many of the photoelectrons to reach the surface on the opposite side and be released. Therefore a semitransparent layer is used as a trade-off. A large fraction of light is consequently transmitted though the cathode, partly explaining the relatively low quantum efficiency Q_E of photocathodes. The most frequently used material is KCsSb, known as the *bialkali* photocathode, which offers high blue and good green response with low dark current. There is also a so-called high temperature bialkali photocathode which can withstand temperatures up to 175°C when used with rugged window and tube materials, such as sapphire and ceramic, respectively.

This is based on NaKSb and is very useful for instance for borehole and oil well logging applications. Further, the *trialkali* or *multialkali* photocathode is also often used; this is NaKSbCs, also known by the code S20. Its sensitivity extends from the UV to the infrared but may require cooling to reduce dark current.

The spectral response of these photocathodes is shown in Figure 4.19. This may also be specified in terms of radiant sensitivity, $R(\lambda)$, which describes the photoelectron current released by the incident radiation power:

$$R(\lambda) = \frac{eQ_E(\lambda)\lambda}{hc} \approx \frac{Q_E(\lambda)\lambda}{123.96} \qquad (4.11)$$

with $R(\lambda)$ in units of mA/W, the wavelength λ in nm and Q_E (λ) in percent.

The number of photoelectrons generated by one γ-ray interaction in the scintillator is far too low to be properly detected by any read-out electronics: For a 100 keV γ-ray event we are talking about less than 1000 photoelectrons, see Equation (4.5). To achieve best possible SNR the photoelectron signal is amplified directly in the second part of the PMT, the electron multiplier. This comprises focusing electrodes, a set of dynodes and an anode or collecting electrode in the end. The focusing electrode voltages direct the photoelectrons towards the first dynode which is held at a potential of several hundred volts. Each photoelectron is thus accelerated to gain sufficient energy to cause emission of several electrons upon impact with the dynode. This process is known as *secondary emission*. These electrons are then accelerated towards the next dynode where the number of electrons is further multiplied. This process is repeated at all dynodes. The number of electrons collected at the anode is typically 10^4–10^8 times that of the number of initial photoelectrons. The gain of the tube depends on the secondary emission yield and the number of dynodes. The latter is typically between 9 and 12 whereas the yield or multiplication factor of each dynode depends on its surface material, the inter dynode voltage and the electron collection efficiency. For conventional dynode materials, such as CsSb, the multiplication factor is typically about 6. There are materials with negative electron affinity, such as GaP, in which the multiplication factor is much higher. The same overall gain may then be obtained with less dynodes. Different types of electron multipliers are presented in the next section.

A scintillation detector for field operation normally has the bias supply built into the unit as illustrated in Figure 4.20. The high voltage stability is very important in PMTs because the overall gain is a sensitive function of applied voltage; it is typically proportional to $(HV)^6$–$(HV)^9$. The low pass filtered high voltage, normally between 700 and 2500 V, is fed into the bias divider (bleeder) consisting of a chain of resistors (see Figure 4.21) in a ring on the socket the tube is mounted in. The high voltage is often adjustable by means of a precision potentiometer or in some cases a low voltage input signal. The latter is convenient for feedback stabilisation of the overall gain; this will be discussed in Section 5.1.2. It is important not to exceed the manufacturer's maximum ratings on high voltage supply: The PMT then becomes unstable with poorer gain linearity, and its lifetime is reduced (see Section 4.6.8). On the other extreme a very low bias extends the PMT lifetime, but with the drawbacks of poor photoelectron collection and gain linearity, restricted dynamic range and slower time response.

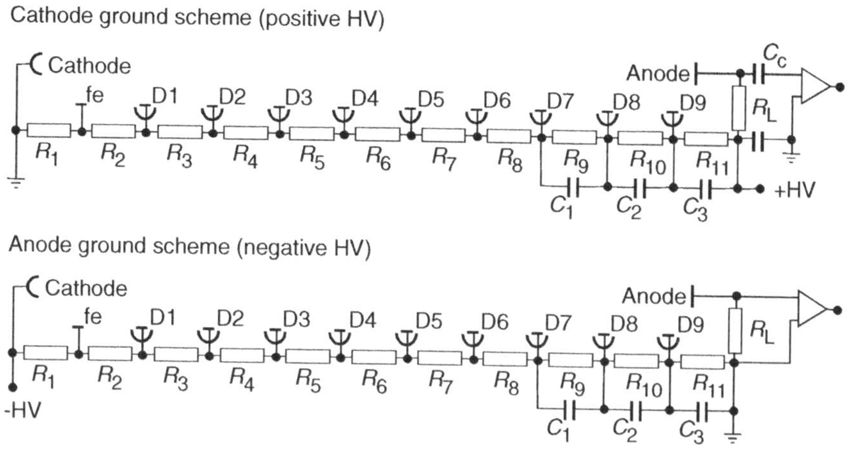

Figure 4.21 Schematic diagrams of voltage divider networks for pulse mode operation [55, 58]. The resistor values are chosen according to the tube in question; those connected to the focusing electrode (fe), the first dynode (D1) and the last dynodes (D8–D9) often differs from those in the centre, which are identical

The voltage divider design is important for the overall performance of the detector. For gain stability the current through the divider (bleeder current) needs to be sufficiently high to maintain stable voltages at the dynodes. On the other hand there is a general demand to have low power consumption. Unstable gain may be a problem at high count-rates where the average (anode) signal current is high. As a rule of thumb the voltage divider current should be at least 10–20 times the average signal current. Using the standard dynode resistor value of 470 kΩ, this is fulfilled for count-rates up to about 50 kc/s. This of course also depends on the energy of the interacting events.

For pulse mode operation of PMTs there are two possible high voltage schemes as shown in Figure 4.21: Cathode ground scheme (positive high voltage) and anode ground scheme (negative high voltage). In both cases PMT manufacturers often prescribe the voltage between the first dynode and the cathode to ensure optimal electron collection efficiency at this dynode. Capacitors are used in parallel with the resistors at the last dynodes to obtain high peak currents and better maintain the dynode potentials at a constant value during pulse durations. The cathode ground scheme is preferred for scintillation detectors because the detector housing and the μ-metal used for magnetic shielding then can be put on the same ground potential as the cathode. The signal is read out through a coupling capacitor (making it impossible to use this configuration for current mode operation). The anode ground scheme requires special measures to be taken to avoid a few phenomena giving rise to noise and photocathode sensitivity deterioration: The grounded exterior of the tube causes leakage currents to flow through the glass into the cathode and electrons to strike the inner wall of the tube.

In the same way as photodiodes all PMTs have some dark current. For a pulse mode operated system, this is more conveniently expressed in terms of *dark count-rate*. Its origin is mainly thermal excitation of electrons in the photocathode although there are other sources such as ionisation of residual gases (ion feedback), glass envelope scintillation and ohmic leakage currents from imperfect insulation. The high temperature bialkali photocathode has the lowest dark count-rate of those mentioned here, closely followed by the bialkali cathode. Typical dark count-rates are between a few hundred and a few thousand per cm^2s of cathode area at room temperature, whereas it may be one or two orders of magnitude higher in the trialkali cathode. It increases with temperature and the area of the photocathode, but is essentially independent of the high voltage or gain. Dark counts do not contribute much to the signal noise in a pulse mode system. This is because they, unlike a scintillation signal, originate from one electron and therefore can be rejected by the low amplitude of their PMT output signal. Normally, the probability that a dark count event adds to a scintillation signal is very low, unless the signal count-rate is very high.

Ionisation of residual gases, so-called *ion feedback*, may also contribute to the dark count-rate. This happens when the accelerating electrons strike and ionise residual gas atoms. These are then positively charged and start drifting towards the cathode. An ion impinging on the cathode gives rise to electron emission and causes so-called *after pulses* around 0.3 μs after the initial pulse. This is normally not a problem because the residual gas concentration is very low. However, there will be some permeation of helium into tubes operated in an environment with helium present. The degree of permeation depends on the helium concentration and the type of glass used. With quartz windows there will be sufficient helium permeation for this to become a significant problem, but not so with standard borosilicate glass or Pyrex glass. This is further discussed in Section 4.6.8.

PMT scintillation detectors are available in many sizes and shapes; however, detectors with circular crystals and entrance windows are the most common. The sort of 'standard' detector is a 2 in. diameter by 2 in. thick NaI(Tl) crystal coupled to a 2 in. diameter PMT with bialkali photocathode. Although there are rugged scintillation detectors based on metal ceramic PMTs, most PMTs are vacuum tubes and need to be treated accordingly. This and the desire for more compact detectors are the reasons for the search for alternative light detectors.

4.6.4 Electron Multiplier Types

The high current amplification and high SNR of PMTs are due to the low noise cascade multiplication of electrons in the electron multiplier. There are many electron multiplier types of as shown schematically in Figure 4.22. A summary of their most important attributes is shown in Table 4.4.

The *linear focused* type is one of the most commonly used and features very fast response time and is widely used in applications where time resolution and pulse linearity is important. It also provides large output current. The *box and grid* type is characterised by very good electron collection efficiency and excellent uniformity. The *venetian blind* type has a large dynode area and is primarily used for large area cathode tubes. It offers better uniformity and a larger output current, but is not the first choice when time response is important. Being one of the first PMTs it has found widespread use for a long time, but it is less used today. The *circular cage* type is compact and provides fast response and high gain at a relatively low high voltage.

The *Micro channel plate* (MCP) is a thin disk with a large number of small diameter (Ø ~ 15–50 μm) tubes fused in parallel with each other. Each tube has a continuous,

Table 4.4 Comparison of the most important attributes of electron multiplier types

Type	Size	Gain	Timing	Linearity	Magnetic immunity	Position sensitivity[a]
Linear focused	*	* * * *	* * * *	* * * *	*	–
Box and grid	* *	* * *	*	*	* *	–
Venetian blind	* *	* * *	* *	* *	* *	–
Circular cage	* * * *	*	* * *	* * *	* * * *	–
MCP	* * * *	* *	* * * *	*	* * * *	* * * *
Mesh	* * * *	*	* *	*	* * * *	* * *
HPD[b]	* * * *	*	* * * *	* * * *	* * * *	* * * *

Note. Four stars represent the highest ranking.
[a] See Section 4.7.
[b] The properties of the hybrid photon detector are, due to its special multiplication mechanism, not directly comparable to the other electron multipliers.

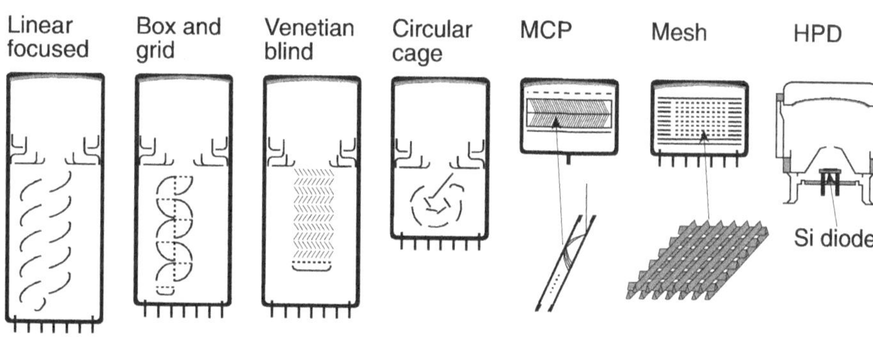

Figure 4.22 The principal types of electron multipliers in use today. Here MCP and HPD are abbreviations for microchannel plate and hybrid photon detector [58, 59]

resistive channel which acts as a continuous dynode chain. Two or three disks are stacked on top each other. The primary advantage of the MCP is its excellent timing properties; it is faster than any of the conventional tubes presented above.

Then we have the *mesh* type multipliers which have a structure of mesh dynodes stacked in close proximity and in several layers as illustrated in Figure 4.22. This provides good pulse linearity. There is also a miniature PMT, the metal package PMT, which uses a mesh like dynode structure called the metal channel dynode type (not shown in Figure 4.22). This uses a TO-8 metal can (Ø = 15 mm) with an entrance window on the top, and is thus very compact.

The *hybrid photon detector* (HPD), also known as the hybrid PMT (HPMT), is shown to the right in Figure 4.22. It differs from the others in the mode of electron multiplication: There are no dynodes, only a set of focusing electrodes forcing the photoelectrons towards a silicon diode, the anode, placed in the centre of the tube. This is normally held at ground potential whereas the cathode is held at very high negative bias, typically in the range of 15–20 kV. This means that every electron acquires a relative high energy that is deposited in the silicon diode. A planar oxide-passivated PIN silicon diode like the one shown in Figure 4.16 is used, but with a very thin surface oxide layer to minimise the dead layer energy loss of the bombarding electrons. There will be some energy loss, but the majority of the energy is deposited in the active volume of the diode, giving rise to a large number of electron hole pairs. This number multiplied with the number of photoelectrons is the total number of charge carriers sensed by the read-out electronics. Gains in the excess of 10^3 are obtained with this multiplication mechanism [59, 60]. The HPD has excellent linearity over many orders of magnitude and very good timing properties. Its gain is less sensitive to variation in the bias voltage and it has reduced signal fluctuations compared to the conventional PMT. The very high voltage required is not really a drawback since the only current flowing is the signal current. The power consumption is much higher in a conventional PMT because this in addition to the signal current requires a certain current through the bias divider to obtain stable operation. So far it has mainly been used in high-energy physics research, but it definitely has potential for other applications. Its fabrication is based on image intensifier technology and it is therefore more expensive than other PMTs. There are some scintillation detector versions available in which the photocathode is deposited on the back of the scintillator and no window is used. There are two HPD types: electrostatically focused like the one shown in Figure 4.22, and proximity focused in which the photoelectrons are accelerated straight onto one or several diodes with total area equal to that of the cathode. The latter allows for a very compact design which is very favourable with regard to its timing properties.

4.6.5 Photodiodes for Scintillation Light Read-Out

The PIN photodiode has several properties making it an attractive scintillation light read-out detector: Compared to the conventional PMT its quantum efficiency is roughly three times higher (see Figure 4.19), it is very compact and can virtually be made in any shape, it has stable low voltage operation and low power consumption, and it is very rugged. On the other hand the PMT by far is the most common scintillation light detector. This is because of two interconnected disadvantages of the photodiode; it has no internal gain and it has limited surface area.

These disadvantages limit the signal-to-noise ratio (SNR): The lack of internal gain means that a γ-ray scintillation event leads to a relatively low number of charge carriers in the photodiode. A rough calculation using $Q_C = 2$–12%, $Q_E = 60$–80%, and in addition some loss [see Equation (4.5)], predicts that the signal energy detected in the diode is only between 1 and 10% of what it would be if the γ-photon interacted directly in the diode. Secondly, because the noise in the diode is proportional to the diode capacitance, which in turn is proportional to the area of the diode (see Section 5.1.4), any increase in the diode area would increase the noise level. The consequence is a poor SNR and a very good illustration of a case where the low energy detection threshold is noise limited as illustrated in Figure 4.2.

The consequence of this is that the CsI(Tl) scintillation crystal which has high scintillation efficiency and good spectral matching to the photodiode, is the most popular crystal for this purpose. A ^{137}Cs γ-ray spectrum acquired with a photodiode, whose data is presented in Figures 4.17 and 4.19, is shown in Figure 4.23. This configuration is a good alternative for low count-rate applications requiring modest area crystals. In this

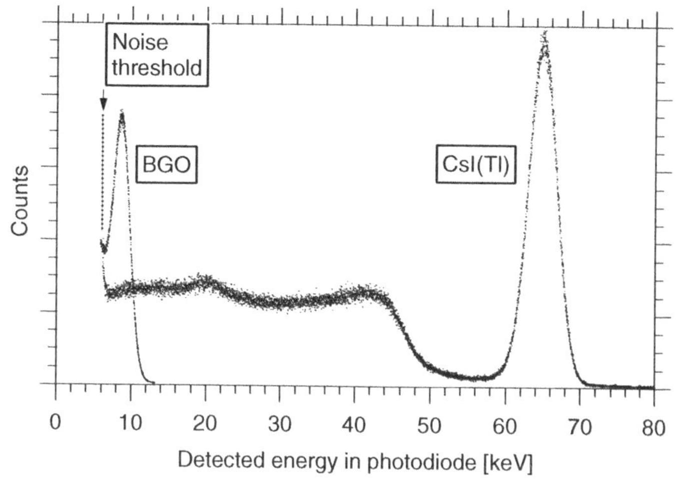

Figure 4.23 Room temperature pulse height spectra acquired by $10 \times 10\ mm^2$ PIN photodiode read-out of ^{137}Cs γ-ray (661.6 keV) scintillation signals in $10 \times 10 \times 25\ mm^3$ BGO and CsI(Tl) crystals. The energy axis is calibrated in terms of direct γ-ray absorption in the diode. The detected energies relative to that deposited in the crystals are about 1.3 and 10% for BGO and CsI(Tl), respectively. Data are taken from [53]

case the low energy threshold is about 50 keV; however, increasing the diode area (or the temperature) would also increase the noise level and consequently the low energy threshold. For BGO, which is a faster crystal with less favourable spectral matching and scintillation efficiency, the threshold for this configuration is about 600 keV as can be seen from Figure 4.23. It is worthwhile noting that in comparing PMT and photodiode read-out, it is possible to achieve better energy resolution with the photodiode at high γ-ray energies.

Silicon *drift diodes* facilitate reduction in the capacitive load at the preamplifier input without reducing the active diode area. This achieved using photolithographic techniques to define a multiple electrode pattern as shown schematically in Figure 4.24. The area of the anode (n^+), which determines the noise level, is now significantly reduced. The cathodes (p^+) are held at increasing potentials towards the anode (n^+) making the electrons drift laterally until they reach the anode where they are collected. Typical drift times are in the order of microseconds; however, the collection efficiency is still good because of the high mobility-lifetime product of electrons in silicon. On the other hand the long drift time limits these systems' speed of response and consequently the maximum detection rate. This is the main drawback with this configuration and also the reason why there has been a considerable effort put into developing detectors where the signal can be increased, rather than just reducing noise level.

The only way to increase the signal level significantly is by charge multiplication. With solid state detectors this a complex task compared to gaseous detectors and vacuum tubes. Research has been conducted on the *avalanche photodiode* (APD) for almost three decades, but stable devices of practical sizes have not been around until lately. Devices with a few hundred mm^2 area and gain of about 1000 are available today. This does not match the specifications of the PMT. However, considering the general advantages of

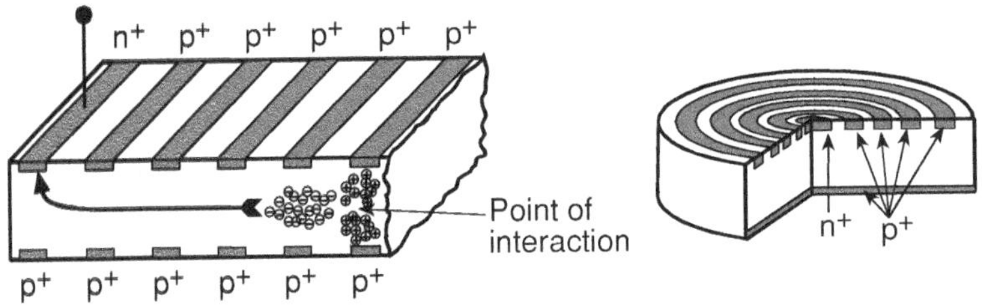

Figure 4.24 Schematic illustrations of two silicon drift diode geometries: linear and cylindrical. Electrons are initially drawn to a potential minimum near the centre of the wafer. There they are transported parallel to the surface by a potential gradient between the p^+ surface strips towards the anode (n^+) where they are collected

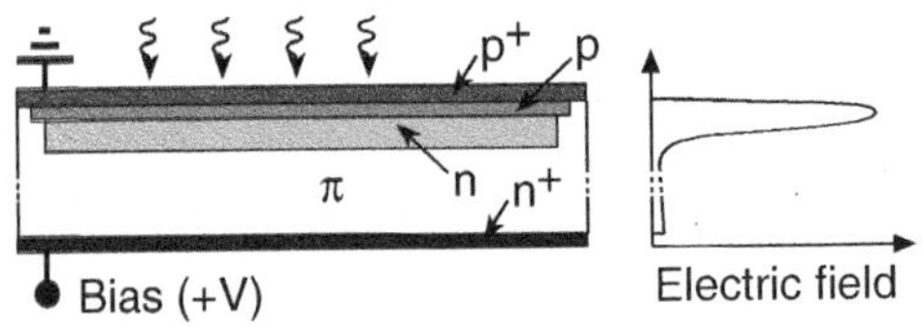

Figure 4.25 Schematic view of the buried junction silicon APD and its electric field distribution. The entrance window of the diode is on the top. The term 'π' means high purity p-type material

photodiodes, APDs are believed to be increasingly used, particularly for scintillation light detection.

The latest APD developments have been on technology tailored for scintillation light detection: This is the so-called 'buried junction' or 'reverse' APD. A schematic cross section of this device is shown in Figure 4.25 alongside the electric field distribution through it. The basic factor of any APD is to use multiple layers doped in such way that the electric field in one region becomes sufficiently high to enable electron multiplication. The diode is illuminated from the top and because the penetration depth of the scintillation photons is very shallow, all are fully absorbed ahead of the multiplication zone. This means that all the created electrons undergo full multiplication, in spite of the fact that the multiplication peaks just a few μm into the diode. Behind the multiplication zone there is a drift region with sufficient field strength for the electrons to migrate to the anode, but without further multiplication. This means that electrons created in this region, whether it is from thermal excitation or ionising radiation, will not undergo multiplication. Holes generated in the drift region, on the other hand, will drift through the multiplication zone, but with a gain which is only a few percent of that of electrons [61, 62]. These buried junction APDs are still being further developed and optimised, several manufacturers fabricate them in different configurations, and they have proven to be very efficient for scintillation light read-out [56, 61–63]. However, the drawbacks with APDs are their strong gain dependency on high voltage and temperature. Even though there are methods for gain stabilisation (see Section 5.4.7), APDs are not immediately suitable for applications in poorly controlled environments.

For the sake of completeness the 'reach-through' APD needs mentioning. Regarding scintillation light detection this may be considered as the previous generation APD. Compared to the reverse APD the multiplication and drift regions are swapped so that the multiplication region is in the end of the electron's drift region. The main disadvantage of this is that most of the leakage current generated in the device also undergoes multiplication. On the other hand, the relatively deep 'sensitive' region, the drift region, is advantageous for detection of low energy X-rays and γ-rays whose energy is below the noise threshold in PIN diodes.

Using photodiodes as scintillation light detectors has an undesirable effect called *nuclear counts*: Some of the nuclear radiation interacts directly in the diode creating signals with higher amplitude than identical interactions in the crystal would produce. These events may to some extent be discriminated by pulse height analysis, but not always. This is less of a problem for γ-ray detection, whereas for detection of penetrating charged particles it may produce significant errors. One solution is to attach the diode sideways onto the crystal so that the diode area facing the beam is considerably less. Finally, also note that the reverse APD is advantageous with respect to nuclear counts because the creation depth for electrons contributing to multiplication is very shallow.

4.6.6 Scintillation Detector Assembling

Scintillation detectors are most often purchased as one unit from the manufacturer: For optimum performance it is recommended to leave the assembling of a scintillation crystal to its light detector to experts. If you like to experiment with this on your own, there are a few considerations you need to make:

The first is the choice of scintillator and light detector. This is primarily done according to the requirements of the application, but with the restriction that there is spectral matching of the scintillation emission and the light detector response. The latter is determined by the inherent spectral response of the light detector and the light absorption in its entrance window. For PMTs borosilicate glass is the standard and lowest cost window material. This is suitable for wavelengths above 300 nm. For better UV-sensitivity for instance for read-out of BaF_2 scintillation light, UV glass or quartz may be used.

The second issue is to ensure efficient reflection properties in all scintillator walls not covered by the light detector. For hygroscopic crystals such as NaI(Tl) this is taken care of by the manufacturer as an integral part of the crystal assembly. Other crystals may be purchased from the manufacturer without assembly. The read-out face of the crystal often is specially polished for optimal light transmission. The other faces are covered with diffuse reflector materials, such as paints based on aluminium, magnesium or titanium oxides. Teflon tape or even correcting fluid may be used for trials or temporary assemblies. Specular reflectors are less efficient.

Thirdly, there must be an efficient coupling between the crystal and the detector to avoid so-called Fresnel losses. This is achieved by refractive index matching; however, the refractive index is normally of secondary concern when choosing the type of scintillation crystal. The refractive index of the crystal is in most cases larger (1.5–2.3) than that of the read-out detector window (1.4–1.6), see Table 4.3. Further, placing the crystal on top of the detector is not efficient because there will be microscopic air gaps (with unity refractive index) in between. For these reasons some type of optical coupling compound is used to facilitate the transmission of light. The refractive index of this should be in-between that of the crystal and detector. Normally high viscosity silicone oil or grease is used, but epoxy resins and clear silicon adhesives are also applicable. A very thin layer is applied carefully, preferably by placing a blob at the centre of the light detector and then pressing and screwing the crystal onto this to avoid formation of air bubbles. When using a photodiode as light detector epoxy resin has turned out to be very efficient: It provides a very efficient attachment of the diode and the scintillator into one rugged unit, it serves as optical coupling and it protects the diode surface from contamination.

For optimal scintillation light collection it is important that the sensitive area of the detector is equal to, or preferably slightly larger than the end face area of the crystal. If it is less, the loss of light will be higher and also exhibit larger variations to the interaction position in the crystal. In such cases a light guide with conical or trapezoidal shape may be used to reduce the dependence of loss on interaction position. But the average loss will then often increase, partly because an additional interface is introduced, and partly because the guide walls favour reflections back into the crystal. Generally, light guides or fibres may also be used for signal transmission over short distances, but normally only for very high energies. This is the only situation where the loss, which always is involved, can be accepted. Modern communication technology using optical fibre transmission is a different world: It is based on continuous and digital signals with power levels several orders of magnitude higher.

So far we have considered the light transmission and collection properties. These are of little value unless the total assembly is rugged, light proof and preferably screened for EMI (Electromagnetic interference). Many crystals are brittle and may cleave if they are exposed to mechanical or thermal shocks. Such cleavage planes act as reflection interfaces in the crystal disabling efficient light collection. A fractured crystal is beyond repair. It is therefore important to ensure that the total assembly is designed for the application and environment. The crystal and light detector is normally assembled in a metal body for best possible protection, see Figure 4.20. For charged particle and low energy γ-ray detection the low energy response of the detector must be considered simultaneously, see Section 4.2.1. When using PMT read-out the cathode ground scheme using positive high voltage makes assembly a lot easier and reduces the chance of electric shock. Proper assembly of a detector for operation in a harsh or demanding environment is probably the best reason for leaving the whole thing to the manufacturer.

4.6.7 Temperature Effects

The temperature is an important parameter for the performance of radiation detectors, particularly for that of scintillation detectors. The scintillation efficiency of most scintillators exhibits strong temperature dependence as shown in Figure 4.26. For most scintillators it peaks around room temperature and then decreases with increasing temperature. The very pronounced negative temperature coefficient of BGO is the largest drawback with this crystal. Experiments show that the temperature dependence of the scintillation efficiency also depends on the volume of the scintillator: Large volume scintillators exhibit the strongest temperature dependency, probably due to changes in surface reflectivity. Declining scintillation efficiency means the 'gain' of the detector system decreases so that the energy resolution consequently increases. The scintillation decay time of several crystals are also temperature dependent, for NaI(Tl) for instance, the decay time decreases with temperature and approaches 100 ns for high temperatures.

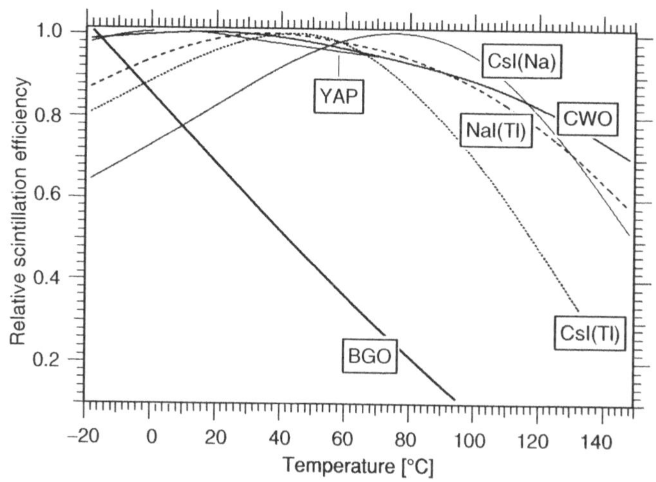

Figure 4.26 The temperature dependence on the scintillation efficiency of some scintillators plotted relative to their individual maximum efficiency in this temperature range [58, 64, 65]. There is some discrepancy in data reported in the literature on this subject

The CsI(Na) crystal is often preferred rather than NaI(Tl) above room temperature and up to about 100°C. This is partly because of the higher quantum efficiency in this range, and partly because of its better mechanical properties. At higher temperatures NaI(Tl) is the preferred choice. At 225°C the quantum efficiency of NaI(Tl) is around 45% of that of its maximum emission, whereas for that of CsI(Na) and CsI(Tl) is around 10% or less at 175°C. Also the inherent energy resolution of NaI(Tl) is relatively stable up to 225°C, but this depends very much on crystal size. For CsI(Na) and CsI(Tl) this is about four times higher at 150°C than at room temperature [66].

For scintillation detectors with PMT read-out it is the temperature dependency of the scintillation efficiency that affects the SNR the most. Photocathode dark current and the dark count-rate increases with temperature, but this is less of a problem for pulse mode systems as explained in Section 4.6.3. Photodiodes are very stable with respect to amplification, but here the noise increases dramatically with temperature, thus affecting the SNR and disqualifying these detectors for operation above about 50–70°C. The temperature influence on energy resolution and noise will be further discussed in Section 5.1.4.

4.6.8 Ageing

While operating a photomultiplier tube continuously over a long period, the anode output signal of the tube may vary slightly over time, even though the operating conditions are not changed. Short-term changes are referred to as *drift* and may for instance be caused by temperature variations. Long term changes where the signal amplitude decreases over time, say over more than 10^3–10^4 h, are called the *life characteristic* or *ageing*. This should not be confused with *burn in* which is an initialisation period of new PMTs where the gain drifts in an unpredictable manner although most often it decreases with time. This effect takes place once the high voltage is switched on and is boosted by high count-rates. The length of this period differs from tube to tube and may take anything from a few hours up to some weeks after which it stabilises. If the HV is switched off and back on again during burn in, there should be no change in gain. Users normally need not to think about burn in since it is by default carried out by the PMT-manufacturers.

Ageing is primarily caused by damage of the last dynode by heavy electron bombardment, effectively reducing its emission efficiency. It is shown in Appendix B.3 that it is possible to estimate the time it takes for the gain of the tube to drop to half of its initial value at given conditions. This is, however, not more than a coarse estimate because the total charge throughput required to reduce the PMT gain by 50% is seldom known. Nevertheless, when long-term stability is of prime importance, it is recommended to keep the average anode current below 1 μA.

Exposure of certain tubes to helium atmosphere will also slowly degrade the PMT performance. This is because there is always some permeation of helium atoms through the PMT glass. This increases the possibility of ion feedback causing satellite pulses as explained in Section 4.6.3. The permeation rate basically depends on the helium concentration in the ambient atmosphere and its temperature and pressure, and on the type of PMT glass. For PMTs with borosilicate glass in air where there are about 5 ppm helium, this is seldom a problem. The helium permeation rate of helium through quartz, however, is about 100 times that of borosilicate. If then also the PMT is operated in an environment with higher helium concentration, it is likely that an increasing number of satellite pulses will appear. Helium permeation also takes place when the tubes are stored. Dry nitrogen atmosphere in an air (helium) tight container is therefore recommended for long term storage of quartz tubes.

So far we have discussed ageing of the PMT; however, the scintillator properties may also change gradually with time. For hygroscopic crystals moisture leaks produce hydration on the crystal surface and degrade the energy resolution. In some cases discoloration of the optical coupling compound may appear reducing the light collection efficiency. Some scintillators are susceptible to radiation damage which also alters their properties, but this is rarely a problem with industrial gauges. For high energy physics research where the detectors are exposed to high radiation fluxes, this is one of the primary considerations when designing detector systems. All effects discussed so far cause a gradual change in the output signal properties. Gradual changes in the signal amplitude, most often a reduction caused by a drop in the scintillation light amplitude, the PMT gain or a combination of these, may be compensated by so-called spectrum stabilising methods. This will be discussed in Section 5.4.7. Sudden changes are more likely caused by failure such as crystal fracturing. Asymmetric or multiple peaks for single γ-ray lines and loss of efficiency are typical indications of crystal cracks.

4.7 POSITION SENSITIVE DETECTORS

In many applications it is desirable to know where the radiation originated or where it interacted in the detector. Position sensitive detectors (PSD) may be used for both purposes. These detectors may be position sensitive in 2 dimensions (2D) defining a detection matrix, or in one dimension (1D) defining a detection array. And for the sake of completeness; there are 3D position sensitive detectors where the interaction depth of the events is also determined. Most PSD systems are based on a combination of a PSD and some collimation or shielding. This will be discussed in Section 5.2.4.

All the main detector categories presented in this chapter have variants that are, or can be made, position sensitive. The straightforward approach is to stack several individual detectors in an array, matrix or any desired geometry. This is often preferred, for instance for measurements on large vessels where only a relatively course spatial resolution and/or a large total area is required. Secondly, this is often the solution with highest speed of response since the detectors and their read-out may be operated in parallel. This is also the case for some *inherent position sensitive* detectors, but not all, as we shall see. Possible drawbacks with stacked individual detectors are the cost and the relatively poor ratio between active and total area. The latter, which disqualifies this approach in some applications, is particularly true for detectors with circular cross section such as cylindrical or coaxial detectors. The ratio of active to inactive area is increased in gamma camera applications by the use of hexagonal photomultiplier tubes that can be stacked like a honeycomb.

All inherent position sensitive detectors have to have at least to two signal outputs, from which the position is either directly available or capable of being deduced by hardware or software. The most common read-out principles are shown in Figure 4.27. In the orthogonal configuration 2D position sensitivity is achieved by reading one dimension at one electrode and the other at the other electrode. This principle is also used in multiwire proportional counters as illustrated in Figure 4.11 (right). In this case the position in each direction is most often determined by using delay-line or graded density read-out. With the former a fixed delay line circuit is inserted between each wire so that the position is found from the delay between the signals at the outmost electrodes. In the latter the electrodes in each plane are connected in two groups in a special pattern so that the position is found from the ratio of these two signals [67]. This is possible with proportional counters because the gas multiplication provides sufficient signal amplitude. For semiconductor detectors there is no gain and the signals have to be read by separate electronics connected to each strip as illustrated in Figure 4.27a. For silicon detectors the multi-strip configuration is achieved by doping. The strip width and the pitch, which is separation between the strips, may then be in the μm region. For CdZnTe detectors the electrodes are realised by evaporation of electrodes through masks. However, in this case the orthogonal configuration is less popular because this also depends on the signal generated by the holes which have poor mobility. For these detectors the pixel configuration is often preferred, see Figure 4.27b [68]. This requires more read-out channels, but at the advantage of higher speed of response since the signal processing is then parallel. For small pixel detectors it has also been shown that the hole contribution to the signal can be removed [32]. However, CdZnTe detectors with strips on one side only may of course be used for 1D position sensitive detectors. For large signal detectors such as scintillation detectors using MCP or mesh electron multiplier tubes (Figure 4.22), position sensitivity is achieved by using resistive anode read-out as shown in Figure 4.27c. The position is determined from the signal amplitude distribution at the four corners.

Position sensitive scintillation detectors often use large crystals with multiple photodetector read-out as shown in Figures 4.27d–4.27f. The interaction position is determined from the signal amplitude distribution in the different photodetectors. This principle is used in many positron emission tomography detector systems (Figure 4.27d) and in the so-called *Anger camera* (Figure 4.27e). The latter requires use of collimators on the detector side as will be shown in Section 5.2.4, and may also be applied with scanning to obtain 2D position sensitivity on static objects. Depending on the exact geometry the amplitude

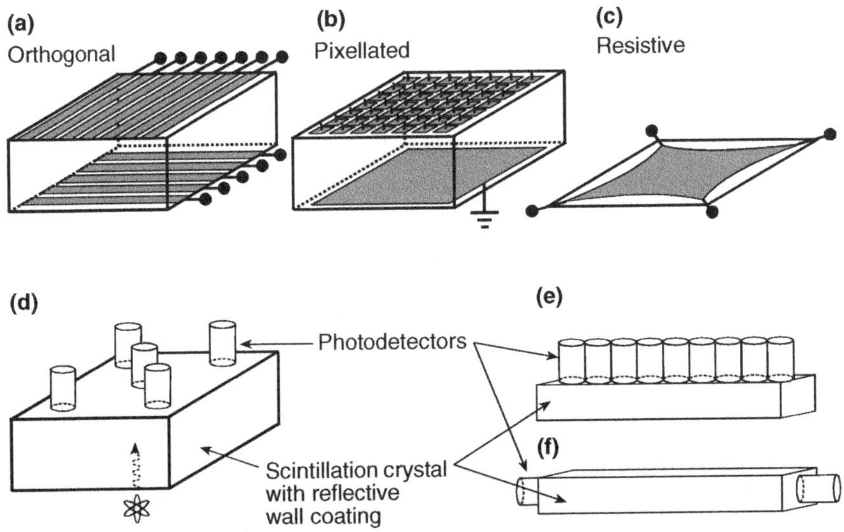

Figure 4.27 The most common principles used in inherent position sensitive detectors. For strip and pixellated semiconductor detectors guard electrodes are often used

distribution principle typically gives spatial resolution in the mm range. It is also dependent on the radiation energy because photodetectors peripheral to the interaction position will be exposed to a small fraction of the signal, which for low energies may be below the noise level. On the other hand the beauty of this concept is that relatively good spatial resolution is achieved with few detectors. For higher spatial resolution scintillation crystals may also be used with position sensitive photodetectors such as silicon strip or pixel detectors, the MCP or the mesh detector as illustrated in Figure 4.22. Even charged coupled devices (CCDs) are used for high spatial resolution read-out of scintillation crystals.

4.8 THERMOELECTRIC COOLERS

Operation of detector systems at low temperatures is beneficial for the SNR, sometimes because of the higher signal strength at lower temperatures, but most often because the noise level is lower. This will be discussed in Section 5.1.4. Cryogenic cooling systems are rarely an option for permanently installed industrial gauges. They are too cumbersome and require liquid nitrogen refilling on a regular basis. Whilst there are no alternatives to achieve their cooling capacity with more compact systems, the *thermoelectric cooler* (TEC), also called *thermoelectric module* or *Peltier element*, may be used for modest cooling down to around −50°C. This is basically a compact heat pump utilising the phenomenon that when an electrical current passes through a closed circuit made up of two dissimilar metals, heat energy is absorbed at one dissimilar metal junction and discharged on the other. This property is quantified by the Seebeck coefficient which in the ideal thermoelectric cooler should be high. For best possible performance this material should also have low electric resistance to minimise heat dissipation, and low thermal conductivity so little heat is transferred from the hot junction to the cold junction. Semiconductor materials have proven to be very appropriate for this purpose, and particularly bismuth telluride for

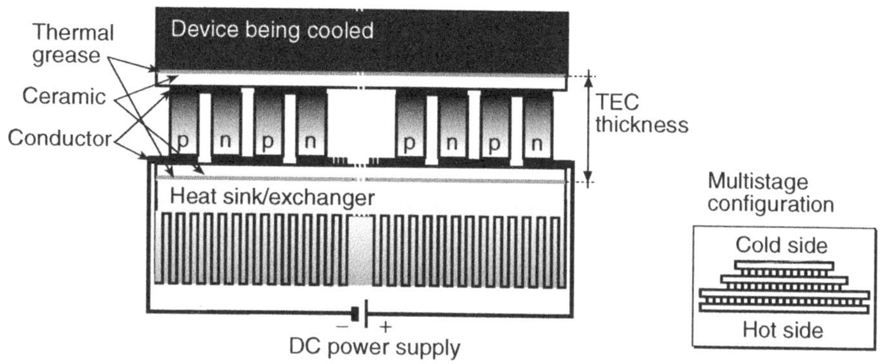

Figure 4.28 Schematic cross section of one of several rows of pn-cells in a matrix constituting a typical single stage thermoelectric cooler. The single stage thickness is typically between 2 and 6 mm. The insert shows a typical multistage configuration

cooling to moderate temperatures. A TEC consists of a number of cells connected in series and driven by a DC power supply as shown in Figure 4.28.

Each cell consists of one n-type and one p-type semiconductor connected by an electrical conductor such as a copper plate. These cells are connected in series in a matrix with a DC power supply. The whole arrangement is then connected thermally in parallel between two heat conducting and electrical insulating ceramic plates. The device to be cooled is attached on the cold side of the TEC by thermal grease, adhesive bonding or soldering. Likewise the hot side needs to be attached to some sort of heat sink. The total heat dissipation needed is the sum of the element heat transfer and the resistive heat generated by the power supply. The heat exchanger may be a finned sink as illustrated in Figure 4.28, possibly in combination with a fan for rapid heat dissipation into air, or it may be another type of heat exchanger based for instance on liquid heat dissipation.

The cooling capacity of a given TEC device is dependent upon the mass to be cooled, the operation temperature and the applied power. There is an optimum current at which the maximum cooling capacity is achieved. Above this current the resistive heating overwhelms the Peltier cooling which otherwise increases linearly with current. At a constant applied current, the cooling capacity increases as the cold side temperature increases. The maximum temperature difference over an element changes with the applied current. If the cooling capacity of a one stage element in insufficient, a multistage element [33] may be applied as illustrated in Figure 4.28.

TECs are very reliable provided they are installed and applied correctly. Efficient thermal contact on the hot side is important to avoid overheating and failure. Moisture inside a TEC can also lead to reduced performance and permanent failure. When using TECs for radiation detector cooling, this also applies to the detector where condensing moisture for instance may increase leakage currents and noise. For this reason sealed dry nitrogen atmosphere is recommended for cooling of detectors to about −30°C. Vacuum encapsulation is recommended for further cooling [33]. Finally, the lifetime of the TEC is reduced significantly when it is exposed to temperature cycling. Even ripples on the DC power supply voltage should be avoided. This means that temperature control systems using pulse width modulation (on/off) must be avoided.

4.9 STOPPING EFFICIENCY AND RADIATION WINDOWS

Detector window attenuation and stopping efficiency were introduced in Section 4.2.1. In this section some examples are given for typical detectors.

4.9.1 Stopping Efficiency

In Figure 4.29 the stopping efficiency of various γ-ray detectors are plotted. These are worked out using Monte Carlo simulations (see Section 8.5) which include secondary interactions by for instance scattered photons. It is also possible to estimate the stopping efficiency of broad beams using Equation (4.1) and ignore the secondary interactions. The error in such estimations increases with the attenuation coefficient and volume, particularly the thickness, of the detector. From the plots in Figure 4.29 it is clear that high-density scintillation crystals provide the best stopping efficiency at high γ-ray energies. At lower energies where photoelectric absorption is dominant and the atomic number determines the stopping efficiency, all attenuation coefficients are higher. The relative difference in the stopping efficiency is thus less in this region.

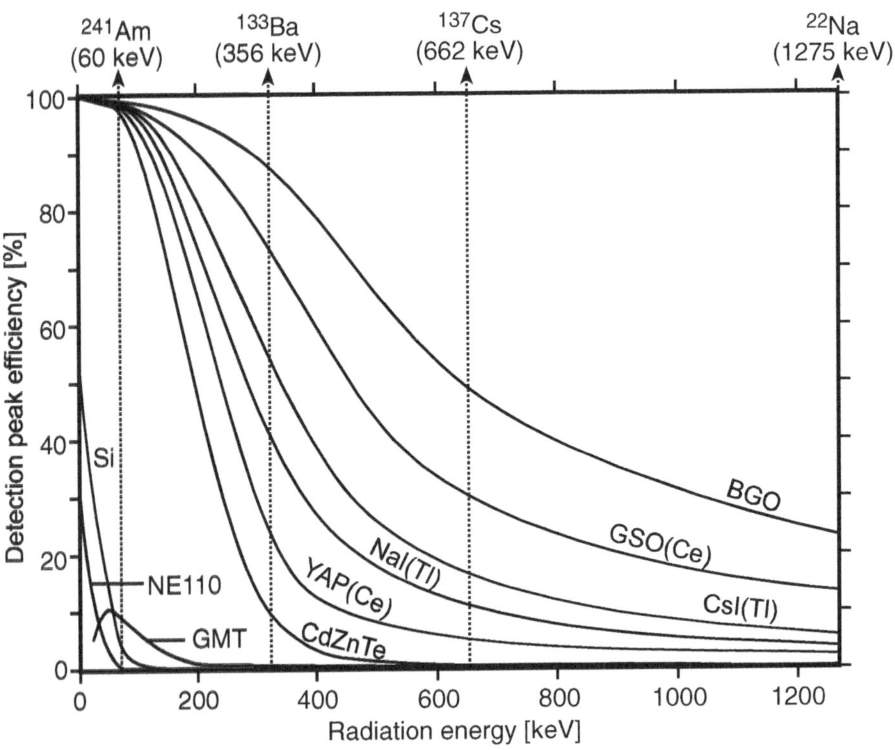

Figure 4.29 Monte Carlo simulations of full energy stopping efficiency of several detectors as function of γ-ray energy [53]. Parallel and monoenergetic 10×10 mm^2 beams are incident to 10×10 mm^2 detectors with thickness 25 mm (all scintillators), 2 mm (CdZnTe) and 1 mm (Si). The GMT data is for a 5-mm diameter tube with a $\sim$90 mg/cm^2 chrome iron cathode, and without energy compensation and mica entrance window [69]. The low energy window drop-off is shown for the GMT, but not for the other materials

The γ-ray stopping efficiency of GMTs is mainly a function of wall detection efficiency and thickness as discussed in Section 4.4.2. In the Compton dominant region this is only about 1% as shown in Figure 4.29. For tubes without energy compensation it is higher in the photoelectric dominant region. For very low energies the main fraction of the beam is attenuated close to the outer surface of the tube wall causing most secondary electrons to be stopped before they reach the gas. There will thus be a stopping efficiency peak whose position depends on the wall thickness of the tube. The stopping efficiency of other gaseous detectors is not shown. However, for a 3.75 cm thick high pressure (40 atm) Xenon ionisation chamber it is about the same as that of a 1.25 cm thick NaI(Tl) scintillation detector [27].

The term stopping efficiency does not apply to charged particles such as α- and β-particles because these loose energy continuously along their track in the detector. Normally all their energy is absorbed and they come to a full stop in the detector. The exception is high energy β-particles which may penetrate the detector and still have energy left. This is particularly true in gaseous detectors where they, according to the plot in Figure 3.2, have relatively long range.

4.9.2 Radiation Windows

For charged particles the main issue in this context is how much energy is lost in the entrance window. For low energy β-particle detection light tight, thin aluminised mylar windows with thickness down to about 25 μm are used on (non hygroscopic) scintillators. Gaseous detectors such as GMTs typically use thin mica windows, about 5 μm thick. In semiconductor detector systems the entrance window may either be part of the detector housing or a layer deposited or evaporated onto the semiconductor surface. The latter is most common for low energy particle detection. Detectors for α-particles are either used without any window, or with extremely thin windows which are transparent to light. Such systems consequently need to be operated in darkness.

Window attenuation is also crucial for detection of low energy γ-rays. The γ-ray window transmission is calculated using Equation (4.1) and plotted in Figure 4.30 for typical windows. Several of these are non-contacting windows; Mica, beryllium and to some extent aluminium, whereas the others may be part of for instance a process vessel wall.

Mica can be made in extremely thin sheets with even surfaces. It has excellent thermal, mechanical and chemical properties and can withstand rigorous shock and vibration. Beryllium is another frequently used low energy window. In addition to low attenuation its wear resistance provides very good protection and its high strength makes it useful for protecting pressurised gaseous detectors. Beryllium is poisonous requiring special machining precautions. Aluminium has relatively low attenuation and is also a low cost alternative satisfactory for many applications. It is for instance used as the entrance window in many proportional counters.

The other windows whose attenuation properties are presented in Figure 4.30 are not typical detector windows, but typical materials used as radiation windows in process vessels. For transmission measurements though a vessel for instance, these windows are used on the source side as well as the detector side. Many process vessels are made in

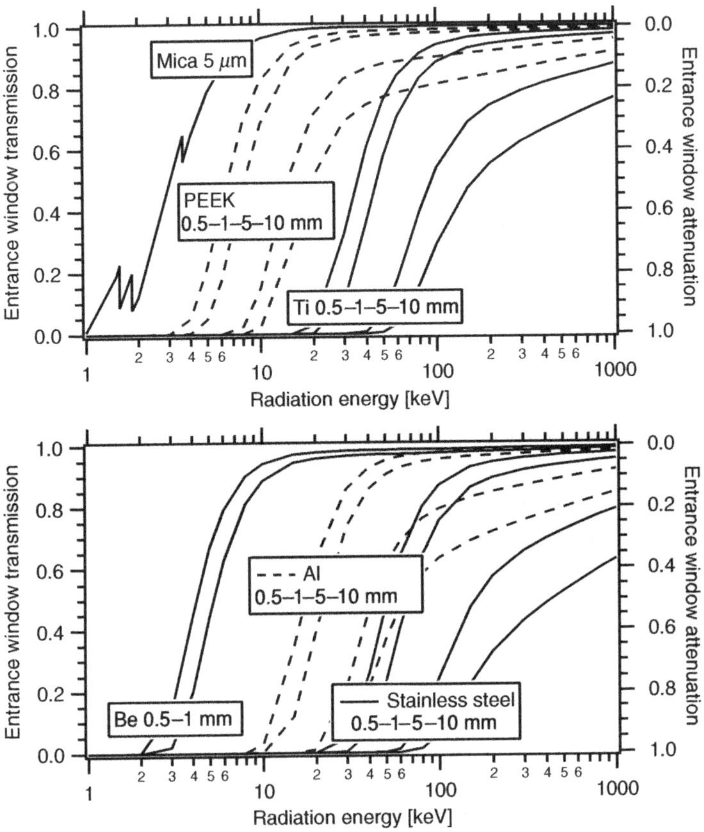

Figure 4.30 Narrow beam γ-ray attenuation properties of typical radiation entrance windows. Data from [12]. Note that the low energy threshold for γ-rays may, as shown in Section 4.2.1, be limited by the noise level rather than the window attenuation

stainless steel which, depending on the process pressure, may be several centimetres thick. As indicated in Figure 4.30 this results in a substantial attenuation, even at higher γ-ray energies. For this reason, radiation windows are often installed in the process vessel. Window materials need to be compatible to the process with respect to chemical and physical properties. Small area titanium windows are often used since titanium, as shown in Figure 4.30, has less attenuation than stainless steel, but otherwise exhibits comparable properties. For even lower attenuation PEEK (Polyetheretherketon) may be used. This is an excellent low attenuation material with outstanding wear and abrasion resistance. It is resistant to attack by most organic and inorganic chemicals. Particularly significant is PEEK's ability to retain its flexural and tensile characteristics at temperatures beyond 250°C, and its high resistance to radiation damage. Carbon fibre reinforced epoxy is also used in some cases. Similar to PEEK this combines low attenuation with excellent mechanical properties.

Note that the use of radiation windows in process vessels where there are temperature fluctuations needs particular consideration with respect to thermal expansion properties.

4.10 NEUTRON DETECTORS

The principle of operation of neutron detectors is similar to other detectors, except that one additional process has to be involved. This is normally nuclear reactions emitting prompt energetic particles which in turn produce secondary electrons, as can be seen from Table 3.1.

Gaseous detectors are most common for detection of slow neutrons, and of these the proportional counter is the most popular, all though ionisation chambers may also be used. The detection principle relies on a nuclear reaction producing secondary radiation as outlined in Section 3.6. The (n, α) reaction is very convenient because α-particles are easily absorbed with relatively high energy deposition in the gas. In proportional counters the $^{10}B(n, \alpha)$ reaction is common, either with boron contained in a coating on the detector wall or more commonly by using boron trifluoride as the counter gas. This reaction releases kinetic energies of either about 2.3 MeV or 2.8 MeV. These relatively high energies make it possible to distinguish neutron interactions from lower energy γ-ray interactions, even though some of the α-particle energy may be absorbed in the counter wall. The $^{3}He(n, p)$ reaction is an alternative which is realised by using ^{3}He of sufficient purity as the fill gas. This reaction releases about 760 keV kinetic energy. The atomic cross section for thermal neutrons is about 3800 barns and 5300 barns for the $^{10}B(n, \alpha)$ and $^{3}He(n, p)$ reactions, respectively (see Figure 3.14).

Scintillation detectors may also be used for neutron detection. Plastic scintillators are loaded with elements such as ^{10}B, ^{6}Li and ^{157}Gd for increased thermal neutron sensitivity. Ce activated ^{6}Li glass or Eu activated LiI are also used as scintillation materials. These utilise the $^{6}Li(n, \alpha)$ reaction in which kinetic energy of nearly 5 MeV is released. Finally, semiconductor detectors may also be used for neutron detection provided they are doped with high neutron cross section elements such as ^{10}B or ^{6}Li. Alternatively a coating containing such elements may be put on the detector surface.

The purpose of slow neutron detection is in most applications to measure the neutron flux or intensity. It is therefore necessary to count only detector output pulses originating from neutron reactions, and reject others such as those from γ-ray interactions. Neutrons are very often accompanied by γ-rays making γ-ray discrimination particularly important. Here ^{6}Li glass and $^{6}LiI(Eu)$ differ in the so-called gamma equivalent energy (about 3–3.5 MeV) compared to where the neutron peak shows up (about 1.8–2 MeV). This is important for neutron gamma pulse height discrimination. Liquid scintillators can every efficiently discriminate neutrons and gammas, especially with modern digital signal processing techniques. These detector cells can be made large (0.5–1 m). In (n, α) detectors discrimination is more easily achieved since the detected α-particle energy is higher than most γ-ray energies.

All slow neutron detectors can be used to detect fast neutrons if these are first moderated in a sheath of hydrogen rich material such as paraffin wax or polyethylene. For energy measurement of fast neutrons other methods must be applied; however, this is beyond the scope of this book. A review of recent developments in neutron detection is given in reference [70].

5 Radiation Measurement

The radiation detector presented in the previous chapter is the core of the radiation measurement system. When operated in pulse mode the amplitude of the output signal of most detectors is proportional to the detected energy. In the previous chapter we also saw that the detected energy is not necessarily equal to the full energy of each interacting photon (event). In summary, a pulse mode detector system is capable of measuring the energy deposition and arrival time (and in some cases the position) of every detected event. The beam intensity is then also readily available. In this chapter we shall focus on how these quantities are measured, and equally important, the *accuracy* with which this can be done. Further, we shall look at the most common measurement methods or modalities that are the foundation of radioisotope gauges for industrial measurement.

Some detectors have built-in gain with excellent signal-to-noise ratio (SNR) whereas others require sensitive low-noise electronics for optimal performance. The scintillation detector with photomultiplier tube (PMT) read-out is a typical example of the former. The PMT may actually be considered to be a high-gain (up to 10^9), wide-bandwidth (up to 1 GHz) amplifier system with extremely low noise, no offset, and low and constant output capacitance. On the other hand semiconductor detectors and ionisation chambers, which have no gain, are good examples of the latter. In either case band-pass filters and amplifiers, so-called shaping amplifiers, are often used in cases where optimal SNR is important. We will pay particular attention to low-noise preamplifiers because these are the most demanding.

In addition, we will also study those signal analysers and methods most applicable for permanently installed gauges. Here the trend is that many features realised by hardware solutions a few years ago, to an increasing extent are implemented as software code on personal computers, microcontrollers and equivalent digital circuitry. There are now rugged industrial personal computers running stable, real-time operating systems available. Further, their computing power approximately doubles every 18 months (known as Moore's law) whereas at the same time the prices drop. Another consequence of this is that functionality that used to be reserved for sophisticated laboratories can now be implemented in field systems. It also facilitates the use of more complex measurement systems where the measurement result, for instance, is based on several independent measurement principles, so-called multimodality systems. Typical examples of this will be given in Chapter 7.

The use of radioisotope gauges is often marketed by the clamp-on or non-contacting advantage most often based on neutrons or high-energy γ-rays. Even though this is a must

Radioisotope Gauges for Industrial Process Measurements. Geir Anton Johansen and Peter Jackson.
 ISBN 0-471-48999-9

in some situations, there now is a trend to also use less penetrating radiation, particularly low-energy γ-rays. The consequence often is that low attenuation radiation windows must be applied as discussed in Section 4.9.2. The advantage of using lower energy is basically higher sensitivity, more efficient collimation and shielding, and less radiated dose to the surroundings. This should also be kept in mind when planning and designing radioisotope gauges. Generally, the installation of the so-called *measurement head* of a measurement system, in our case radiation source and detector, may be categorised either as

1. *Non-contacting*, also called *non-intrusive*, such as for clamp-on gauges.
2. *Intrusive*, also known as *non-invasive*, as when using radiation windows.
3. *Invasive* Here the sensor head is, fully or partly, internal to the process vessel, for instance, inside one or several dip pipes. It may or may not disturb the process or object being investigated.

The second, and particularly the third point, can only be applied when compatible with the process or object being investigated. There are many examples where intrusive as well as invasive systems are fully accepted because the benefits overshadow any disadvantages. We will study some examples of this in Chapters 7 and 8.

5.1 READ-OUT ELECTRONICS

The principal components of a pulse mode read-out system are shown in Figure 4.1: The detector output is most often connected through a preamplifier to a so-called shaping amplifier in which the signal is amplified and filtered for optimal SNR. Then some sort of pulse analyser is used, depending upon the application. All radiation detector applications, with the exception of some using GMT or scintillation detectors with PMT read-out, require a preamplifier.

5.1.1 Preamplifiers

The major task of the *preamplifier*, also known as the *front-end electronics*, is not really to amplify the charge signal, but rather to interface the detector and provide a low-impedance source for the main amplifier. In other words its primary function is to extract the signal from the detector without significantly degrading the intrinsic SNR. Therefore, the preamplifier is located as close as possible to the detector, and the input circuits are designed to match the characteristics of the detector. The low-impedance output of preamplifiers also makes them suitable for driving a long cable to the main amplifier. There are three types of preamplifiers; the voltage sensitive, the current sensitive and the charge sensitive. Which one to choose, depends on the radiation detector.

The PMT has very high output impedance and is close to an ideal current generator. Its equivalent circuit is shown in Figure 5.1, with $R_0 \sim 1\ \mathrm{G}\Omega$ and C_0 typically between 3 and 10 pF. By adding an external resistance (R_1) and capacitance (C_1) the input resistance and capacitance to the preamplifier will become a parallel connection of these components.

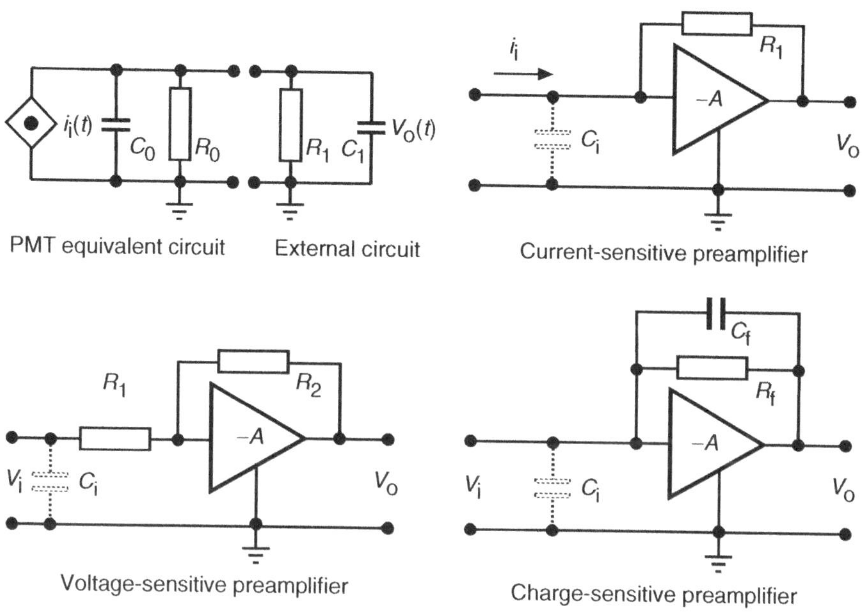

Figure 5.1 Equivalent circuit of a PMT and simplified diagrams of current-, voltage- and charge-sensitive preamplifiers used for radiation detectors

For fast timing the load resistance is typically $R_l = 50\ \Omega$ so that $R_i \approx R_l$, making the time constant of the preamplifier input, $\tau = R_l C_i \sim 1$ ns, depending on the input capacitance C_i. The *current-sensitive preamplifier*, also known as the *transimpedance amplifier* or simply the *current-to-voltage converter*, has a low-impedance input and is used to convert fast current pulses to voltage pulses. A simplified diagram of this type of preamplifier is also shown in Figure 5.1. A time constant of about 1 ns is often much less than that of the current signal from the PMT, which for a scintillation detector is basically the decay constant (τ_D) of the scintillation signal. The output voltage is in this case very much a reproduction of the signal current at the input, i_i:

$$V_o(t) = \frac{GN_{pe}eR_L}{\tau - \tau_D}\left[e^{(-t/\tau_D)} - e^{(-t/\tau)}\right] \approx -\frac{GN_{pe}eR_L}{\tau_D}e^{(-t/\tau_D)} \tag{5.1}$$

provided $\tau \ll \tau_D$. Here G is the gain of the PMT and N_{pe} the number of photoelectrons liberated at the photocathode. On the other side, for short decay time signals, such as those from plastic scintillators, τ and τ_D are in the same range. The approximation in Equation (5.1) then cannot be made and the shape of the output signal depends on the exact values of C_i and R_i.

The current-sensitive preamplifier with $R_l = 50\ \Omega$ is used for fast timing and for fast scintillators. For inorganic scintillators with longer decay time, either the current-sensitive amplifier is used with higher input resistance ($R_l \sim 10$–100 kΩ) or the *charge-sensitive preamplifier* is often preferred because this integrates or smoothes out the PMT signal. The charge-sensitive preamplifier uses a feedback loop with relatively large resistance, R_f, and a small capacitance, C_f. The latter adds to the total cold input capacitance such that this equals $C_i + C_f$. The effective input capacitance of the circuit when operating is then $C_i + (1 + A)C_f$ (the Miller effect). Assuming R_f is very large the input voltage is $V_i = Q_i/[C_i + (1 + A)C_f]$. The output voltage is then expressed as

$$V_o = -AV_i = -A\frac{Q_i}{C_i + (1 + A)C_f} \approx -\frac{Q_i}{C_f} \quad \text{provided } AC_f \gg C_i + C_f \tag{5.2}$$

This means that the amplitude of the output voltage is virtually independent of the input capacitance and variations in this. For an ionisation-sensing detector with unity charge collection efficiency the charge collected in the detector equals $E_{det}e/w$. Here E_{det} is the deposited energy and w is the average energy required to create one charge carrier pair, as explained in Section 4.2.5. The *sensitivity* of a detector system using a charge-sensitive preamplifier is thus defined as

$$\text{Sensitivity} = \frac{V_o}{E_{det}} = \frac{e}{C_f w} \tag{5.3}$$

This is usually expressed in units of mV/MeV and is typically about 50 mV/MeV for a semiconductor detector without gain.

In most cases the signal amplitude of the charge-sensitive preamplifier output is proportional to the total charge collected from the interacting event; that is, the charge at the input is integrated. The rise time of the output signal is thus limited by the charge collection time, which typically is in the nanosecond range for semiconductor detectors. The preamplifier input is discharged through R_f so that the time constant of the output signal (τ) equals $R_f C_f$. This is typically in the millisecond region. The situation is different for some gaseous detectors where the charge collection time also may be in the millisecond region, and in worst case comparable to the discharge time constant. This introduces an error since the output signal amplitude now becomes dependent on the charge collection time, which in turn depends on the interaction position in the detector.

The value of R_f has to be matched to requirements of the application. For high count-rate operation R_f needs to be sufficiently low ($R_f \sim 10\ \text{M}\Omega$) so as to avoid the preamplifier running into saturation, so-called *lock-up*. This can be seen from the illustration in Figure 5.2, where the count-rate is below the critical level. On the other hand, to achieve optimal energy resolution R_f should be highest possible ($R_f > 1\ \text{G}\Omega$) because of SNR considerations as we shall see in Section 5.1.4. There is thus a trade-off between count-rate capability and energy resolution.

For very high energy resolution systems the feedback resistor is replaced with a so-called reset discharge system [71, 72]. This comprises a control circuit that allows the preamplifier output signal to accumulate in steps without discharging until a trigger level is reached (see Figure 5.2). The control circuit then triggers the reset process on the basis of either pulsed optical feedback or transistor reset. Both methods operate on the input stage of the preamplifier: the field effect transistor (FET). With pulsed optical feedback the drain–gate junction of the FET is illuminated by a short light pulse from a LED (light emitting diode), causing current flow and discharge. One type of transistor reset uses a specially designed FET with two gates, where the additional gate controls the reset [73, 74]. With a few exceptions active reset is used only with very low-noise detectors such as cryogenic germanium detectors. The drawback of the active reset methods is that the reset period introduces some *dead time* during which interacting events will be lost. The dead

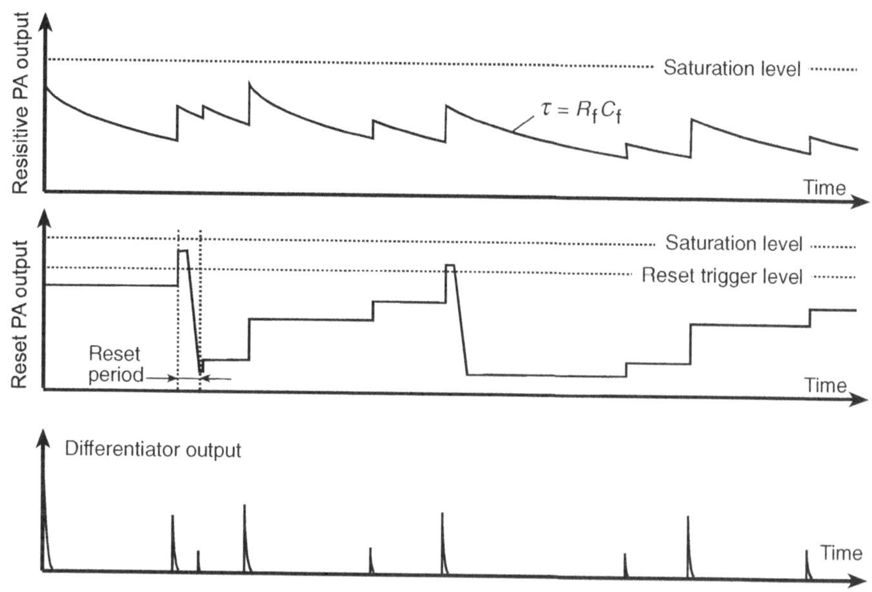

Figure 5.2 Illustration of the output signal from charge-sensitive preamplifiers with resistive feedback (top) and reset feedback (middle) with identical input. At the bottom the output of the differentiator is shown. This is normally the first stage of the main amplifier. The reset preamplifier also creates negative pulses on the differentiator output at reset. These are not shown in this illustration

time is typically a few and up to tens of microseconds for optical feedback and transistor reset, respectively [30].

The charge-sensitive preamplifier is the most widely used for radiation detectors, including those with built-in amplification, such as PMT scintillation detectors. In this case it may be realised relatively simply and at low cost. For detection systems where good energy resolution is necessary, charge-sensitive preamplifiers require more careful design to optimise the SNR. This is particularly true for detectors without gain where the collected charge is, as shown in the previous chapter, as low as 10^{-15} C. We shall take a closer look at the practical side of this, including important design rules, in Sections 5.1.4 and 5.1.5.

The *voltage-sensitive preamplifier* needs mentioning even though this is seldom used today. It has a high-impedance input and its output voltage is expressed as

$$V_o = \frac{-1}{\frac{1}{A} + \frac{R_1}{R_2 A} + \frac{R_1}{R_2}} V_i \approx -\frac{R_2}{R_1} V_i = -\frac{R_2}{R_1} \frac{Q_i}{C_i} \tag{5.4}$$

provided that $R_2 \ll R_1 A$. Here Q_i is the charge collected in the detector. It is also assumed that the charge collection time, which is the time constant of the signal, is much less than the time constant of the input circuit. We see that the output signal amplitude depends on the total input capacitance C_i. This is a drawback because C_i need not be constant, for instance if there are variations in parasitic capacitances. Furthermore, the capacitance of some semiconductor detectors also varies with the applied bias, as can be seen from Figure 4.17. This is also why the charge-sensitive preamplifier is preferred to the voltage-sensitive one.

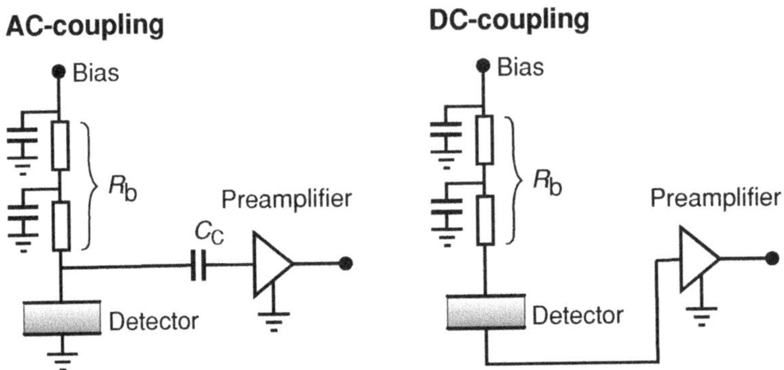

Figure 5.3 The AC- and DC-coupling schemes for connection of bias supply to the detector. The AC-coupling requires a coupling capacitor C_C to disconnect the DC bias voltage from the preamplifier input

5.1.2 Bias Supply

All electronic radiation detectors require a *bias supply* spanning from a few tens of volts for silicon detectors to several thousands of volts for some scintillation light detectors. The latter is particularly demanding because it requires the combination of high voltage (~3000 V) and relatively high current (~10 mA) for stable operation of the bleeder (see Section 4.6.3). The stability of the bias voltage is critical for all detectors with built-in gain because of its strong dependence on the applied voltage. Scintillation detectors using silicon PIN (p-type–intrinsic–n-type) diode read-out are probably the less demanding because of low voltage (~50 V), low current (~100 pA) and relatively low sensitivity to bias variations.

As part of their load resistor R_b, most bias supplies have built-in low pass filters (as shown in Figure 5.3). This is to reduce noise, which otherwise would propagate to the amplifier input. This filter also protects the detector and front-end electronics from voltage spikes when the bias supply is switched on and off.

DC-coupling is used only in high-energy-resolution systems because it shows slightly better noise performance. The leakage current through the detector then has to be accommodated by the preamplifier whose input is at virtual ground potential. AC-coupling uses a coupling capacitor, C_C, to isolate preamplifier from the bias voltage. For high-voltage biasing this of course needs to be a high-voltage capacitor. We will see in Section 5.1.4 that the bias resistance R_b should be as large as possible from an SNR point of view. In practice, however, it is limited by the voltage drop caused by the current flowing through the detector. As a rule of thumb this voltage drop should not exceed 10% of the bias voltage. Typical values range from 10 to 100 MΩ for most detectors and maybe a few gigaohms for detectors with very low leakage current.

Special biasing considerations for the different detector types were discussed in the previous chapter; see for instance Section 4.4.5 for the GMT (Geiger–Müller tube) and Section 4.6.3 for the PMT. The optimal bias voltage for a detector is always a compromise between several important performance parameters, particularly speed of response and energy resolution, as will be shown in Section 5.1.4 in the case of semiconductor detectors. The detector manufacturers always have recommendations regarding detector biasing in the data sheets: for instance minimum and maximum values, which you should always observe.

The output voltage of most bias supplies can be tuned, at least within a limited range. In most supplies this is done through adjustment of a potentiometer. Others allow the high voltage to be controlled through a low voltage input, typically in 1:1000 ratio, or more seldom through a digital input. This allows the bias supply to be used in a feedback loop for gain stabilisation (see Section 5.4.7).

5.1.3 The Shaping Amplifier

We shall now focus on signal processing for detectors using charge-sensitive preamplifiers where the output signal is a so-called *linear tail pulse* (see Figure 5.2). In most cases this signal needs amplification and filtering to achieve the desired system performance. This is particularly true for unity gain detectors where the low-signal level dictates a low-noise level to obtain the required SNR. On the other hand there are situations where there is virtually no need for pulse processing, for example detection of energetic radiation with PMT scintillation detectors. In this case a preamplifier may be sufficient to interface the detector and provide a low-impedance output suitable for driving a signal cable.

The output of the charge-sensitive preamplifier is fed into a so-called *shaping amplifier*, also known as a *linear amplifier* or simply *shaper*. This has a twofold mission: to amplify and filter the signal so that its characteristics suit the subsequent signal analyser. This may, for instance, be an analogue-to-digital converter (ADC) where it is advantageous if the range of the signal amplitude at the shaping amplifier output matches the input range of the ADC. The shaping amplifier has a band-pass filter, which for historical reasons is characterised by the *peaking time* of the output signal, τ_0, rather than the centre frequency and the pass bandwidth, which are the common filter terminology. There are many alternatives as to how this filter is implemented; one of the most common is shown in Figure 5.4. This is referred to as the CR–RC^n shaping amplifier because it comprises one differentiator (C_dR_d – high pass filter) and a ladder of n integrators (R_iC_i – low pass filters). This is also referred to as *semi-Gaussian unipolar shaping* because the shape of

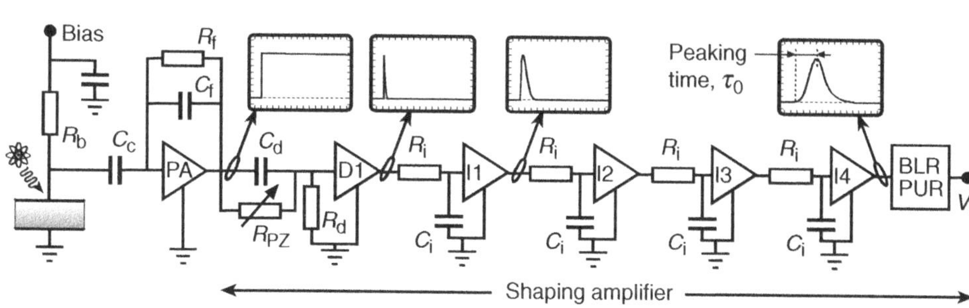

Figure 5.4 Outline of the frequently used CR–RC^4 shaping amplifier constituting one differentiator (D1) and four integrators (I1–I4). The pole-zero cancellation is tuned by adjusting R_{PZ}. The baseline restorer (BLR) and pile-up rejector (PUR) are two separate circuits. Both are optional and not required for low count-rate applications. The shape of the output signal (V_o) is now ideal for subsequent measurement circuitry

the output signal is close to a Gaussian curve: The more integrators that are used, the closer one gets to a symmetric Gaussian shape. Optimal SNR is achieved by using equal time constants in the differentiator and the integrators so that $\tau = C_d R_d = R_i C_i$. The peaking time of this filter is then given as $\tau_0 = n\tau$.

The peaking time (τ_0) is also shown in Figure 5.4: this is the time from 1 to 100% of full amplitude of the pulse at the shaping amplifier output. Sometimes the *shaping time* is used; this is equal to the pulse width and is somewhat more than twice the peaking time for a CR–RC^4 filter. We shall see in Section 5.1.4 that the choice of peaking time, and through that the properties of the band-pass filter, is critical for the SNR. The peaking time is equally important for the count-rate capability of the system since it dictates how close pulses can arrive in time.

The input stage is similar for most shaping amplifier types. This is the differentiator, a high pass filter, which shortens the width of the preamplifier output pulse as illustrated in Figures 5.2 and 5.4. In the timescale of the shaping amplifier where τ_0 typically is a few microseconds, the preamplifier output looks like a step signal. But it is not; it decays with a time constant defined by the preamplifier feedback as $\tau = R_f C_f$. Without a so-called *pole-zero cancellation* (PZC), this causes an undershoot on the output pulse of the differentiator, which will propagate all the way to the output of the shaping amplifier. The principle of PZC is seen from the transfer function of the differentiator output to the preamplifier input. There will be a pole in this function defined by the time constant $\tau = R_f C_f$, and a zero defined by the time constant $\tau = R_{PZ} C_d$. Pole-zero compensation is carried out by examining the output on an oscilloscope and carefully adjusting R_{PZ} until the undershoot disappears. This means that $R_f C_f = R_{PZ} C_d$ so that the zero overshoot is cancelled out in the transfer function. This procedure is demonstrated in Figure 5.5; however, real signals will be a lot more fuzzy because of noise. The value of R_{PZ} is critical: if it is too low we have undercompensation and the undershoot will still be there, whereas if it is too high there will be an overshoot or a tail on the output signal.

Undershoot and overshoot must be avoided because it effectively increases the pulse width of the output signal. In turn this increases the probability of pulses to arrive on the tail of the previous pulses. For high count-rates the effect of this, which is known as *pile-up*, may cause large errors in the amplitude measurement because then there is a high probability that the pulses arrive more or less on top of each other, as illustrated in Figure 5.5. The effect of overcorrection is easily seen in a pulse height spectrum as a high-end tail on peaks such as the full energy peaks of γ-ray lines caused by successive pulses standing on the tails of the preceding pulses, thus increasing their amplitude. Conversely, undercorrection of the overshoot gives rise to pulses with lower than true amplitude and shows up in a spectrum as a lowering of the apparent energy of the peak. For this reason correct pole-zero compensation is important, especially when operating at high count-rates. Likewise short peaking times are preferred in high count-rate applications. To keep the probability of pile-up low, the inverse of the average count-rate should, as a rule of thumb, be more than 10 times the pulse width because of the random radiation emission. That is, if the pulse width is 5 μs, the average pulse separation should be at least 50 μs, corresponding to a maximum average count-rate of 20 kc/s. For systems where pile-up cannot be avoided it is possible to use a pile-up rejector. Any pulse arriving on the tail of the previous one is then rejected. Alternatively, both are rejected if the second arrives on the rising tail of the first, that is before it has peaked. Pile-up rejection may be done

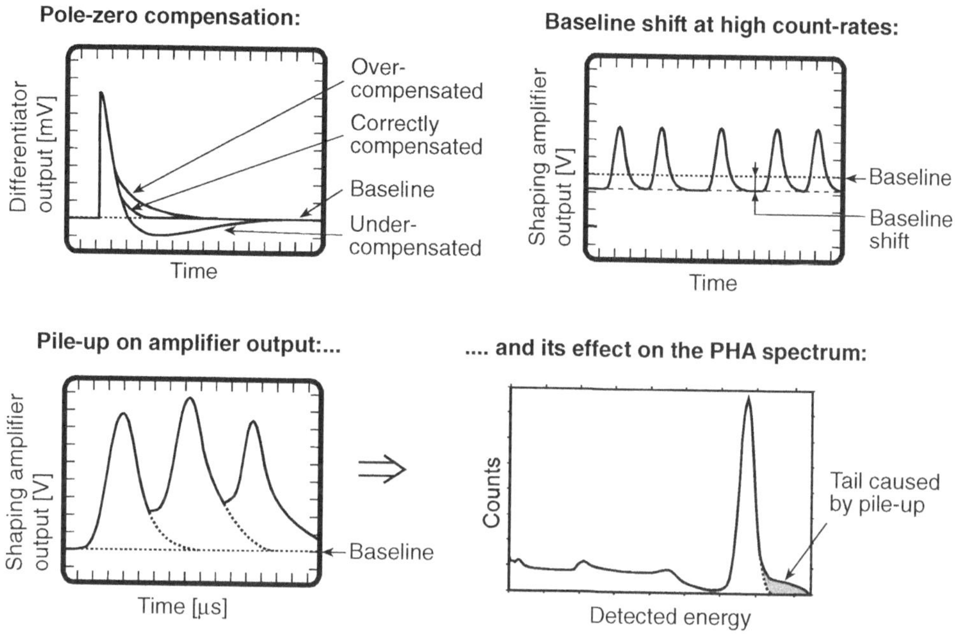

Figure 5.5 Illustration of PZC at the output of the differentiator, baseline shift and pile-up at the output of the shaping amplifier, and the latter's effect on a typical pulse height spectrum. In the case where the output pulse of the shaping amplifier has an undershoot, for instance caused by undercompensated PZC or at bipolar shaping, pile-up would cause a tail on the low end of the full energy (or any) peak

on the basis of abnormal pulse shape at the amplifier output, as indicated in Figure 5.4. Alternatively, a second parallel shaping amplifier with high temporal resolution may be used to measure the separation between succeeding pulses. Pile-up rejection is then carried out by a comparison of peaking time and pulse width changes. Pile-up rejection is important for accurate measurement of radiation energy, such as spectrometry. It is also important for radiation intensity measurement, but here the recorded count-rate may be compensated on the ground of peaking time, pulse width and dead-time models. This will be explained in Section 5.4.11.

For some detectors the choice of peaking time is a consideration of not only count-rate capability and noise filtering, but also so-called *ballistic deficit* or *ballistic error*. For scintillation detectors with long scintillation decay time (τ_D) part of the signal is lost if the peaking time is made too short. The *ballistic error* is defined as the amplitude ratio of the output signal to that of a system with infinitely long peaking time. This is less critical for scintillation detectors because constant decay time means constant ballistic error. For ionisation-sensing detectors the poor mobility of positive charge carriers causes large variation in the collection time at the cathode. Depending on the anode–cathode separation and the electric field, these variations may be of the same timescale as the peaking time. The ballistic error is then dependent on the interaction position in the detector and will no longer be constant. From this point of view, long peaking times are preferred. It is also possible to some extent to use pulse shape analysis to reject events with ballistic error because this will influence the rise time of the peak.

In some shaping amplifiers there will be a *baseline shift* at high count-rates (see Figure 5.5). Any series capacitor in a shaping amplifier, such as in the differentiator, prevents transmission of the DC-component of the signal. There will thus be a baseline shift to make the overall transmitted charge equal to zero. The baseline shift thus increases with increasing count-rate. Needless to say, this results in serious errors in the measured signal so as amplitude and count-rate in the case of a counting system. The action of the *baseline restorer* (BLR) is basically to connect the signal line to ground in absence of a signal so as to establish correct baseline for the arrival of the next pulse. There are passive BLRs based on diode clamping; however, today so-called gated BLRs are mostly used because of their better performance. The BLR also helps to suppress low-frequency noise such as microphonics and power line disturbance.

The problem with baseline shift is avoided by using *bipolar shaping* (e.g. CR^2–RC^2) This is achieved by introducing a second differentiator in the shaping amplifier, for instance one replacing the second integrator in Figure 5.4. The positive lobe of the output signal is then followed by a negative lobe slightly longer in duration, but with a lower amplitude. The areas of these lobes are equal so that the baseline inherently is preserved and the DC level always is zero. Bipolar shaping is the simplest circuitry to implement with regard to baseline preservation. It is useful as long as the pile-up probability is kept low; however, it has a drawback in that its filtering properties are not as good as those of unipolar shapers (see Table 5.2).

Unipolar *delay line shaping* amplifiers have among the best filtering properties. The pulse shaping or bandpass filtering is here achieved by splitting the output of the differentiator into two branches, delaying one of them and feeding this to the inverting input on the next amplifier stage. The other branch is fed to the non-inverting input so that the output signal is ideally a square pulse with pulse length equal to the delay. This output is then integrated so that the shaper is known as DL-RC. There is also a so-called *double delay line shaper* denoted DDL-RC, whose filtering properties are even better. The noise filtering properties of the different shapers will be dealt with in more detail in the next chapter. Design examples of CR–RC^2 and DL-RC shaping amplifiers are given in Section B.5.

5.1.4 Electronic Noise

Electronic noise is an important contributor to the overall energy resolution performance of some radiation detection systems, primarily those using detectors without internal gain. Because of the relatively low signal from these detectors, low noise is essential to obtain a high SNR. Because of the high impedance of the detector and FET, some sources of noise become significant, which are never even considered in other applications of low-noise amplifiers. In this section we will use a silicon PIN detector connected to a charge-sensitive preamplifier as a case study, partly because this is unity gain detector, and partly because of its excellent properties with respect to creation and collection of charge carriers. The latter will be explained in more detail in Section 5.3.6. Nevertheless, the considerations made in this section also apply to other detector types.

The main amplifier noise contribution to the total system noise is negligible in most detector/amplifier systems. The dominant electronic noise sources are those at the input stage of the preamplifier, and those of the detector. Figure 5.6 shows the equivalent electronic

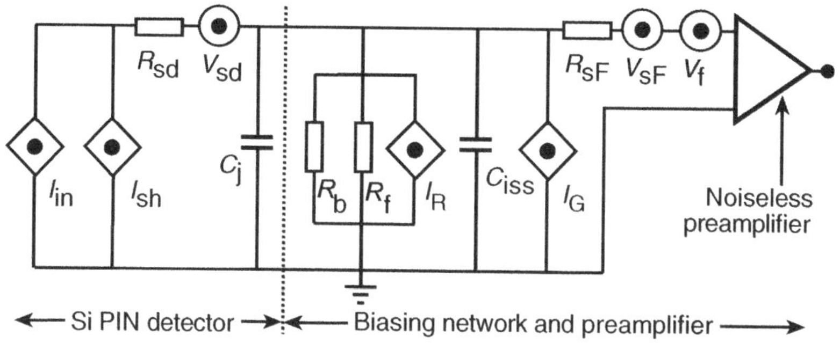

Figure 5.6 Equivalent electronic diagram of the detector, biasing and preamplifier circuitry shown in Figure 5.4. The detector is assumed to be a silicon PIN diode. Here I_{in} is the signal current generated by the interacting radiation

diagram of the charge-sensitive preamplifier shown in Figure 5.4 connected to a silicon PIN detector. All intrinsic noise sources giving a measurable or significant contribution to the total electronic noise are included in this equivalent circuit.

The different noise sources are explained in Table 5.1 and expressed in terms of ENC (equivalent noise charge). The function of the bandpass filter in the shaping amplifier is included by the so-called noise coefficients: Here $\langle N_S^2 \rangle$ is the *step noise* coefficient, which is proportional to the peaking time (τ_0), and $\langle N_\Delta^2 \rangle$ is the *delta noise* coefficient, which is proportional to the inverse of the peaking time. The other coefficient, $\langle N_{f^{-1}}^2 \rangle$, is independent of the peaking time. The dependence of the noise coefficients on the shaping amplifier is summarised in Table 5.2.

ENC is the charge introduced in a step at the input of a noise-free preamplifier, which gives an output pulse with amplitude equal to the root mean square (rms) noise of a real preamplifier. It is often preferred to express the noise in terms and units other than ENC:

- In terms of ENC, numbers of electrons rms: $\text{ENC}(\#\text{e}^- \text{ rms}) = \frac{\text{ENC(C)}}{e}$
- In terms of noise voltage rms: $E_{\text{NO}}(\text{V}) = \frac{\text{ENC(C)}}{C_f}$
- In terms of energy line width: $\text{FWHM}_{\text{DET}}(\text{eV}) = \frac{2.35 w_{\text{DET}}}{e}\text{ENC(C)}$

The line width (FWHM; see Section 4.3.1) is expressed in units of energy (often keV) relative to energy deposition in the detector. Here w_{DET} is the average energy required to generate one charge carrier pair in the detector in question (see Section 4.2.5). For silicon used in our case study, $w_{\text{Si}} = 3.61$ eV such that the conversion factor from ENC to FWHM_{Si} keV is $k_c = 5.29 \times 10^{16}$ V. For the terms in Table 5.1 we then define $E_{sh} = k_c Q_{sh}$ and so forth. Because none of these noise contributions are correlated the total diode noise (E_D), the total preamplifier and biasing network noise (E_A), and the total electronic noise (E_E) can be expressed as (see Section 5.3.3)

$$
\begin{aligned}
E_D &= \sqrt{E_{sh}^2 + E_{sd}^2} \\
E_A &= \sqrt{E_R^2 + E_G^2 + E_{sF}^2 + E_{f^{-1}}^2} \\
E_E &= \sqrt{E_D^2 + E_A^2} = \sqrt{E_{sh}^2 + E_{sd}^2 + E_R^2 + E_G^2 + E_{sF}^2 + E_{f^{-1}}^2}
\end{aligned}
\qquad (5.5)
$$

Table 5.1 Explanation of the different noise sources in the equivalent circuit shown in Figure 5.6[a]

Symbol	Source	Expression [ENC^2]
I_{sh}	Shot noise caused by the diode leakage current (I_l) and its random generation and recombination of charge carriers in the diode pn-junction.	$Q^2_{sh} = eI_l\langle N^2_S\rangle$
V_{sd}	Thermal noise generated in the inherent series resistance (R_{sd}) in the undepleted region of the diode. Here C_j is the diode junction capacitance.	$Q^2_{sd} = 2kTR_{sd}C^2_j\langle N^2_\Delta\rangle$
I_R	Thermal noise generated in the bias and feedback resistors. Here R_T is the resultant resistance of R_b in parallel with R_f.	$Q^2_R = \frac{2kT}{R_T}\langle N^2_S\rangle$
I_G	Noise due to lossy dielectrics, with D as the effective dielectric dissipation factor of materials in the vicinity of the FET gate. Here C_d is the dielectric capacitance.	$Q^2_G = \frac{kTDC_d}{\pi}\langle N^2_{f-1}\rangle$
V_{sF}	Intrinsic thermal noise in the conducting channel of the FET. The equivalent channel resistance $R_{sF} = \frac{2}{3}g_m^{-1}$, where g_m is the transconductance of the FET.	$Q^2_{sF} = 2kR_{sF}C^2_{in}\langle N^2_\Delta\rangle$
V_f	Flicker noise, also known as 1/f noise, in the FET. Here A_f is the flicker noise constant. For a typical FET this is about 10^{-14} V^2.	$Q^2_{f-1} = \frac{A_f C^2_{in}}{2}\langle N^2_{f-1}\rangle$

[a] The noise is expressed in terms of ENC. Here k is Boltzmann's constant, T the temperature (in K), C_{in} the total capacitance at the preamplifier input; that is, $C_{in} = C_j + C_{iss}$ + stray capacitances, where C_{iss} is the gate-drain capacitance of the FET [71, 75–79].

Table 5.2 Noise coefficients and figure of merit (FM) for the different shapers[a]

Shaper	$\langle N^2_S\rangle$	$\langle N^2_\Delta\rangle$	$\langle N^2_{f-1}\rangle$	FM
CR–RC	$1.85\tau_0$	$1.85/\tau_0$	7.54	1.36
CR–RC^2	$1.28\tau_0$	$1.71/\tau_0$	6.90	1.22
CR–RC^3	$1.04\tau_0$	$1.87/\tau_0$	6.67	1.18
CR–RC^4	$0.90\tau_0$	$2.05/\tau_0$	6.56	1.17
CR^2*–RC*^{4}				1.38
DL–RC				1.10
DDL–RC				1.08

[a] The latter expresses the filter performance relative to the optimal theoretical filter ('Cusp' FM = 1) [75].

The line width of avalanche photodiodes (APD) also has contributions from avalanche multiplication noise in addition to electronic noise [61, 80]:

$$\mathrm{ENC_{APD}}(\#e^-\ \mathrm{rms}) = N\sqrt{\left(\frac{\mathrm{ENC_E}(\#e^-\ \mathrm{rms})}{MN}\right)^2 + \frac{(F-1)}{N}} \tag{5.6}$$

where M is the APD multiplication gain, N the number of primary photoelectrons generated in the APD and F the excess noise factor due to APD amplification.

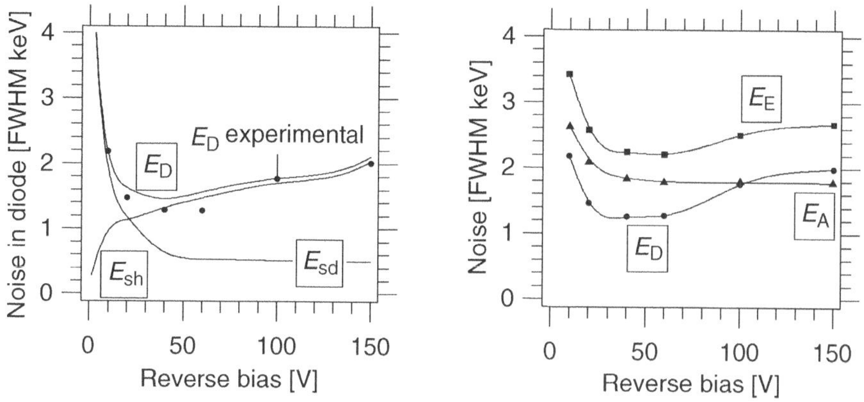

Figure 5.7 Room temperature noise properties of a silicon PIN detector and a charge-sensitive preamplifier using a CR–RC[4] shaping amplifier (Tennelec TC244) where $\tau_0 = 8\ \mu s$. The detector is identical to that whose leakage current and capacitance are plotted in Figure 4.17. The preamplifier is an Amptek A250 with a Sony 2SK152 input FET with $g_m = 50$ mS. Further $R_b = R_f = 2\ G\Omega$, i.e. $R_T = 1\ G\Omega$

Increasing the reverse bias of the silicon PIN diode increases its depletion region and decreases the junction capacitance all the way through until full depletion. According to the plot of the diode junction capacitance (C_j) shown in Figure 4.17, this happens at about 50 V reverse bias for this diode. At the same time the equivalent series resistance (R_{sd}) of the undepleted region of the diode also decreases. This affects the noise by a significant drop in E_{sd} with the reverse bias, as can be seen from the expression in Table 5.1 and the plot in Figure 5.7 (left). Also seen here is that the shot noise (E_{sh}) increases with increasing reverse bias, simply because the leakage current increases as can be seen from the plot in Figure 4.17. The decrease in junction capacitance also causes the total input capacitance of the preamplifier (C_{in}) to decrease. The total amplifier noise (E_A) consequently decreases because E_{sF} as well as $E_{f^{-1}}$ decrease (see Table 5.1). All together this means that regarding noise there is an optimal reverse bias for this type of detector. This is basically a balance between junction capacitance and leakage current.

This is less complex for other ionisation-sensing detectors, such as MSM (metal–semiconductor–metal) detectors and gaseous detectors, where the capacitance is fixed and independent of the bias. The capacitance can be calculated, provided the geometry and the dielectric properties are known:

$$C = \varepsilon_0 \varepsilon_r \frac{A}{d}$$
$$C = 2\pi \varepsilon_0 \varepsilon_r \frac{L}{\ln\left(\frac{r_C}{r_A}\right)} \tag{5.7}$$

for planar and cylindrical detectors, respectively. Here A and d are the area of the electrodes and the cathode–anode separation of the planar detector, respectively, while r_C is the inner diameter of the cathode (cylinder), r_A is the anode diameter and L the length of the cylindrical detector. Further, ε_r is the dielectric constant (relative permittivity) of the detector.

The dependency of noise on the peaking time (τ_0) of the shaping amplifier is equally important for all types of detectors [81, 82]. This is plotted in Figure 5.8 for the detector

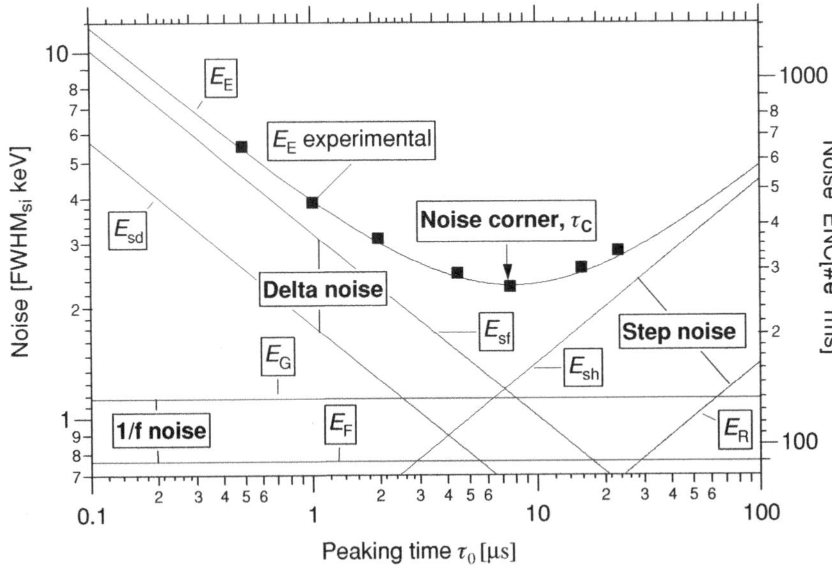

Figure 5.8 The dependency of room temperature (295 K) noise on peaking time for the system described in Figure 5.7. The silicon diode is operated fully depleted so that $I_l \approx 500$ pA, $R_{sd} \approx 30\ \Omega$ and $C_{in} \approx C_j + C_{iss} = 55$ pF. The contribution from the different noise sources has been calculated using the expressions in Table 5.1 and measurements of leakage current, junction capacitance, etc.

system used in our case study. All noise sources listed in Table 5.1 are included in this plot. The *step noise*, also referred to as *parallel noise*, increases with τ_0, whereas the *delta noise*, also known as *series noise*, decreases. There is thus always an optimal peaking time where the noise is at minimum. This, the so-called *noise corner* (τ_C), explains why there is a trade-off between optimal energy resolution ($\tau_0 \approx \tau_C$) and count-rate capability (smallest possible τ_0). In high count-rate systems it is consequently necessary to accept higher noise, provided the SNR is sufficient for triggering the counting system. Photodiode read-out of scintillation light is a good example of a case where the latter is a problem. In the BGO (scintillation crystal) spectrum displayed in Figure 4.23 the shaper has to be operated near τ_C for the signal not to be buried in noise.

There are a few other interesting observations that can be made from the plot in Figure 5.8: The so-called *1/f noise*, which is independent of the peaking time, is in most systems important only in the vicinity of the noise corner. The main contributor to 1/f noise is losses in the dielectric; however, in many cases E_G and E_F are presented as one noise source, partly because E_G is difficult to quantify. Furthermore, it is evident that the step noise is dominated by the diode shot noise (E_{sh}), which is proportional to the leakage current. Even in this case with relatively low leakage current, the thermal noise in the resistor (E_R) is relevant only at long peaking times. Moreover, the thermal channel noise in the input FET (E_{sF}) is the dominant delta noise source. In optimising a system for high count-rates this is the major enemy, and from the expression in Table 5.1 it is clear that E_{sF} is kept low by keeping the input capacitance and temperature low, and using a FET with the highest possible transconductance. Additionally, keep in mind that low capacitance also means that stray capacitances should be kept at a minimum. This also means that the preamplifier should be located as close as possible to the detector. For this reason the preamplifier is integrated in the same housing as the detector in many designs. We will take a closer look at this in the next section.

The noise expressions in Table 5.1 also show the importance of the temperature, given that some of the variables, such as the diode leakage current (I_l), have strong temperature dependence. This explains why even moderate cooling significantly improves the SNR. Note that it is equally important to cool the input FET as the detector, particularly for operation with short peaking times.

5.1.5 Electronics Design

The design of radioisotope gauges involves several stages, and the first experimental ones are trials in the laboratory. This is also true for the read-out electronics. Even though all circuitry may be modelled using, for instance, PSpice or MathLab, it is important to carry out experiments, especially for design of low-noise electronics where physical properties with major importance may not be taken into account in the models. It is then very handy and efficient to use so-called *nuclear instrument modules* (NIM) where the various parts of the detector read-out system are available as separate modules that fit into a rack, so-called NIM-bin, with a common power supply. Typical modules are

- Bias supply with adjustable voltage, polarity switch and current limiter.
- Shaping amplifier with adjustable peaking time and gain, selectable shaping filter, various options for BLR and PUR, etc.
- Multichannel analyser, e.g. pulse height analyser (PHA), interfaced to a personal computer where a variety of functions are available through software.
- Preamplifiers are most often separate units in order that they may be located as close as possible to the detector. These preamplifiers also have separate test inputs that allow controlled amounts of charge to be fed to the preamplifier input stage. The bias filter and decoupling capacitor are also normally integrated into the preamplifier box.
- Precision pulser with adjustable pulse amplitude and repetition rate. The output pulses are step signals fed through a high pass (RC) filter at the preamplifier test input, creating constant charge pulses equivalent to those of detectors.

This system, which has become a very much used standard, allows for great flexibility when carrying out laboratory experiments and determining the performance of different configurations. For industrial field operation NIM is less used, except for situations where back-end units such as multichannel analysers, counters, etc., are placed in the vicinity of the process in control rooms. Most often permanently installed gauges use dedicated electronics as an integral part of the gauge housing. Current loops or fieldbus technology is then used for communication to the process control system.

The most straightforward way of designing read-out electronics for radiation detectors is to use hybrid circuits based on surface mounted technology. There are also VLSI silicon chips available, but these are mainly used for multiple channel systems such as cameras. The design of low-noise front-end electronics can be quite a challenge. By using hybrid circuits optimal performance is achieved at a relatively low cost. There is only one practical way to achieve impedance matching between the preamplifier input stage and the detector without introducing additional noise components: This is to select a FET whose input capacitance (C_{iss}) is in the same range as the capacitance of the detector. Some hybrid circuit preamplifiers are designed to use external input FET (and feedback resistor) so that it can be matched to different detectors. Other hybrid preamplifiers are manufactured in different versions with different input FETs and feedback resistors. To make selection easier data sheets often have plots of amplifier noise (E_A) as a function of input capacitance for different FET versions. For low-noise and compact design, so-called flip-chip mounting of the preamplifier chip on the back of the detector chip is used. Some silicon detectors even have the preamplifier integrated on the detector chip to reduce noise [83, 84, 231]. Hybrid circuits are also available for shaping amplifiers, BLRs, etc., although these, compared to preamplifiers, are more easily realised by using operational amplifiers without loss of performance. Some design examples of the latter are given in Section B.5.

Generally in measurement science grounding and shielding of the sensor and its read-out electronics are important to achieve optimal performance. This is critical for charge-sensitive preamplifiers because of the low signal level and the high impedance at its input. It is beyond the scope of this book to go into this in detail; however, there are certain precautions that should be taken: So-called earth loops should be avoided near the preamplifier input. This implies careful grounding of coaxial cable screens and detector housing. The latter should be made in a material with high electrical conductivity, such as aluminium. Radioisotope gauges are often operated in a noisy environment. For the power and bias supply transformer shielding is therefore recommended to eliminate stray common-mode coupling, as well as other sources of pickup. There should be separate analogue and digital ground to reduce the influence of high frequency digital switching noise. There is excellent literature available on low noise design and grounding and shielding [85–88]. All nucleonic equipment on the market must be tested to ensure that it is both immune to interference from and does not emit electromagnetic fields.

We mentioned in the introduction to this chapter that there has been a trend where computers and software replace hardware solutions. This is generally true for measurement science, but also for radiation detector read-out systems. In some laboratory instrumentation, analogue electronics of pulse mode systems are being replaced with digital signal processors (DSPs), which sample and digitise the preamplifier output directly at high speeds. Noise filtering, BLR, PUR, PHA and so forth are then efficiently implemented by software with the added advantage of high flexibility [85, 89–93]. A careful extrapolation into the future suggests that this will also be the case for permanently installed gauges. DSP techniques will play a key role in any future instrument development.

Often radioisotope gauges also have to comply with legislation for operation in explosive atmospheres. The two common methods of achieving electrical safety are intrinsic safety and explosion proof. Intrinsic safety is achieved by design of the electrical circuits. The values of components that can store electrical charge, such as capacitors and inductors, must be below certain limits. All currents are also limited so that no spark with sufficient energy to ignite an explosive mixture can be produced, even under fault conditions. Explosion proof is achieved by designing the detector housing in such a way that even if the electronics develop a sparking fault, and the housing is full of explosive gas, the resulting explosion will be contained within the housing. The containment is accomplished by firstly designing the housing to withstand the pressure of the explosion without bursting and secondly ensuring that any leakage paths are long enough to ensure that the escaping gases are too cold to ignite any surrounding explosive mixture [94, 95] (gas environment) and [96] (dust environment).

5.2 DATA PROCESSING ELECTRONICS AND METHODS

The output from a shaping amplifier generally carries information about five properties of the radiation, which may be used for different purposes. One of these, *the pulse shape* (including rise time), is primarily used inherently for optimising the detector system for measurement of one of the other properties: for instance discrimination of signal contribution from slow positive charge carriers in systems requiring high-energy resolution. This is less of interest in the context of this book and we shall not dwell upon it. The other four properties are used, as listed in Section 4.3, to determine

- The number of output signals per time and through that radiation beam intensity.
- The pulse height and through that the detected energy.
- The arrival time of each pulse and through that the time of interaction. This is in most cases used relative to some other event, for instance from another detector.
- The interaction position and through that some sort of geometrical information such as the origin of the radiation.

5.2.1 Intensity Measurement

The signal analysing functionality required for radioisotope gauges is in most cases fairly simple. Very often it is a question of measuring the *radiation intensity* as a function of time. This is done by counting amplifier output pulses with amplitude above a certain threshold or within a certain amplitude range. The counting is carried out continuously in time intervals with fixed length, known as the *integration, measurement* or *counting time* (or *interval* or *period*), τ_I. If now the number of counts recorded in this interval is n_C, the count-rate is simply given as

$$n = \frac{n_C}{\tau_I} \tag{5.8}$$

Its unit is thus counts per second, c/s or sometimes cps. Note that n often is referred to interchangeably as count-rate and number of counts. The *radiation intensity* is in many situations set equal to the count-rate in the detector ($I = n$) simply because the intensity is relative to another intensity measured with the same detector, for instance I_0, as illustrated in Figure 3.5. The stopping efficiency of the detector needs to be known to determine the true incident intensity. In some cases the *flux* (Φ), which specifies the intensity per beam area or solid angle relative to a point source, is used instead of intensity.

The *discriminator* is used to count shaper output pulses with amplitude above a certain threshold. Electronically it may be realised by using a comparator with two analogue inputs

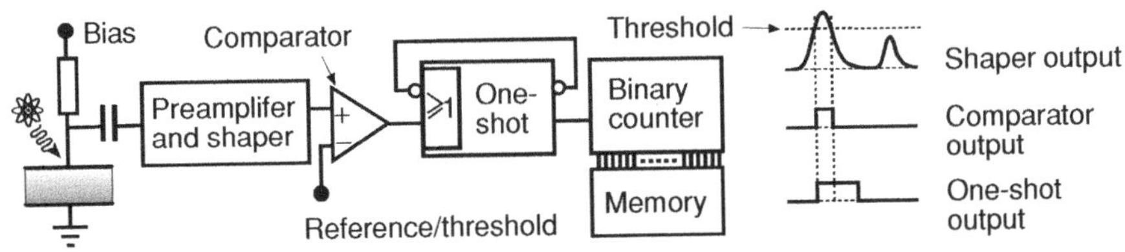

Figure 5.9 Outline of a discriminator used for pulse counting. The pulse length of the one-shot (monostable multivibrator) output is set by an external RC time constant

and one digital output: The output is false as long as the input from the shaper output is below the reference level, which is the other input, and true otherwise. The reference level, the *counting threshold*, is either set by a resistive voltage divider or provided by a DAC-(digital-to-analogue converter) output, which allows the threshold to be programmed. An outline of this circuitry is shown in Figure 5.9. Positioning of the threshold will be discussed in Section 5.4.6.

There is always noise superimposed on the shaper output signal. This, which is not shown in the illustration in Figure 5.9, may be sufficient to cause multiple triggers in the comparator on each input signal and produce several spikes on its output. For this reason a monostable multivibrator, a one-shot, is used between the comparator and the binary counter to ensure that each qualified shaper output pulse is counted only once. The one-shot provides an output signal with fixed duration, typically set equal to the pulse width of the shaper output signal. Normally it is also configured to be non-retriggerable so that the length of its output signal is constant and independent of possible multiple false trigger pulses. Often a Schmitt trigger is used instead of the comparator and the one-shot. It provides very much the same functionality because of its built in hysteresis. The data on the output of the binary counter is transferred to a memory, in its simplest form a shift register, at the end of each counting interval. The counter is then reset for a new interval at the same time as the data in the memory is transferred further to the processing unit.

The *single channel analyser* (SCA) is used to count shaper output pulses within a certain amplitude range, a so-called *energy window*. We shall see that this is very useful for a variety of purposes. This is realised very much as illustrated in Figure 5.9, but by duplicating the circuitry behind the shaper output so that this is fed into two comparator inputs. This therefore requires two trigger thresholds, denoted with subscripts 1 and 2 (e.g. H_1) for the lowest and highest, respectively. In some NIM SCAs this is set by H_1 and ΔH (where ΔH is the width of the window), i.e. $\Delta H = H_2 - H_1$. The window count is found either by subtracting the number of counts in counter 2 from that of counter 1, or by introducing a logic circuitry behind the one-shots and using only one counter. Some applications require the use of multiple windows. In this case the use of separate counters for each trigger threshold gives the highest flexibility.

The *scaler* is traditionally associated with radiation intensity measurement. This is basically a counter as outlined above, originally with a display, but also available with interface to other data recording equipment. The count-rate meter, or simply *rate-meter*, is an analogue display instrument showing the average measured intensity. It has a built-in RC circuit that smoothes the random emission and detection of nuclear radiation. However, permanently installed radioisotope gauges rarely use rate-meter devices. These gauges normally use binary counting systems as a part of their signal processing circuitry, as explained above.

The electronics associated with intensity measurement is apparently not very complex. The greater challenge lies in the interpretation of the data with respect to measurement errors caused by the random emission of radiation, background radiation, etc. We will deal with this in Section 5.3.

5.2.2 Energy Measurement

The measurement of radiation energy is closely related to the counting of radiation events dealt with in the previous section. The *pulse height analyser* (PHA) may be regarded as a large number of continuous SCAs where the number of counts in each channel is recorded continuously and displayed. The result is a spectrum that may be regarded as a histogram where a large number of output signals from a pulse mode system are sorted according to their amplitude (which is proportional to energy deposition). A more correct name would thus be *pulse height distribution analyser*. This is a very important instrument, not necessarily as part of a gauge in the field, but as a tool in the laboratory. A pulse height spectrum reveals many properties of radiation detector systems that are otherwise difficult to discover just by studying the shaper output pulses on an oscilloscope. We have seen several examples of this in this and the previous chapter. It may, for instance, be used to measure line width (FWHM) and SNR, so as to reveal distortions such as pile-up and unwanted radiation from collimators and shields.

Actually, the PHA is just one of several operational modes of the *multichannel analyser* (MCA) whose general function is to sort and count pulses, and to store and display the result. In PHA mode the sort criterion is the pulse amplitude. The basic components of a typical PHA are shown in Figure 5.10. The output signal from the shaper is supplied to an SCA where it is determined whether its amplitude is within the window defined by the adjustable lower and higher level discriminators, LLD and ULD respectively. If so, the linear gate will be kept open to accept the delayed shaper output signal. Depending on the type of ADC this signal is fed through a pulse stretcher which detects the pulse amplitude and holds this sufficiently long for the ADC to finish the conversion. Once it has the content of the memory location equal to the digital number on the ADC, it is incremented by one, i.e. the event is counted in this channel.

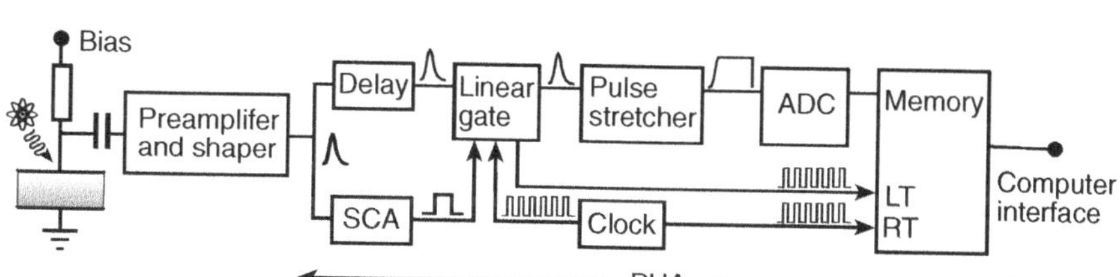

Figure 5.10 Outline of a PHA using the computer for analysis, display and storage of data. Not shown are the computer programmable control unit and all its control signals. The pulse stretcher is also known as peak-find-and-hold unit

Different types of ADCs are used in commercial PHAs: The *Wilkinson* or *linear ramp* ADC has been very popular because of its excellent linearity. On the other hand the linearity of modern *successive approximation* ADC is also very good, making it an attractive alternative because it is normally faster than the Wilkinson ADC. The processing (conversion) time of the PHA depends on the ADC type, number of channels, clock frequency, etc., but is typically in the range of 10–50 μs. Depending on the pulse rate this means there is a probability for a pulse to arrive while the system is busy processing the previous event. The *dead time* of the system is the difference between the *real time* (RT) and the *live time* (LT). The latter is the time the gate is open and the system is not available for new pulses to be processed (see Figure 5.10). The dead time is displayed (in %) while the system is running. If it is high, adjusting LLD and ULD should be considered, particularly the former so as to avoid spending time analysing noise pulses in the lower end of the spectrum. The discussion in Section 5.4.6 is very relevant in this context. The data acquisition time in a PHA may in most cases be preset, either as RT or as LT. The number of memory locations or channels varies from 256 (2^8) to 16,384 (2^{14}). These may be configured in groups so that a 16,384-channel PHA, for instance, may sequentially acquire and store 8 spectra each with 2048 channels.

Today, computer-interfaced PHAs are very common, either as stand-alone or as NIM units. The PHA has a variety of useful functions for spectral analysis. The user may, for instance, mark a peak in the spectrum as a so-called region of interest (ROI). The centroid and FWHM of the peak are then reported as well as integral or *gross counts* and *net counts* in the ROI. Net counts equal gross counts with the background continuum subtracted. These numbers are often also given with uncertainties (see Section 5.4.8). The peak centroid and FWHM are given in number of channels, or in energy, provided the PHA energy calibration feature is used. In addition to the LLD and ULD, the zero offset of the PHA may also be adjusted. Note that once an energy calibration is performed, this is valid only for the present settings of offset and gain (in the PHA and in the shaper if this has adjustable gain).

PHA units are primarily laboratory tools and seldom used for industrial field gauges. These also often need energy analysis functionality, but in most cases the energies to look for are known. For this reason it is more common to use multiple windows covering the pulse height (energy) ranges in question, rather than a full multichannel analyser. We will see examples of this in the succeeding chapters. Using multiple windows is faster allowing higher count rates unless a PHA with a *flash* ADC is used.

5.2.3 Time Measurement

The essence of time measurement, or *timing*, is to determine the time of interaction in a detector with high precision, and very often relative to another event, for instance in another detector. We know that in some detectors large charge collection times and dependency on interaction position may cause variations in the measured time. Also, using the comparator output of the circuit shown in Figure 5.9 for measurement of interaction time would be highly dependent on the amplitude of the output signal. This is because the pulse width is constant and independent of amplitude. A pulse with large amplitude would thus trigger the comparator earlier than one with small amplitude. The error caused by this effect is referred to as *walk*. *Jitter* is another error produced by noise, which is always superimposed on the signal.

Nuclear physics experiments often require timing in the sub nanosecond region. This is commonly referred to as *fast timing* and performed on unshaped pulses from fast detectors. *Slow timing* is in the range of one to a few tens of nanoseconds. Timing is rarely relevant for field-based industrial gauges; however, two examples of laboratory methods illustrate the usefulness of timing: positron emission tomography (PET; see Section 5.5.4) and Compton suppression systems (see Section 5.4.9). The fundamental principle of PET is to use two or more detectors and determine whether simultaneous interactions in any two of these detectors are caused by two back-to-back photons emitted from the same annihilation process. These detectors are thus operated in *coincidence*. Two events are accepted as coincidental if they arrive at the electronics within the resolving time τ_R of the system. The requirement for PET timing is typically within the slow timing range [97, 98]. Inevitably some events will be registered accidentally as in coincidence. If now the count-rates of two detector systems operated in coincidence are n_1 and n_2, the accidental rate equals $2\tau_R n_1 n_2$ [97, 99]. A Compton suppression system has a spectrometry (high-energy resolution) semiconductor detector surrounded by several scintillation detectors. The system performance is improved by rejecting as many Compton interactions as possible, accepting only full energy interactions. So if any two events are simultaneously detected in any of the surrounding detectors and in the spectrometry detector, this is interpreted as a Compton interaction and thus rejected (see Section 5.4.9). These detectors are said to operate in *anticoincidence*, and again a time resolution in the slow-timing range is acceptable.

Although best timing resolution requires unshaped signals, slow timing is achieved by the use of shapers with very short peaking time. In many cases, such as pile-up rejection as discussed in Section 5.1.3, fast shaping optimised for timing is combined with slower shaping optimised for energy measurement. This means there are two parallel shaping amplifiers connected to the preamplifier output.

5.2.4 Position Measurement

The position-sensitive detector (PSD) presented in Section 4.7 is the cornerstone in applications where the origination or interaction position of the radiation needs to be determined. However, in many cases shielding and collimation of the source, the PSD or both play an equally important role. Collimation is discussed further in Section 5.4.3.

Imaging of remote γ-ray and X-ray emitters, for instance in astrophysics and space physics, are typical examples of the first category where the source position and distribution are to be determined. The *pinhole camera* is used for this purpose, as illustrated in Figure 5.11. Its principle is based on the camera obscura introduced in Chapter 1: The radiation can enter only through the pinhole on top of pyramidal- or conical-graded shielding,* producing a mirrored image of the projected distribution of the radiation source in the 2D PSD. The second example of the first category is the principle of PET (see Section 5.5.4). Here the position and distribution of a β^+-source is found by detecting the interaction

* A so-called *coded aperture* arrangement is often used instead of the pinhole concept so as to increase the number of detected events from the source (the geometrical factor) [67].

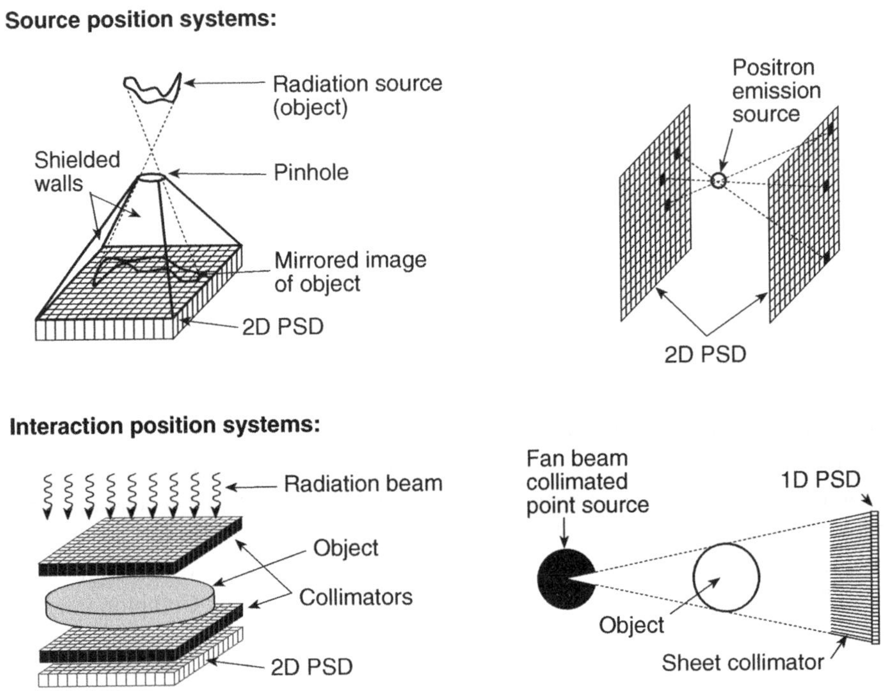

Figure 5.11 Typical systems using PSDs for determining the source position (top) and the interaction position (bottom). The interaction position may also be determined by using systems combining 1D position sensitivity with mechanical scanning, such as the Anger camera

positions of several back-to-back annihilation photon pairs in two or more 2D PSDs operated in coincidence. This and related imaging principles are powerful tools for medical diagnostics and for analysis and investigation of industrial processes and their dynamics. It has, as the pinhole camera, little relevance for permanently installed gauges. For these the second category is of greater interest.

In radiography the attenuation of a collimated radiation beam is measured in a 2D PSD, as shown in Figure 5.11. This is projection imaging where the information lies in the attenuation measured at each interaction position in the detector. However, the most appropriate methods for permanently installed gauges are those based on a fan beam collimated radiation source. The examples shown in Figure 5.11 have all the detectors in the same plane. It is often more convenient to use other geometries like in X-ray computed tomography where the 1D PSD is a ring around the cross section of the object (see Section 7.7.4).

From all the examples given here it is clear that position-sensitive measurements require a combination of a PSD and some kind of collimation or shielding. The latter is to reduce the impact of scattered or unwanted radiation, and the geometry of this is often equally as important as the detector to the performance of the total system (see Sections 5.4.2 and 5.4.3).

For instantaneous position measurement we have to use an approach similar to or based on those presented in Figure 5.11. In some cases, however, the same information may be acquired by scanning, although this is not preferred in industrial gauges because any mechanical movement increases the need for maintenance and risk of failure. Nevertheless, it needs mentioning as a low-cost alternative: Position sensitivity may be obtained by using only one source and one detector in a scanning system. Such a system is based on moving the

- source,
- detector,
- collimator,

or any combinations of these. The movement of a small detector (or one using narrow collimation) can yield position information like that from the systems in Figure 5.11. But it is also possible to use one or several large area detectors and move a narrow collimator slit in front of each. Or, as already mentioned, a combination of these methods may be used.

Scanning may be realised by manually moving the involved parts, but then with a limited accuracy. Automatic scanning systems are often based on stepper motors with appropriate gear and position feedback mechanisms. Combined with precision machining of the system a spatial resolution of less than 1 mm is achievable, depending on the actual system. Precision scanning is most readily realised with low-energy sources because these require less shielding and thus have less mass to be moved.

Except for the drawback of mechanical moving parts, scanning systems are limited to measurements on processes with time constants typically 1 order of magnitude longer than the scanning time. On the other hand, it may also be used on such processes where temporal averaging of the process dynamics is sufficient.

5.3 MEASUREMENT ACCURACY

5.3.1 The Measuring Result

All *measuring results* are twofold: the *measured value* and its *uncertainty* [100]. The first may be regarded as an estimation of the *true value* of the *measurand*, also known as the *input quantity*, whereas the second is an expression of how accurate this estimation is. This is important but every so often subject to negligence. In this context it may also be useful to clarify some other concepts: *Accuracy* is defined as the closeness of the agreement between the result of a measurement and the true value of the measurand. Accuracy is a qualitative concept [24]. The *error* of measurement is defined as the result of the measurement minus a true value of the measurand [24]. Error is thus a quantitative concept, but in practice impossible to specify because the true value by nature is indeterminate. In practice, error is used as an expression of the measurement uncertainty, which, to close the circle, is an estimation of the error.

The final measuring result (*output estimate*) is very often a function of several individual measurements (*input estimates*) of input quantities [100, 101]. For instance, when measuring the area (number of counts) of a peak in a PHA spectrum the spectral background has to be measured and subtracted (see Section 5.4.8). Very often the final measuring result includes measurements by totally different measurement principles such as pressure and temperature. In either case the total or final measuring result (output estimate) of the output quantity is often a function of several input estimates $x_1, x_2, x_3, \ldots, x_n$ and their associated uncertainties $u(x_1), u(x_2), u(x_3), \ldots, u(x_n)$:

$$y = f(x_1, x_2, x_3, \ldots, x_n) \tag{5.9}$$

Finally, if, for instance, the high bias voltage of a scintillation detector is included in y (which often is the case), this is also an input estimate because it has an associated uncertainty and cannot be regarded as a constant.

5.3.2 Estimation of Measurement Uncertainty

The consequence of the two-sided *measuring result* is that providing the measured value is not the end of the job; its *uncertainty* also must be estimated and provided. This is done either by experiments, by using knowledge about the nature of the measuring system, or by using data provided, for instance, by the instrument manufacturer. The ISO Guide *Guide to the Expression of Uncertainty in Measurement* [100] is internationally recognised in this context. Further, EA-4/02 *Expression of the Uncertainty of Measurement in Calibration* [101], which is based on the ISO Guide, is also very useful. There are two approaches as to how the uncertainty is estimated:

- *Type A evaluation of uncertainty* by statistical analysis of a series of observations.
- *Type B evaluation of uncertainty* by means other than statistical analysis of a series of observations.

Statistical analysis may be performed by a series of many independent measurements or observations $(q_1, q_2, q_3, \ldots, q_n)$ of an input quantity under the same conditions; that is, the true value of the input quantity is kept constant. If there is sufficient resolution in the measurement process, there will be an observable scatter or spread in the values obtained. We can display this spread graphically by first sorting and counting all the measurements after their value. We then make a histogram with the measured value along the x-axis and the number of observations at each value along the y-axis. Then by generating the curve of an envelope around this histogram, we most often end up with a plot as shown in Figure 5.12. This is generally known as a *probability distribution*, and the particular one in Figure 5.12 is the *Gaussian* or *normal distribution* $p(q)$.

The *standard uncertainty* $u(x_i)$ of an input estimate is most often defined as one standard deviation (σ) of the average of all observed values ($\bar{q}$): $u(x_i) = \sigma$. This means, according to the table in Figure 5.12, there is a 68.3% probability that a measured value (input estimate) of an input quantity (x_i) will be within the *confidence interval* from $-\sigma$ to σ. Or in other words, 68.3% of all observations will be within this range. The *coverage factor* (k) is unity in this case. In some situations it is preferred to use the *expanded uncertainty* of the measurement. For a Gaussian distribution this is defined at the coverage interval with $k = 2$, that is -2σ to 2σ, corresponding to 95.5% coverage probability.

The distribution in the measured energy of a γ-ray emission line in a PHA spectrum is Gaussian, as we discussed in Section 4.3.1. Counting statistics is another very important

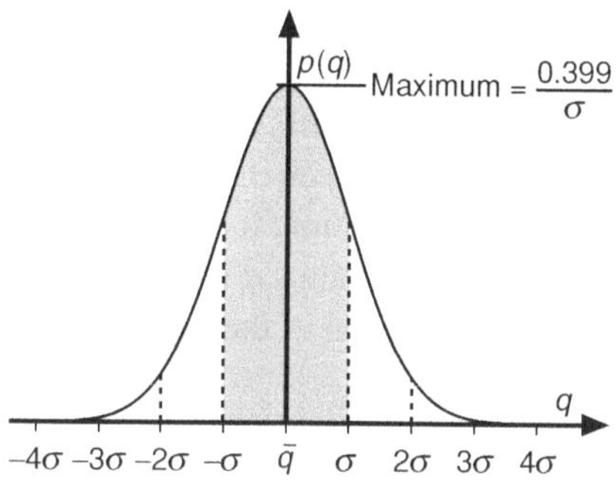

Confidence limits	Coverage factor k	Probability
$\pm\sigma$	1	68.3%
$\pm 1.645\sigma$	1.645	90.0%
$\pm 1.96\sigma$	1.96	95.0%
$\pm 2\sigma$	2	95.5%
$\pm 2.326\sigma$	2.326	98.0%
$\pm 2.576\sigma$	2.576	99.0%
$\pm 3\sigma$	3	99.9%

Figure 5.12 The Gaussian distribution where σ is the standard deviation and q is the observed (measured) value. The shaded area within $\bar{q} \pm \sigma$ is 68.3% of the total area of the envelope curve

application of statistical analysis to radioisotope gauging. Because of the random nature of the radiation emission the time between succeeding radioisotope disintegrations is not constant. This means that even if the number of counts (n_C) in a limited time is measured very accurately by a pulse mode counting system, it is only an estimate of the true average number of emissions in the counting period. Hence it also has a corresponding uncertainty. The random emission of radiation follows the so-called *Poisson distribution*, provided the observation time is small compared to the half-life of the source. For a large number of counts the Gaussian distribution also adequately describes the emission process. This is normally the case for applications of radioisotope gauges. Both the Poisson and Gaussian distributions make the fundamentally important prediction:

$$\sigma = \sqrt{\bar{q}} \tag{5.10}$$

This means that in the case of radioisotope pulse counting the standard deviation, or standard uncertainty, may be estimated as the square root of the number of counts: $\sigma = \sqrt{n_C}$ for large number of counts, say $n_C > 100$. This is because the single count n_C is the best estimation of the true average number of counts. This relationship is not valid when the counting time (τ_I) is long compared to the half-life of the source isotope. This is, however, seldom the case for radioisotope gauges.

Type B evaluation of uncertainty is based on a scientific judgement using all available information predicting something about how accurate the measurement will be. Such information may be derived from

- Previous measurement data.
- Experience with or general knowledge of the behaviour and properties of relevant materials and instruments.
- Manufacturer's specifications.
- Data provided in calibration and other certificates.
- Uncertainties assigned to reference data taken from handbooks.

This is then used to estimate the standard deviation, which is at 68.3% coverage probability. For some instruments the manufacturers say the true value will always be within $\pm a$ of the measured value. We may regard this as a rectangular probability distribution where the probability for a measured value (q) to occur anywhere between $\pm a$ is equal; i.e., it is not highest for the average value ($\bar{q}$) as for the Gaussian distribution. It can then be shown that the standard uncertainty $u(x_i) = \surd(a/3)$. For most instruments a is given relative to the *actual reading*, or relative to the *full range*, or a combination of these.

5.3.3 Error Propagation and Uncertainty Budget

When all the standard uncertainties of the input estimates, $u(x_1)$, $u(x_2)$, $u(x_3), \ldots, u(x_n)$, are known, we can calculate the standard uncertainty of y expressed in Equation (5.9). Provided the input quantities are not correlated this *combined uncertainty* is expressed by the error propagation formula:

$$u_c(y) = \sqrt{\left(\frac{\partial f}{\partial x_1}\right)^2 u^2(x_1) + \left(\frac{\partial f}{\partial x_2}\right)^2 u^2(x_2) + \cdots + \left(\frac{\partial f}{\partial x_n}\right)^2 u^2(x_n)}$$
$$= \sqrt{\sum_{i=1}^{n}\left(\frac{\partial f}{\partial x_i}\right)^2 u^2(x_i)} = \sqrt{\sum_{i=1}^{n} c_i^2 u^2(x_i)} \tag{5.11}$$

Here the *sensitivity coefficient* is defined as

$$c_i = \frac{\partial f}{\partial x_i} \tag{5.12}$$

In some cases each of the uncertainties of the different input quantities contributes equally to the combined uncertainty, i.e. all $c_i = 1$. It is then relatively straightforward to analyse the relative importance of the different contributions. Such analysis is important when determining which input quantity to deal with in order to reduce the combined measurement uncertainty. Such analysis is a lot more difficult in cases where all $c_i \neq 1$. The uncertainty budget shown in Table 5.3 is then recommended. This may be implemented in a spreadsheet and is very handy for uncertainty analysis. The importance and contribution of the different input quantities are then exposed.

5.3.4 Pulse Counting Statistics and Counting Errors

All radioisotope intensity measurements are subject to statistical fluctuations in the number of counts because of the random nature of photon or particle emission. As we saw in Section 5.3.2, for realistic (say >100) numbers of detected events, n_C, the relative standard deviation is inversely proportional to the square root of the number of detected photons:

$$\sigma_{n_C} = \sqrt{n_C} \quad \text{so that} \quad \frac{\sigma_{n_C}}{n_C} = \frac{1}{\sqrt{n_C}} \tag{5.13}$$

Table 5.3 Uncertainty budget as recommended by ISO Guide EA-4/02[a] [101]

Quantity X_i	Estimate x_i	Standard uncertainty $u(x_i)$	Sensitivity coefficient c_i	Contribution to total uncertainty $u_i(y) = c_i u(x_i)$
X_1	x_1	$u(x_1)$	c_1	$u_1(y)$
X_2	x_2	$u(x_2)$	c_2	$u_2(y)$
$\vdots$	$\vdots$	$\vdots$	$\vdots$	$\vdots$
X_n	x_n	$u(x_n)$	c_n	$u_n(y)$
Y	y			$u(y)$

[a] This has many excellent examples on how the uncertainty budget may be used. Note the importance of specifying all uncertainties as standard uncertainties ($k = 1$). The combined standard uncertainty ($u_c(y)$) may then be expanded ($k = 2$) or specified for any other coverage probability afterwards.

This statistical error will propagate and influence the accuracy by which, for instance, radiation transmission measurements can be carried out. γ-Ray and β-particle transmission is according to Lambert–Beer's exponential decay law, as stated in Equation (3.7). This equation can be solved with respect to the attenuation coefficient or the absorber thickness, as will be discussed in Section 5.5.1:

$$\mu = \frac{1}{x}\ln\left(\frac{I_0}{I}\right) \quad \text{and} \quad x = \frac{1}{\mu}\ln\left(\frac{I_0}{I}\right) \tag{5.14}$$

In Section B.4.1 it is shown how the statistical error, given Equation (5.13), results in relative standard deviations in the measurement functions in Equation (5.14):

$$\frac{\sigma_\mu}{\mu} = \frac{1}{\mu x}\sqrt{\frac{e^{\mu x}}{I_0 \tau_I}} \quad \text{and} \quad \frac{\sigma_x}{x} = \frac{1}{\mu x}\sqrt{\frac{e^{\mu x}}{I_0 \tau_I}} \tag{5.15}$$

when it is assumed that the error in the incident beam intensity, I_0, is negligible. This is realistic since I_0 can be determined with high accuracy through initial calibration measurements of longer duration (τ_0). The relative error given in Equation (5.15) is plotted as a function of the relaxation length (μx) and the total number of incident photons or particles ($I_0 \tau_I$) in Figure 5.13.

These plots reveal some very important properties for pulse counting transmission measurements:

1. The relative error is at minimum when $\mu x = 2.0$ (86% attenuation), that is, $\mu_M \rho x = 2.0$ when using the mass attenuation coefficient. This is established through derivation of Equations (5.15) as shown in Section B.4.1.
2. Increasing the total number of incident photons ($I_0 \tau_I$), by increasing either I_0, τ_I or both, increases the number of counts in the detector count-rate and reduces the error.
3. The relative reduction in the error with increased number of counts is significantly higher for μx-values outside the optimal value, making long counting time or high incident intensity more important in this instance.

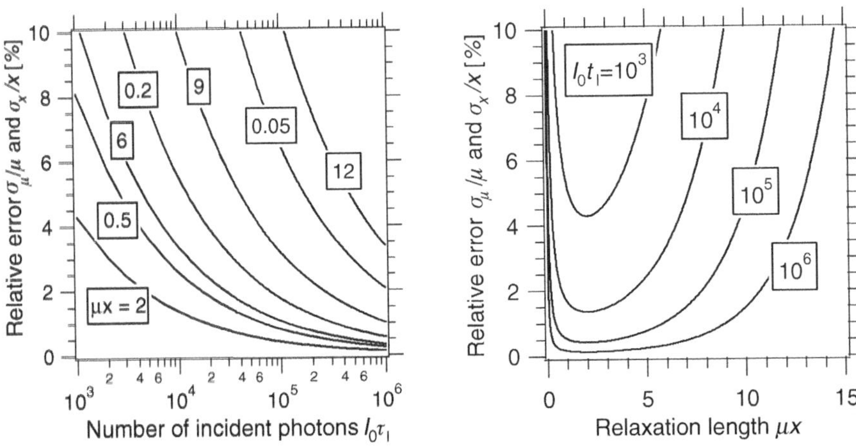

Figure 5.13 Relative error (standard deviation) in the measured average linear attenuation coefficient or thickness as a result of statistical fluctuations in the measured beam intensity, assuming the error in the incident intensity I_0 is negligible

4. The relative error reduction with increased number of counts is highest for low $I_0\tau_I$-values, making long counting time or high incident intensity most important here.

Minimisation of statistical errors is an important design criterion for radioisotope gauges. We will study an example of this in Section 8.3. This applies to any kind of gauges, not only transmission gauges. There is always a trade-off between speed of response (short τ_I) on one hand and accuracy (long τ_I) on the other. For γ-ray gauges we will see in Chapter 6 that, from a safety point of view, the solution is not to increase the incident radiation intensity (source activity). However, here the radiation energy plays a more important role than the intensity.

In the energy region where Compton scattering is dominant, say between 100 keV and 1.5 MeV, the mass attenuation (μ_M) may be regarded as constant (see Section 3.3) for a given energy, and the attenuation is a function of the density (ρ). Equation (5.14) may then be expressed as

$$\rho = \frac{1}{\mu_M x} \ln\left(\frac{I_0}{I}\right) \tag{5.16}$$

The standard deviation in ρ due to statistical errors is (as shown in Section B.4.2)

$$\sigma_\rho = \frac{1}{\mu_M x \sqrt{I\tau_I}} \tag{5.17}$$

5.3.5 Probability of False Alarm

Many nucleonic instruments are configured as high-level switches: if the detector count-rate falls below a preset value, the relay contained within the instrument de-energises to provide an alarm. But we have seen that in normal operation the count-rate is subject to statistical fluctuations. What is the probability that the count-rate fluctuations will cause a false alarm?

From the table in Figure 5.12 you can determine that the probability that any count in a series of counts is less than -3σ away from the mean value is 0.9987. So, the probability that n consecutive count measurements will be less than -3σ from the mean is given by $P = (0.9987)^n$. If the alarm is set to operate 3σ below the count-rate observed during normal instrument operation, P represents the probability that no false alarms will occur, and $(1 - P)$ is the probability that at least one false alarm will be observed. If the instrument operates with a 1-s time constant, then there are about 31,536,000 consecutive count periods in a year. So, in this case, $P = (0.9987)^{31,536,000} \approx 0$ and $(1 - P) \approx 1$. We are thus almost certain to get at least one false alarm in a year. Clearly, the alarm must be set more than 3 standard deviations below the count-rate observed in normal operation if we are to avoid false alarms. If the alarm is set to operate 6σ below the count-rate observed in normal operation, the probability of any reading being $>$(mean -6σ) is 0.999999997. In this case, $P = (0.999999997)^{31,536,000} \approx 0.91$ and $(1 - P) \approx 0.1$ and so there is a 10% chance of getting at least one false alarm per year when a 1-s time constant is used. If a 10-s time constant is employed, the chance of getting a false alarm will reduce to 1% per year.

Whatever time constant is employed, you can see that the alarm setting must be more than 6σ below the mean count-rate observed in normal operation if the risk of false alarms is to be avoided. The risk of getting a false alarm is infinitesimally small if the alarm is set to operate 7σ below the mean count-rate observed in normal operating conditions.

5.3.6 Energy Resolution

In Section 4.3.1 we defined the concept of energy resolution of a radiation detector system to be its ability to resolve radiation energies. In practice, it is not difficult to discriminate peaks whose centroids are separated by three FWHMs or more. For one FWHM separation the use of deconvolution algorithms is required. The total energy resolution or its line width (E_T) has contributions from different sources in the radiation detector system. These are not correlated, and when expressed identically such as FWHM keV (see Section 5.1.4), they add up as given by the error propagation formula in Equation (5.11) with all $c_i = 1$. The total line width including the most relevant contributions is then

$$E_{\mathrm{T}} = \sqrt{E_{\mathrm{S}}^2 + E_{\mathrm{E}}^2 + E_{\mathrm{d}}^2 + \cdots} \tag{5.18}$$

Here E_S is the line width contribution from statistical fluctuations in the signal generation in the detector, E_E the electronic noise contribution and E_d the line width contribution from drift during the measurement. As mentioned numerous times before, temperature variations are the most important source for the latter. Each one of these contributions can also be broken into sub-contributions, such as E_E given in Equation (5.5) in the case of semiconductor detectors. As indicated in Equation (5.18) there may also be further contributions depending on the detector system in question: Many compound semiconductor detectors, for instance, have a contribution from incomplete charge collection. Scintillation detectors have a contribution called intrinsic effective line width (E_I) caused by the nonlinear dependence of scintillation light production to the energy of the secondary electrons.

The line width contribution from statistical fluctuations (E_S) may be estimated by using charge carrier statistics, analogous in Equation (5.8) to the number of counts. If now N is the number of charge carriers, its predicted standard deviation is $\sigma = \sqrt{N}$, which thus is the inherent statistical fluctuation in N. Further N may be expressed in terms of the detected energy. If we now assume that the full energy of a γ-ray photon (E_γ) is deposited in the detector, the number of charge carriers will be $N = E_\gamma/w$, where w is the average energy required to generate one charge carrier pair (as explained in Sections 4.2.5 and 4.2.6). Based on the Gaussian distribution, Equation (4.7) states that the line width equals 2.35σ, and in this case, $2.35\sqrt{N}$. In terms of energy the so-called Poisson prediction of the line width contribution from statistical fluctuations in the charge generation may then be expressed as

$$E_\mathrm{S}^\mathrm{P} = 2.35w\sqrt{\frac{E_\gamma}{w}} = 2.35\sqrt{E_\gamma w} \tag{5.19}$$

This turns out to be fairly accurate for scintillation detectors; however, for semiconductor detectors the observed line width is much less. To cope with this an empirical factor, the *Fano factor* (F), is introduced to account for the difference. The standard deviation is then expressed as $\sigma = \sqrt{(NF)}$, and so Equation (5.19) takes the form

$$E_\mathrm{S} = 2.35\sqrt{FE_\gamma w} \tag{5.20}$$

The physical explanation of this is that the generation of individual charge carriers for some detector types cannot be regarded as an independent process. The Fano factor is approximately 0.1 or slightly less for semiconductor detectors, slightly more for proportional gaseous detectors and about unity for scintillation detectors. Using Equations (4.6) and (5.20) the energy resolution is given as

$$R_\mathrm{S} = \frac{E_\mathrm{S}}{E_\gamma} = 2.35\sqrt{\frac{Fw}{E_\gamma}} \tag{5.21}$$

Equation (5.20) also explains the significance of having lowest possible values of w. Keeping our rule of thumb in mind, w is roughly 3, 30 and 300 eV for semiconductor, gaseous and scintillation detectors, respectively. For the latter we then consider E_S also to include the photoelectron generation process in the PMT. Note that Equation (5.20) does not apply for all scintillation crystals; some are reported to have deviations that are difficult to explain [102, 103]. All the different stages in the signal generation process in scintillaton detectors contribute to the total line width: the scintillation photon generation, the collection efficiency of these photons at the cathode, the photoelectron generation, the collection of these at the first dynode, the multiplication process and the collection efficiency at the anode. Sometimes the statistical line width of scintillation crystal (E_SC) and that of the photomultiplier (E_PMT) are considered separately. These are plotted for a NaI(Tl) detector in Figure 5.14 alongside data for photodiode read-out of the crystal. In this case E_I is included in E_SC, explaining the bend at about 400 keV where E_I peaks. The photodiode and PMT read-outs presented in Figure 5.14 are not directly comparable because of the smaller area of the photodiode detector. For a CsI(Tl) crystal which matches the spectral response of the photodiode better (see Figure 4.19), the line width using a photodiode read-out is then less than that using a PMT read-out for energies above about 500 keV under otherwise identical conditions [104].

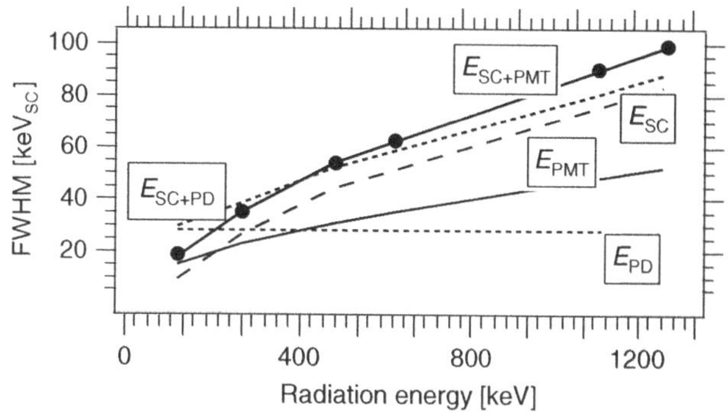

Figure 5.14 Illustration of the composition of the total line width in a NaI(Tl) detector with PMT read-out (E_{SC+PMT}) and with photodiode read-out (E_{SC+PD}) [104]. The contribution of the scintillation process (E_{SC}), the photomultiplication process (E_{PMT}) and electronic noise in the diode ($E_{PD} \equiv E_E$ in Section 5.1.4) are also plotted. The PMT read-out is for 1.5-in. diameter crystal and PMT [105], whereas the photodiode read-out is for a crystal and diode of 1-cm^2 area (presented in Section 4.5.4)

5.3.7 Measurement Reliability

Reliability is the other side of the accuracy coin, and in many cases equally important. This is simply because measurement accuracy has no meaning once a gauge has critical failure. For this reason the process industries are often willing to sacrifice accuracy for increased reliability. The reliability is traditionally expressed as the *mean time between failure* (MTBF), which is a statistical number based on experience, models or most often combinations of these. Laboratory instruments are easily serviced and maintained. For permanently installed gauges this is more difficult, and in some cases even impossible. Sea-bed and down-hole gauge applications are examples of the latter.

We quoted Albert Einstein in the beginning of this book: 'Everything should be made as simple as possible, but not simpler'. If we also keep another saying in mind, 'no chain is stronger than its weakest link', we probably have the two most important design rules to optimise reliability. These imply that the system complexity should be reduced whenever possible, simply because complex systems and components are more likely to fail than simple ones. The practical side of this is to select system components with estimated lifetime longer than the desired MTBF. The next implication of our design rules is to identify the critical elements in the total system (including software) and focus on improving these.

One way of accomplishing better reliability is the use of *redundant* systems. This may be the use of multiple beam, energy and modality measurements (see Section 5.5.6). Even though the goal of the methods presented in this section is to provide complementary information, some degree of redundancy is also often obtained. Or in some cases the complete system is designed with this in mind. The combination of completely different (e.g. nuclear and non-nuclear) measurement principles is then often used because these have the highest probability of responding differently to unforeseen process conditions. One principle may, for instance, be very sensitive to deposits on the process walls and cause complete failure if this happens, whereas another principle is completely undisturbed by this. Redundancy is also achieved by using parallel read-out electronics; most fieldbus standards include redundant communication on the sensor and actuator level. From this discussion it is also clear that the system cost is an important design parameter when balancing accuracy and reliability.

5.4 OPTIMISING MEASUREMENT CONDITIONS

In the majority of industrial radioisotope gauges the actual measurement is carried out by counting pulses, either gross spectrum counting or window counting. In this section we shall study various methods by which optimal measurement conditions are obtained for pulse counting. Broadly speaking, this is achieved by increasing the sensitivity to the measurand and reducing the interference of all other variables, very much as we optimise the SNR for sensors and electronics. The general strategy for reducing interference and noise also applies to radioisotope methods where the interference is commonly referred to as *background radiation*: Preferably remove or rearrange the background radiation source, and if this is not possible shield the radiation source, shield the detector, and if any of this is not or only partly realisable, correct for the background.

5.4.1 Background Radiation Sources

Needless to say cosmic radiation and most naturally occurring background radiation materials (NORMs) cannot be removed or rearranged anyway. For measurement of very low radiation levels, however, special grade materials with very low concentration of radioisotopes must be selected for constructions close to the detector. The only advantage with this background is that it is fairly stable with time, even though it varies with the location. An exception to the former is build-up of radioactive scale inside pipelines and process vessels originating from the oil and gas reservoirs.

Man-made background often represents a more complex problem. This arises when interfering radiation is encountered from such things as radiographers, radioisotope tracer studies or even adjacent nucleonic gauges. When multiple gauges are sited on adjacent vessels careful installation can prevent interference, otherwise the level change in one vessel will show on the nearby vessel as a false level change (see Figure 5.15).

Random rare interference from other uses of radiation in the vicinity of a level gauge is harder to deal with. For short infrequent periods of interference it is sometimes sufficient

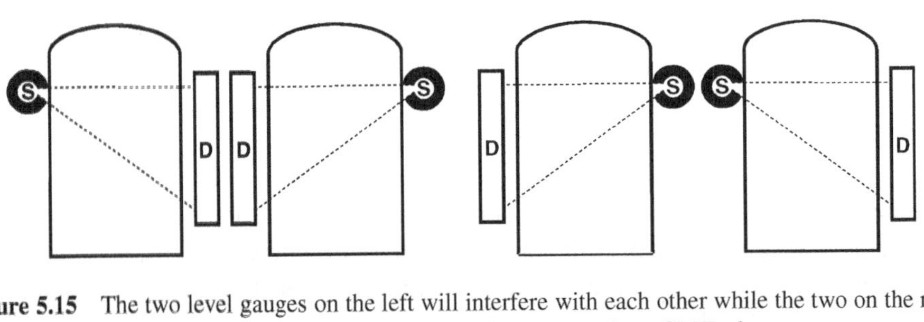

Figure 5.15 The two level gauges on the left will interfere with each other while the two on the right will not (S = source and D = detector). Level gauge detectors are long GMTs that are very unpractical and expensive to shield

to use the work permit system to prevent process upset. If before any exposure the control room is informed and asked to place the affected vessel onto manual control, then a short period of interference will have little or no effect on the smooth running of the process. Some gauges use a separate detector to 'freeze' the gauge output when a high background is detected. This has the same effect as placing the vessel on manual control. Various schemes for using a separate detector to correct the gauge output have been suggested, but correction depends on the direction from which the interfering radiation is arriving and unless this is known the correction is inadequate. In extremely critical installations the detector can be shielded but such shielding is heavy and expensive. Fortunately interference is rare as users of radioactive materials are well used to carefully planning any exposure.

5.4.2 Shielding

The purpose of a shield is to absorb radiation energy in order to reduce the radiation intensity to a desired level. For radioisotope gauges shielding is used for two purposes: The first is to reduce the dose rate emitted by the radiation source to the surroundings to a legislated level, or preferably lower. This will be dealt with in Chapter 6. Secondly, in the context of this chapter the focus is on shielding gauge detectors from background both fixed and variable to improve measurement accuracy. If a system is operating on a low radiation field then it is preferable to reduce the background count-rate to an insignificant level rather than correcting for it in the measurement. The latter will be discussed in Section 5.4.8. A typical density gauge scintillation counter with say a 50-mm diameter by 50-mm-thick crystal will have a background count-rate of about 100 c/s when unshielded. This is from the natural surroundings and will vary from place to place, but will not vary significantly with time. Natural background can be assumed to be constant for a given installation and a single long count on commissioning of the gauge can be used to correct all subsequent counts by subtracting the appropriate background. If the gauge is working on a low count-rate then the random variation in the background may become a significant proportion of the total error and the detector should be shielded. A simple cylindrical lead shield about 2 cm thick will reduce the background of a density gauge scintillation detector to about 10 c/s. There are applications, such as extended detectors used on level gauges, which are very difficult to shield. In these cases shielding is usually not used and the background is subtracted as explained in Section 5.4.8.

Detectors used for spectroscopy and measurement of low radiation levels, such as environmental radiation measurement, need to be very sensitive and may require heavy shielding in order to eliminate most extraneous radiation. Such detectors are used to measure a complete spectrum of radiation including low energies. In this case special *graded shielding* may be employed to eliminate characteristic X-rays, which are produced by the interaction of high-energy radiation in the shield. The design parameters of a typical shield are listed in Table 5.4, and the attenuation coefficients of the materials used are plotted in Figure 5.16. The shielding is graded starting with a layer of a high-*Z* material. Lead is frequently used because of the combination of high stopping efficiency and low cost. The thickness of this layer is determined by the energy and intensity of the incident radiation. To stop the Pb K-line X-rays at about 77 keV a 3-mm layer of tin is used. Its attenuation coefficient below the Pb K-line is about the same as that of lead. However, the

Table 5.4 Typical composition of a graded shield, in this case for shielding a detector from γ-rays from a ^{137}Cs source[a]

Layer	Material	Density	Z	Thickness	Comments
1	Pb	11.4	82	10 cm	Depends on γ-ray energy and intensity
2	Cd	8.65	48	3 mm	Should be avoided in neutron fields
2	Sn	7.3	50	3 mm	Good alternative to Cd
3	Cu	8.96	29	0.7 mm	

[a] The first layer faces the radiation source. For the second layer Sn is often preferred to Cd because the latter is highly toxic and used only if unavoidable.

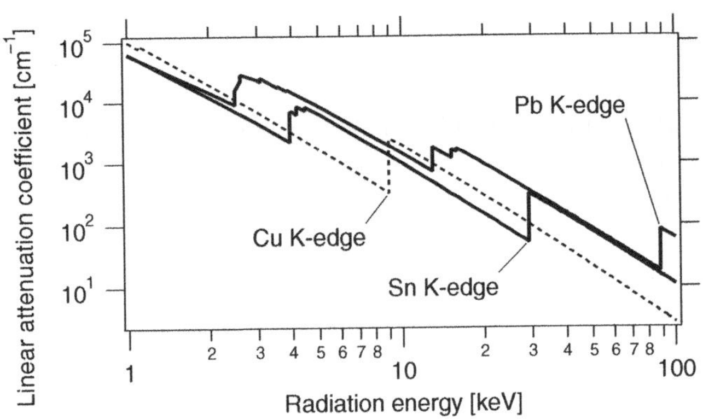

Figure 5.16 The attenuation coefficients of frequently used graded shield materials

main achievement is that tin has no X-ray emissions in this energy region. This is true all the way down to the Sn K-line X-ray emission line at about 26 keV. To stop these a thin layer of copper is used. The K-line fluorescence of copper is at about 8 keV, and this is often below the energy region of interest. If not, a thin layer of another lower-Z material, such as titanium, may be added.

5.4.3 Collimation

Collimation is used at the radiation source to define the beam to illuminate the desired volume of the process or object to be measured. Likewise collimation is used on the detector to define its desired view into the volume irradiated by the source. For transmission measurement the intersection of the source illumination and detector view defines the *measurement volume*. This is the volume where radiation interactions and emissions contribute to the measurement result. For measurement principles where the source and the detector are placed next to each other facing the process or object, the remote boundary of the measurement volume is diffuse and defined by the radiation attenuation properties. We will come back to such measurement geometries, of which backscatter is an example, in Section 5.5.2.

For the radiation source the collimation is part of the shield that reduces radiation leakage from the gauge. *Fan beam collimation* or *focussed collimation* (see Figure 5.17a)

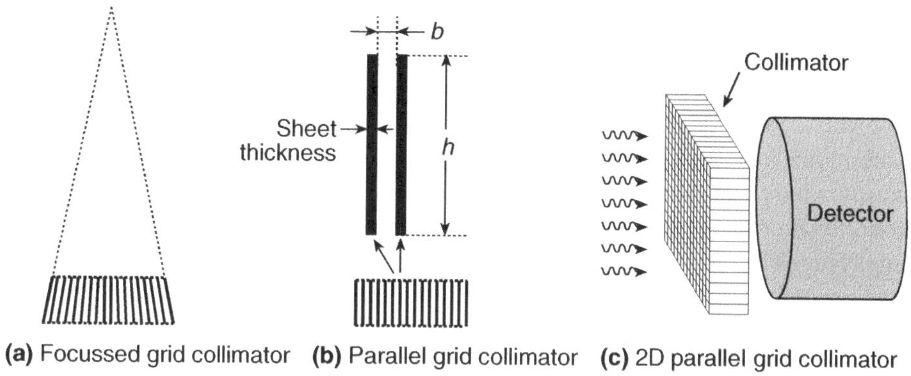

Figure 5.17 Examples of detector collimators. The focussed grid collimator (a) is often used fan for beam collimation of point sources. For the parallel grid detector (b) the so-called *grid ratio* is defined as h/b. A 2D grid collimator (focussed or parallel) (c) is efficient for low-energy γ-rays and X-rays

is frequently used for point sources. We saw some examples of this in Section 5.2.4. For the detector, proper collimation in some cases also improves the performance by reducing the quantity of scattered radiation reaching the detector. The depth and diameter or width of the hole in a collimator depends on the size of the detector and the source/detector separation (see Figure 5.17b), but also depends on the energy of the radiation. The higher the energy, the thicker a collimator needs to be in order to be effective. As with shielding, a detector used to measure low energies will benefit from the use of graded materials in the collimator to reabsorb secondary radiation produced therein.

With level gauges the sole purpose of the source collimation is to prevent the extraneous dose rates as in all but the smallest vessels the beam will inevitably be broad in relation to the detector size. In density or thickness gauge applications the collimator may be narrow enough to direct the beam into the centre of the detector and true narrow beam conditions can be achieved. A collimator that directs the beam into the centre of a detector will improve the resolution of the spectrum and increase the full energy detection fraction (see Section 4.2.3) by reducing the number of incomplete interactions at the edge of the crystal and by eliminating all but the forward scattered radiation, thus reducing build-up (see Section 3.5.1). It is worth noting that a finely collimated beam into the centre of a detector will not obey the inverse-square law with regard to source/detector separation. Similarly on a large vessel with a broad beam level gauge installed the dose rate beyond the detector reduces more quickly with distance than would be expected because a significant proportion is produced within the vessel nearer to the detector than the source. One downside of collimating the detector is that the collimator effectively reduces the useful diameter of the detector and will therefore reduce the count-rate and hence the statistical accuracy of the measurement. For low-energy radiation it is possible to collimate the beam into the whole of the detector using a collimator that is effectively a bundle of small collimators as shown in Figure 5.17c.

The smaller tubes reduce the angle of acceptance of the detector in a much shorter length collimator than could be achieved with a single tube. This arrangement only works for low energies because the walls between adjacent collimators can only be thin. For such precision collimation other materials are preferred to lead because this is too soft, particularly for sheet collimators. These, which are used for 1D collimation (see Figure 5.11), are often so thin that lead may bend by its own weight.

Tungsten is frequently used for shielding and collimation because of its high density (19.3 g/cm^3 – 50% higher than that of lead) and high atomic number ($Z = 74$). But it is difficult to machine; this also has to be done at elevated temperatures. For this reason a *heavy alloy* based on tungsten, nickel and copper is often used for precise and efficient beam collimation on the source and detector sides. It is also used for shielding, for instance internally in radioisotope sources (as shown in Figure 2.6), beam shutters, etc. The tungsten content typically varies between 90 and 95% by weight and the density is between 17 and 18 g/cm^3. These alloys can be machined conventionally and exhibit excellent mechanical properties, but tend to be expensive.

5.4.4 Neutron Collimation and Shielding

The purpose of a neutron shield or collimator is to absorb neutrons with a minimal of γ-ray emission and residual activity. The shield has to be dimensioned so that the total neutron and γ-ray dose-rate emission is below the required level. Collimation and shielding of neutrons is a multistage process analogous to the graded shielding of γ-rays and X-rays. Firstly, fast neutrons are slowed down in an efficient moderator, secondly these slow neutrons are captured in a material with high cross section to (n, α) or (n, p) nuclear reactions, and thirdly any γ-rays or X-rays emitted by the source or the two former stages are stopped in a suitable material, as discussed in the preceding section. The only restriction to the latter is that materials with high cross section to slow-neutron capture reactions, particularly (n, γ), must be avoided. For the first two stages, materials with a minimum probability of induced radioactivity must be used. The first two stages may also be as achieved using one material containing both moderation and absorption elements. Efficient and economic fast neutron moderators are heavy water, deionised water, beryllium, polyethylene, graphite and zirconium hydride of nuclear level purity. Popular and efficient slow-neutron absorbers are ^{6}Li, ^{10}B and ^{113}Cd. The two former elements are ideal because, as we saw in Section 3.6, absorption takes place by the (n, α) reaction. The latter is used because it has a very high thermal neutron cross section, even though this is for the (n, γ) reaction, meaning there will be considerable γ-ray emissions. Borated (1–6% by weight) polyethylene is frequently used as a combined moderator and absorber.

5.4.5 Alternative Transmission Measurement Geometries

We shall see in Section 5.5.1 that the most common transmission measurement geometry is to position the source on one side of the process vessel and the detector diametrically opposite on the other side. To minimise the measurement error we saw in Section 5.3.4 that $\mu x = 2$ is optimal (86% attenuation). This is achieved by accommodating proper values for μ, which in practice means radiation energy (see Figure 3.7), x or both. Very often there are design restrictions, for instance in that the dimensions of a vessel at the measurement position are fixed and cannot be changed. As a consequence it may be difficult to obtain the optimal (86%) attenuation in some situations. The solution may then be to use different

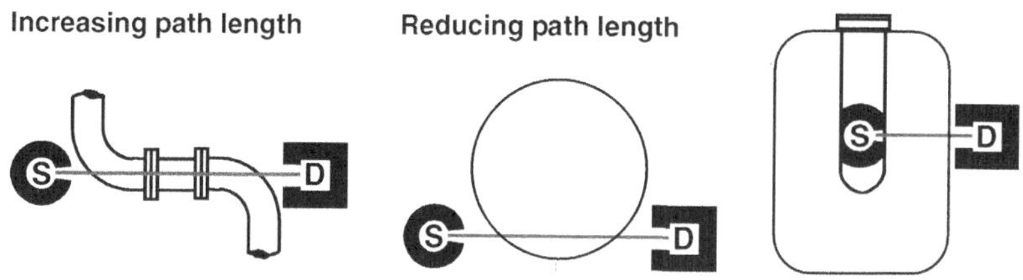

Figure 5.18 Possible transmission measurement geometries to achieve optimal attenuation. Positioning of the source inside dip pipes (right) and other internal parts of the process is a frequently used method

measurement geometries such as those suggested in Figure 5.18. It is often a question of creativity. There are, as we shall see in Chapters 5 and 7, many examples of solutions, with the source inside dip pipes and other internal parts to reduce path length. This may be the only solution for vessels with large diameter and thick walls. In vessels with highly attenuating walls the treatment very often is to use radiation windows, as discussed in Section 4.9.2.

Similar strategies also apply to thickness measurement, where, for example, the beam may be tilted to increase the path length through the measurement object.

For very large vessels where dip pipes cannot be used, backscatter measurements should be considered. This will be presented in Section 5.5.2.

5.4.6 Counting Threshold Positioning

In Section 5.2.1 (Figure 5.9) we studied a typical discriminator circuit for count-rate measurement. One of the last stages in the design of such a circuit is to acquire a pulse height spectrum with the source to be used for the actual measurement, and use this to determine at what pulse height the counter threshold should be placed. A transmission spectrum of a thin window ^{137}Cs source using a scintillation detector is shown in Figure 5.19. The intention normally is to count the number of transmitted 661.6 keV photons, suggesting that the threshold should be positioned just below the full energy peak (position ① in Figure 5.19). There are, however, two other aspects not in favour of this that should be considered: Firstly, most of the spectral background below the full energy peak is the Compton continuum of detector interactions where the scattered photon escapes the detector. We saw in Section 4.2.3 that this is very likely in a realistic detector. Any event in the Compton continuum thus is equally important to the transmission measurement as an event in the full energy peak. The problem with this is that the spectral background may also be build-up caused by events scattered into the detector from structure outside the measurement volume. We discussed this in Section 3.5. Except for the backscatter peak, it is impossible to tell the difference between build-up events and Compton continuum events.* In many cases, however, it is possible to do some sort of calibration measurements, for instance a so-called empty vessel measurement (see Section 5.5.1), which will account for most of the build-up. We also need to keep in mind that part of the build-up is from events that have passed through the measurement volume before or (and) after they are

* This is less of a problem for high-Z detectors where a larger fraction of the interactions are full energy interactions.

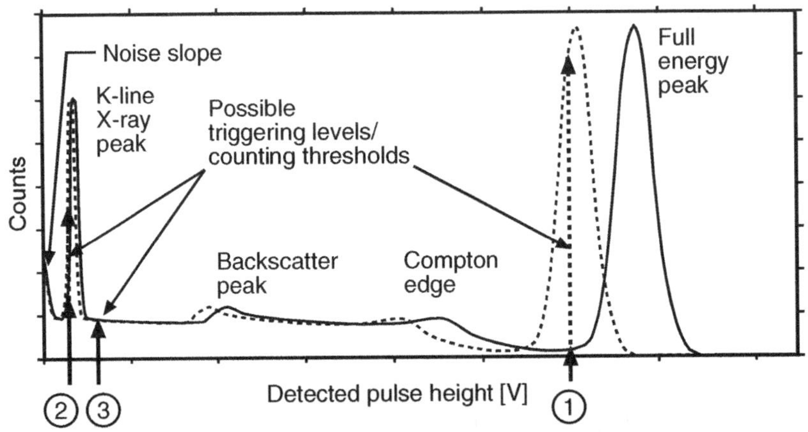

Figure 5.19 Positioning of counter threshold (trigger level) in a typical transmission spectrum using a scintillation detector and a ^{137}Cs source (identical to that presented in Figure 4.5). The dashed line spectrum is acquired under conditions identical to the solid line one, but with a negative gain shift of about 10%, which is realistic under some conditions

scattered. The intensity of such events also depends on the transmission properties of the measurement volume and thus contributes to the actual transmission measurement. With reference to the discussion in Section 3.5.3, we now measure the effective attenuation coefficient (μ_{eff}). The reason for including the Compton continuum in measurement is of course that the statistical counting error decreases (Section 5.3.4) when the count-rate increases. In the case shown in Figure 5.19 the count-rate would roughly be doubled by positioning the threshold level, for instance, in position ③.

The second reason for considering using a low counting threshold is that the absolute count-rate error caused by drift in the gain of the detector system then is less. The influence of a negative gain shift of 10% is illustrated by the dashed spectrum in Figure 5.19. Assuming that the original centroid position of the full energy peak is at about 6.8 V pulse height, this gain shift causes a drop of 680 mV to 6.12 V. The threshold at position ① is fixed at 6 V, meaning that most of the leading edge counts would be lost by the 10% gain shift as can be seen. In the region around threshold position ③ at about 0.6 V the gain shift causes an absolute shift in pulse amplitude of about 60 mV. The shift error in count-rate would thus be 10 times higher with threshold position ① under otherwise identical conditions. This is not the case here where the error is even higher because threshold position ① is on the leading edge of the full energy peak. The change in number of counts with pulse height is thus much higher here, therefore increasing the error. For the same reason threshold position ③ is preferred to position ②. The conclusion is that the threshold should always be placed in a flat area of the spectrum away from peaks in case of gain shift, and preferably in the low-energy end of the spectrum.

There is one more point we can make concerning threshold positioning in the illustration in Figure 5.19: Why not put the threshold in the valley between the noise and the fluorescence X-ray peak? This is a safe choice regarding gain shift errors; however, in doing so the measurement is influenced by the attenuation properties at two different energies. This is less critical for close emission energies. But here the γ-ray peak is in the Compton region, making the transmission measurement sensitive to the density of the medium. The X-ray peak is in the photoelectric region where the composition of the medium (Z_{eff}) influences the measurement. Therefore threshold position ③ above the X-ray peak is recommended. Having said so, this example is not very realistic because normally the ^{137}Cs γ-ray sources used in gauging have sufficient steel encapsulation to absorb virtually all low-energy X-ray emissions.

Gain errors are most problematic with scintillation detectors where the scintillation efficiency is influenced by the ambient temperature, and where the PMT gain is very sensitive to variations in the high voltage bias. For unity gain semiconductor detectors for instance, the signal amplitude is very stable. But on the other hand the noise level in these is very sensitive to temperature variations. As a consequence the counting threshold should be placed above the noise slope with some margin. Failing in doing so may cause devastating count-rate errors. This also applies to the positioning of the LLD in PHAs (see Section 5.2.2), but here with increased dead time as the consequence.

In this section we have used γ-ray transmission measurements as our example, but most of the count-rate error reduction methods discussed here also apply to threshold positioning in β-particle transmission gauges and various types of scatter gauges.

5.4.7 Spectrum Stabilisation

The low threshold method described in the previous section is only applicable to discriminator counting. For SCAs and multiple windows counting a stable spectrum is a requirement for proper operation. Otherwise the errors may be significant, especially for the high end of the spectrum as described in the previous section. Before continuing it is worthwhile appreciating that most radiation detector systems have an initial electronics warm-up time, typically a few minutes, before the gain is stabilised.

There are various methods that can be applied for gain stabilising of detector systems. Traditionally, the most frequently used principle monitors the position of a well-defined peak in the high end of the spectrum where sensitivity to gain shift is highest. An error signal is produced and used to control the PMT high voltage bias if there is a peak shift. Alternatively, for some detector systems it may be more convenient to adjust the threshold level(s) accordingly. This is achieved by comparing the content (integral counts) in a window on the leading edge of the peak to the content of a window on the falling edge (see Figure 5.20).

For this method to be successful it is important to have a well-defined peak, the content of which is not disturbed by other radiation effects such as scatter. In γ-ray transmission applications the full energy peak is used for this purpose. This peak will always be present in transmission measurements using detectors where photoelectric absorption is not negligible. For small low-Z detectors, such as plastic scintillators, and high radiation energy, the spectrum may be dominated by the Compton continuum with only a small full energy peak present. This normally cannot be used for gain stabilisation.

In other cases the spectrum may be too complex with too much peak interference to find a suitable peak to use for gain stabilisation. An *active pulser* method may then be applied to produce a well-defined peak in the top end of the spectrum. For some scintillators, such as NaI(Tl) and CsI(Na), a frequently used and reliable method is the integration of a small ^{241}Am α-source between the scintillator and the PMT. This is a low-activity source,

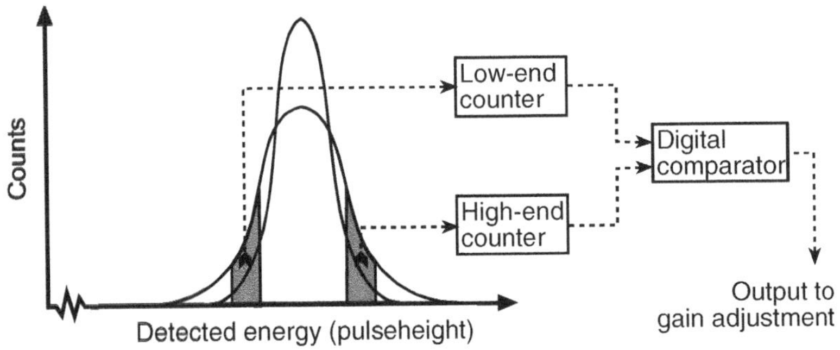

Figure 5.20 Gain stabilisation by comparison of integral number of counts in a low-end window on the leading edge of a spectral peak to that in high-end window on the falling edge. Peak shift towards lower pulse heights gives more counts in the low-end counter fewer in the high-end one, and vice versa. Changes in the energy resolution (line width) will as illustrated not affect this principle because it has equal impact on both windows for symmetrical peaks. The gain can be controlled, for instance, by adjusting the high voltage to a scintillation detector PMT

between 10 and 1000 Bq, on an encapsulated foil. The α-particle energy deposition in the crystal is very stable, but the exact value in each case depends on the thickness of the encapsulation. Typical values are between 1.5 and 3.5 MeV. The drawbacks of this method are that this source also emits low-energy γ-rays and X-rays producing spectral background below 60 keV, and that the temperature changes in the scintillation efficiency are not exactly the same for γ-rays and α-particles. Depending on the measurement spectrum a γ-ray emitting ^{137}Cs source may be embedded the same way, but with the drawback of adding spectral background by the Compton continuum. The advantage of radioisotope pulsers over precision LEDs, which are also used, is that radioisotope pulsers monitor the pulse amplitude shifts in the scintillation crystal in addition to the PMT. A precision LED pulser covers only PMT gain shift, and furthermore the light output from the LED is temperature dependent. On the other hand, changes in the crystal performance with temperature can be characterised well, but the PMT behaviour is still a problem. The latter also applies to drift as a function of count-rate. One therefore tends to move away from using radioisotope sources. The new development is towards controlled and measured light pulses injected into the crystal.

So far we have discussed only drift in the system gain because this is the most common problem. But in some cases there may also be drift in the zero offset. This is tackled by monitoring the drift of two spectral peaks, one in the high end and one in the low end of the spectrum. Finally, software programmable techniques are becoming increasingly popular for spectrum stabilisation. These, often referred to as digital stabilisers, are based on algorithms continuously identifying peak positions in full PHA spectra.

5.4.8 Background Correction

In this section we will study the last strategy in combating background: accepting its presence, estimating its magnitude and correcting for it in the measurements. There are two approaches as to how background estimation can be done, and these can also be combined:

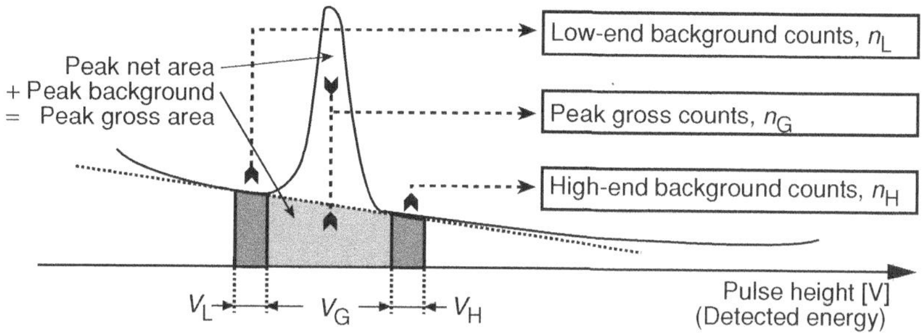

Figure 5.21 Method for estimation of spectral peak background based on linear interpolation between the count levels just below and above the peak. This could, for instance, be the fluorescence X-ray peak superimposed on the Compton continuum, as shown in Figure 5.19. The numbers of counts within the three windows are in practice found by using four thresholds, and subtracting the number of counts between these as appropriate

1. Assume the background is constant and count it with the source shutter closed or preferably with the source removed.
2. Use spectral analysis to estimate the background of peaks in the detection spectrum continuously during measurement.

The latter only applies to window counting (SCA) and requires the detector system to be energy sensitive. Only the first method is thus applicable to measurement systems using GMTs. Although the background often varies, its contribution from several of the sources discussed previously may be regarded as constant for a particular set-up on a location.

The traditional approach to the second method is to estimate the peak background in a spectrum, assuming that it is on average linear between the count levels just below and above the peak. This is illustrated in Figure 5.21. The gain of the detector system needs to be stable for this background correction method to work. Spectrum stabilisation as described in Section 5.4.7 is therefore often required. It is now necessary to use a PHA to establish the properties of the spectrum in the vicinity of the peak. To find the count levels just below and above the peak we need to establish two counting windows (SCAs) with widths V_L and V_H in which the integral counts are n_L and n_H, respectively. Keeping in mind that a PHA spectrum is a differential spectrum (dn_C/dV), the average count levels in these two windows are n_L/V_L and n_H/V_H, respectively. The number of background counts (n_B) is then expressed as

$$n_B = \left(\frac{n_L}{V_L} + \frac{n_H}{V_H}\right)\frac{V_G}{2} = (n_L + n_H)\frac{V_G}{2V_L} \tag{5.22}$$

The simplification to the right is made assuming equal widths of the two windows ($V_L = V_H$). Here V_G is the width of the peak window where the number of gross counts (n_G) is recorded. The number of net counts in the peak is then

$$n_N = n_G - n_B = n_G - (n_L + n_H)\frac{V_G}{2V_L} \tag{5.23}$$

In PHA applications it may be more convenient to express the window widths in number of PHA channels rather the voltage.

Equation (5.23) states that the peak area or number of counts is a function of three variables: n_L, n_G and n_H, which all are subject to statistical errors. The statistical error in n_E in terms of one standard deviation (σ_{n_E}) is found using the error propagation formula given in Equations (5.11) and (5.23).

$$\sigma_{n_N} = \sqrt{\left(\frac{\partial n_N}{\partial n_G}\right)^2 \sigma^2_{n_G} + \left(\frac{\partial n_N}{\partial n_L}\right)^2 \sigma^2_{n_L} + \left(\frac{\partial n_N}{\partial n_H}\right)^2 \sigma^2_{n_H}} = \sqrt{n_G + \frac{V_G^2}{4V_L^2}(n_L + n_H)}$$

$$= \sqrt{n_G + \frac{V_G}{2V_L} n_B} = \sqrt{n_N + n_B + \frac{V_G}{2V_L} n_B} = \sqrt{n_N + n_B \left(1 + \frac{V_G}{2V_L}\right)} \tag{5.24}$$

where $\sigma_{nN} = \sqrt{n_N}$, $\sigma_{nL} = \sqrt{n_L}$, and so forth, and the sensitivity coefficients are given as

$$\frac{\partial n_N}{\partial n_G} = 1 \quad \text{and} \quad \frac{\partial n_N}{\partial n_L} = \frac{\partial n_N}{\partial n_H} = -\frac{V_G}{2V_L} \tag{5.25}$$

Indeed, the statistical error decreases with increasing V_L and V_H, but on the other hand this may increase other errors such as interference with other peaks or the noise slope. Therefore, in determining V_L and V_H it is necessary to evaluate the spectral neighbourhood of the peak. These need not of course be equal, and in some cases it may even be better to estimate the peak background using extrapolation of the spectral properties on one side of the peak only. This may, for instance, be the case when there is interference by another peak or noise so that the linear interpolation approach fails. In cases where the background is fairly constant longer counting times may be used for n_L and n_H to reduce the error in n_B.

From Equation (5.24) we see that the statistical error in the calculated peak net counts not surprisingly is less for narrow peaks, i.e. for small values of V_G. The peak width is determined by the energy resolution or the FWHM, as discussed in Section 5.3.6. Attention should be paid to the effect of variations in the line width during measurement. The line widths in semiconductor detector spectra are, for instance, very sensitive to variations in the temperature. If this increases, so does the line width. With reference to Figure 5.21, this causes the background to be overestimated because an increasing fraction of valid peak counts is now recorded in n_L and n_H. The net peak area will thus be underestimated unless the positions of the counting windows are adjusted accordingly.

A generally accepted definition for the *lower limit of detection* (LLD) is that the number of counts in the peak should be equal to 2 standard deviations of the background number of counts. Based on this it can be shown that the LLD, expressed in number of counts, [106] is given as

$$\text{LLD} = 3\sqrt{n_B} \tag{5.26}$$

The first option presented at the beginning of this section was to assume that the background is constant and count it with the source shutter closed or with the source removed. Suppose we now count the full spectrum by using a low counting threshold. We use a count time τ_B for the number of background counts (n_B), and thereafter a count time

τ_G for the number of gross counts (n_G). The net number of counts in the full spectrum is given as $n_N = n_G - n_B$. In most cases this is more conveniently expressed in terms of count-rates. We then have

$$n_{r_N} = n_{r_G} - n_{r_B} = \frac{n_G}{\tau_G} - \frac{n_B}{\tau_B} \tag{5.27}$$

where the subscript r denotes count-rate. If we further assume that the counting times are known without error, we can use the error propagation formula to calculate the statistical error in the net count-rate:

$$\sigma_{n_{r_N}} = \sqrt{\frac{n_G}{\tau_G^2} + \frac{n_B}{\tau_B^2}} \approx \sqrt{\frac{n_G}{\tau_G^2}} \tag{5.28}$$

This approximation is valid when a long τ_B ($\gg \tau_G$) is used for the number of background counts. In some cases there is a limited time available for the total measurement ($\tau_T = \tau_B + \tau_G$). To minimise the statistical error ($\sigma_{n_{r_N}}$) in the net count-rate, the optimal ratio between these times then is [6, 8]

$$\frac{\tau_B}{\tau_G} = \sqrt{\frac{n_{r_B}}{n_{r_G}}} \tag{5.29}$$

5.4.9 Compton Anticoincidence Suppression

We saw in Sections 4.2.2 and 4.2.3 that there is always a possibility of radiation leakage through Compton scattered photons escaping the detector, particularly with operation in the Compton dominant energy region (see Figure 3.6). In such cases Compton interactions contribute to the Compton continuum and not to the full energy peak. This is a problem in spectroscopy, or generally when the contents of two or more peaks are to be measured, because the Compton continuum of high-energy peaks adds background for the low-energy peaks. Even though this may be corrected for as described in Section 5.4.8, the best result is obtained if the background is reduced. *Compton anticoincidence suppression* may be applied for this purpose: The detector, most often a high-resolution spectroscopy detector, is surrounded by one or several scintillation crystal detectors that normally are shielded from the radiation beam. All these detectors are operated in anticoincidence with the spectroscopy detector in such way that all coincidental events are rejected from the latter and not added to the acquiring spectrum. These events are thus interpreted as originating from the same incident event that undergoes Compton interaction. The surrounding detectors also behave as an active shield in the sense that all events scattered from any of these into the spectroscopy detector are also rejected. Compton anticoincidence suppression is very efficient and provides significant improvement of the detection spectrum. Because of the high cost, however, it is primarily a tool for spectroscopy laboratories.

A novel coincidence detection system using two NaI(Tl) detectors has been developed to improve the detection of high (MeV) energy γ-rays. A well-type crystal with separate PMT read-out is surrounding the primary crystal so as to detect annihilation radiation emitted in the latter [107].

5.4.10 Source Decay Compensation

The activity decay of radioisotopes is another source of error for relatively short-lived isotopes unless it is corrected for. In contrast to the other count-rate corrections we have discussed previously, this one is an easy task. Equation (2.12) expresses how the activity of a radioactive source decays with time. The radiation intensity emitted in any direction decays with an identical rate. Therefore the decay of the incident intensity (I_0) in a measurement system is also expressed as

$$I_0 = I_0^{\mathrm{cal}} \, e^{-\left[\frac{\ln(2)}{T_{1/2}}\right]t} \tag{5.30}$$

where $T_{1/2}$ is the half-life listed in most nuclide indices and I_0^{cal} is the initial incident activity. Although all radioisotope sources are provided with a certificate stating the activity at a specific point of time, the most reliable method is to define the value of I_0^{cal} by calibration in the actual measurement system at any point of time. Actually, often it is I_E^{cal} rather than I_0^{cal} that is determined as explained in Section 5.5.1.

5.4.11 Dead Time Correction

In high count-rate gauges there is a risk for measurement error caused by the detector or its read-out electronics loosing events because the system is busy processing previous events. The *dead time* of a detector system may be defined as the time the system is busy processing an event. This error can be corrected for provided a model of the dead time losses is available. For a discriminator or SCA system the dead time is basically limited by detector response time or the pulse width of the shaper output signal. If the peaking time of the shaper is made very short or if no shaper is applied as discussed in Section 5.2.3, then the dead time is defined solely by the detector response. When using a PHA this normally is the dominant dead time contributor because the input SCA, the ADC and the memory accessing all take time, typically several tens of microseconds in total. At high pulse or count-rates the dead time losses will be appreciable, and large dead time corrections are inevitable. There are two general categories of dead time: *non-paralysable* and *paralysable dead time*, also known as *non-extending* and *extending dead time*, respectively.

A non-paralysable detector system is one where any new interaction in the detector within the dead time of the preceding one, is ignored. Hence, it does not give rise to a new pulse that extends the dead time accordingly. This can be modelled by assuming n is the interaction rate, m measured rate and τ_D the dead time. The fraction of all time the detector is dead is then $m\tau_D$ so that the rate at which true events or interactions are lost is $nm\tau_D$. The relationship between true and measured rates with the non-paralysable model is then expressed as

$$n - m = nm\tau_D \quad \Leftrightarrow \quad n = \frac{m}{1 - m\tau_D} \tag{5.31}$$

In a paralysable detector system any new interaction in the detector within the dead time of the preceding one gives rise to a new pulse that extends the dead time accordingly. The model now has to be derived from the probability that this will happen. This is because the duration of the dead time now varies and the approach used for the non-paralysable model cannot be applied. The relationship between measured and true rates with the paralysable model is now expressed as [25, 108]

$$m = n\, e^{-n\tau_D} \tag{5.32}$$

This has to be implemented using an iterative scheme because the true rate cannot be solved explicitly. These are idealised models with certain limitations. If we look at the temporal development of a GMT pulse as presented in Figure 4.13, all interactions happening within the dead time of the GMT will be ignored and not produce any pulse that would extend the dead time. This is thus a non-paralysable case. However, interactions happening when the tube starts to recharge, i.e. at the end of the period defined as the resolution time in Figure 4.13, will extend the effective dead time of the tube and not be counted. Hence this is also a paralysable case. A new hybrid model combining the two idealised models has been developed and found to be accurate within 5% for count-rates up to 70 kc/s in a GMT with 300-μs dead time [110]:

$$m = \frac{n\, e^{-n\tau_{DP}}}{1 + n\tau_{DN}} \tag{5.33}$$

where τ_{DN} is the non-paralysable dead time, i.e. the one referred to just as dead time in Figure 4.13. Further, τ_{DP} is the paralysable dead time equal to the difference between the resolution time, as defined in Figure 4.13, and τ_{DN}.

These dead time models are fairly good approximations as long as the dead time losses are below, say 30% [25]. For higher losses one should generally consider changing the measurement conditions, e.g. by reducing the source activity or using a detector with smaller dead time.

Dead time can be measured quite easily by two simple methods: firstly using two sources, the activities of which need not be accurately known. Each source is counted individually at a fixed distance from the detector and then the two sources are counted together at the same distance. With careful choice of sources and distance the counts from each source individually will be low enough for dead time to be insignificant while the two together will produce significant dead time. Now the true count n is given by the sum of the two individual counts and the combined count is the measured count m. Substitution into Equation (5.31) gives the dead time τ_D.

The second method uses the inverse square law to produce two count-rates of a known ratio. A single source is placed at two measured distances from the detector. When the source is furthest away the count-rate will be low and the dead time will be negligible. The source is then moved to the closer position and another count rate is recorded; now dead time will be significant. If for convenience the longer distance is double the shorter one, then the true count for the shorter distance, n, is four times the count taken at the longer distance. The measured count m is the actual count taken at the shorter distance, and substitution into the equation reveals the dead time τ_D.

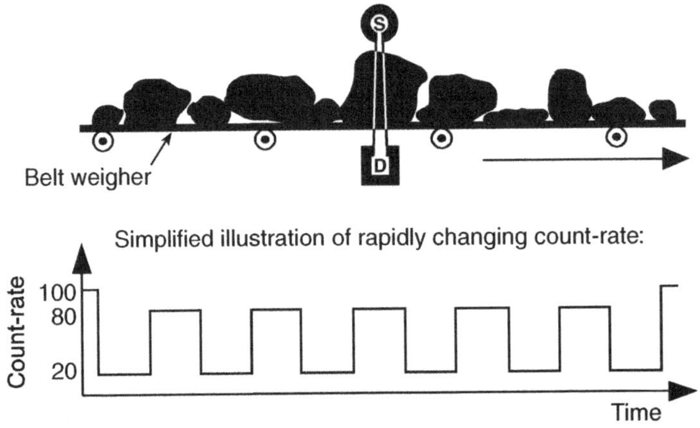

Figure 5.22 Illustration of rapid variation in detector count-rate for a transmission system applied, for instance, on a quarry conveyor belt with blocks of stone on it

5.4.12 Data Treatment of Rapidly Changing Signals

In systems where the material between source and detector is changing rapidly such as on a mine belt weigher (Figure 5.22) or a slugging multiphase flow from an oil well, it is important to count in short intervals. The accuracy of each individual count is low and so the data are combined after each count is logged. This does not give the same result as logging the average count taken over a longer time period.

A 10-s count through the blocks would give an average count-rate of 50 c/s. The density or mass on the belt is proportional to $\ln(I/I_0)$, which here is $\ln(50/100) \approx -0.693$. If we had taken 1-s counts, which luckily coincided with the block edges, we would have 20, 80, 20, 80, 20, 80, 20, 80, 20, 80 counts in each count period. This time, $\ln(I/I_0)$, for each period equals -1.61, -0.223, -1.61, etc., the average of which is -0.92 and a very different result as that derived from the average count-rate.

Of course we cannot depend on the blocks or slugs coinciding with our count periods and so the count period must be small compared to the transit time of the objects. This is analogous to sampling and digitising of an analogue signal. Nyquist's sampling theorem states that the sampling frequency must be at least twice the highest frequency component of the signal. Transferred to our case this means the counting period must be at least half the smallest time constant in the process.

5.4.13 Dynamic Time Constants

Most digital counting systems used on radioisotope gauges have the benefit of a dedicated microprocessor. This enables the application of numerical techniques not available in analogue systems. Firstly a simple *moving average* can be applied; this involves placing the counts accumulated in a given time period into a FIFO (first-in first-out) memory. When a pre-set number of memory locations are full their contents are averaged. One time period later, the number in the first memory is discarded, the new number is added onto the total and a new average is calculated. This process is repeated at each time interval and allows a frequent update of the average output while still maintaining the accuracy of a longer count time. A large step change of input will show a response in a single update time although the initial response will be smaller than the input change. This method is sometimes known as the *bucket brigade*, which is a useful descriptive analogy of a line of people fighting a fire; when the last bucket is filled the first one is emptied. The data is the water. Often a fast response to change is required, coupled with an accurate density reading when in a steady state, these two things are almost mutually exclusive when a long counting time is required to achieve the desired accuracy. One device used to achieve this is to run a moving average as above for the density reading, but to calculate a new control output signal with each successive input count. Each new input count is compared to the moving average and if the latest input varies from the mean by more than a pre-set amount the control output is taken over a smaller number of buckets. A typical example would be a gauge where the density measurement is taken over 30 1-s intervals in order to achieve the desired accuracy. Full response to a step change in the input would take 30 s but it may be important to respond by, say, closing a valve in a shorter time. If each 1-s input is compared to the mean and its deviation from the mean is calculated, then, say, if the count was within 1 standard deviation of the mean the output would be the average of the previous 30 counts. If the input varied by 2 standard deviations from the mean then the probability of the step change being real and large is increased and so the average can be weighted by disposing of, say, the oldest 15 1-s counts. If the input varied by say 3 standard deviations then the certainty is even greater and so the average could be weighted even more by dumping, say, the oldest 25 input counts.

5.4.14 Errors in Scaler Measurements

Suppose we have recorded N repeated counts from the same source for equal counting times. These counts can be designated $n_{C1}, n_{C2}, \ldots, n_{CN}$ and their sum is n_{CT}. If we apply the error propagation formula [Equation (5.11)] to find the expected error (standard uncertainty) in n_{CT}, we find that

$$\sigma_{n_{CT}}^2 = \sigma_{n_{C1}}^2 + \sigma_{n_{C2}}^2 + \cdots + \sigma_{n_{CN}}^2 = n_{C1} + n_{C2} + \cdots + n_{CN} = n_{CT} \tag{5.34}$$

because $\sigma_{n_{Ci}} = \sqrt{n_{Ci}}$ for each independent count. This result shows that the standard deviation expected for the sum of all the counts is the same as if the measurement had been carried out performing a single count extending over the entire period represented by all the independent counts.

Now if we proceed to calculate a mean value from these N independent measurements, $\bar{n}_C = n_{CT}/N$. Since N is a constant,

$$\sigma_{\bar{n}_C} = \frac{\sigma_{n_{CT}}}{N} = \frac{\sqrt{n_{CT}}}{N} = \frac{\sqrt{N\bar{n}_C}}{N} = \sqrt{\frac{\bar{n}_C}{N}} \tag{5.35}$$

We can use Equations (5.34) and (5.35) to clarify a common misunderstanding: that the accuracy of a measurement based on scaler counting is different depending on whether we use a large number of short counts or a single long count. Suppose we take 10 10-s readings and the average count in a reading is 1000. Equation (5.35) gives that

$\bar{n}_C = \sqrt{(1000/10)} = 10$, and we would claim that the best 10-s measurement is 1000 ± 10 counts. Now, instead of taking 10 10-s readings, take only 1 100-s reading. Equation (5.34) gives that $\sigma_{n\text{CT}} = \sqrt{10{,}000} = 100$ and we would claim that the best (and only) 100-s reading is $10{,}000 \pm 100$ counts.

Clearly, the measurements are different in each case, and the standard deviations are different in each case. However, we can see that $\sigma_{n\text{CT}}/n_{\text{CT}}$ ($=100/10{,}000$) equals $\sigma\bar{n}_C/\bar{n}_C$ ($=10/1000$). Similarly, if we express the results as count-rates, then in each case the best measurement is 100 ± 1 c/s. Consequently, the *accuracy* of the measurement is the same in both cases. The accuracy of a measurement depends only on the total counting time of the scaler. Unless we think that counting conditions are going to change during the course of the measurement (e.g. change in background radiation level) there is no advantage in taking a large number of short counts over a single long count. For fixed source/detector conditions, the only way to improve the accuracy of our measurement is to increase the total number of counts accumulated at the scaler (i.e. to count over a longer total period), and even here we have to bear in mind that accuracy will improve by a factor of 2 only if we quadruple the total count (see Section 5.3.4).

5.5 MEASUREMENT MODALITIES

The most common radioisotope *measurement modalities*, or *measurement principles*, are summarised by the illustrations in Figure 5.23. These may be categorised in different ways, such as transmission, scattering and emission systems, or as those using sealed sources, which most often are external to the process, and those using tracers or sources internal to the process. We discussed this in Section 1.2, where we also concluded that some principles are more applicable for so-called process diagnostics than for permanently installed gauges and nuclear control systems. In this section we will make some general considerations about the different modalities; in Chapter 7 a variety of application examples are given, whereas the design of selected systems will be treated in Chapter 8.

5.5.1 Transmission

Transmission measurements are the most used and most straightforward modality of all: A radiation source is placed on one side of the process vessel and a detector on the other side, most often diametrically opposite to the source. For γ-rays it is the narrow beam attenuation in the process material that is measured according to Lambert–Beer's exponential decay law as stated in Equation (3.7):

$$I = I_0\, e^{-\int_0^x \mu \mathrm{d}l} \quad \Rightarrow \quad I = I_0\, e^{-\mu x} \tag{5.36}$$

where it is assumed that the process material is a homogeneous mixture throughout the process vessel (see Section B.1 for the derivation). The intensity is measured by pulse counting, as explained in Section 5.2.1. The attenuation thus depends on the beam path length through the absorber (x) and its linear attenuation coefficient (μ). Equation (5.36) may thus be solved with respect to either of these variables. Furthermore, the latter is

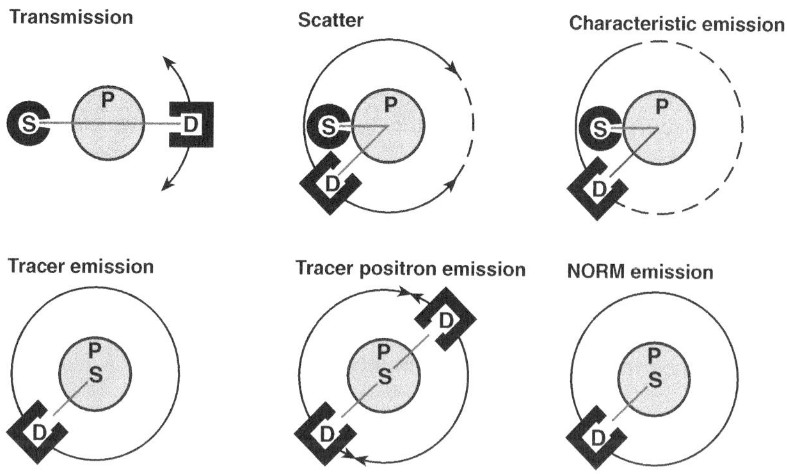

Figure 5.23 Cross-section illustrations of measurement modalities applicable to industrial gauging systems. The process (P) here is represented by a circular vessel or pipe, but of course these modalities also apply to other geometries. The radiation source (S) is shielded and collimated except in the case of tracers and NORM (natural occurring radioactive materials), where the source is an integral part of the process material. Likewise collimation and shielding are applied to the detector(s) (D). As demonstrated in Section 5.2.4, this is an essential part of many radioisotope gauges

dependent on the atomic composition (Z) and the density (ρ) in the photoelectric and Compton dominant energy regions, respectively (see Figures 3.6 and 3.7). γ-Ray transmission may consequently be used to measure either

- Thickness (μ constant).
- Average density, known as γ-ray densitometry (x constant – operation in the Compton dominant energy region).
- Effective atomic number (see Section 3.7) (x constant – operation in the photoelectric or pair production dominant energy regions).
- Component fractions or interface positions in processes with two components such as gas/liquid, gas/solid, liquid/liquid or liquid/solid, provided the components have sufficient difference in attenuation properties (ρ or Z). It may also be applied on processes with more components, so-called multiphase processes or systems, as will be discussed in Section 5.5.6.

The latter accounts for the majority of industrial applications of γ-ray gauges and is based on one of the former three methods. We will study several such applications in Chapters 7 and 8.

Transmission measurements are often made through a vessel with wall thickness x_w and attenuation coefficient μ_w. The total attenuation is then

$$I = I_0\, e^{-\mu_w x_w}\, e^{-\mu x}\, e^{-\mu_w x_w} = I_0\, e^{-2\mu_w x_w}\, e^{-\mu x} = I_E\, e^{-\mu x} \quad \text{where} \quad I_E = I_0\, e^{-2\mu_w x_w} \tag{5.37}$$

Here I_E is the empty vessel intensity with constant beam attenuation properties ($\mu_w x_w$) on the entrance and exit sides. In practice this usually means that the intensity measured with air at atmospheric pressure inside the vessel since the attenuation is then virtually zero. Very often I_0 is used in the sense of I_E for this type of transmission measurements.

In practice a real γ-ray transmission gauge seldom has what may be defined as narrow beam attenuation as defined in Section 3.5. The build-up from scatter can seldom be completely ignored, and this is most commonly accounted for by introducing effective attenuation coefficients, also defined in Section 3.5. Further, calibration measurements are often used to relate the transmission measurement directly to, for instance, the density when operation in the Compton dominant energy region. In this case the mass attenuation coefficient as defined in Section 3.3 is commonly used:

$$I = I_0 \, e^{-\mu_{M\,\mathrm{eff}} \rho x} \tag{5.38}$$

where $\mu_{M\,\mathrm{eff}}$ is the effective mass attenuation coefficient. This is independent of the atomic composition (Z) and thus constant for a given energy in the Compton dominant region.

We saw in Section 5.3.4 how to optimise a γ-ray transmission gauge to obtain lowest possible measurement uncertainty. In general, γ-ray transmission is applicable for path lengths in the range between about 2* and ~200 cm. The high-end cut-off is limited by type of material and the thickness of the process as well as the vessel wall. For measurement on process vessels the low-energy threshold most often is window limited if we regard the vessel wall as a radiation window. When measuring on open processes we most often have noise limitation as explained in Sections 4.2.1 and 4.9. Typical radioisotope sources used for γ-ray transmission are listed in Table 2.3.

The availability of a tuneable X-ray source, like those mentioned in Section 2.3.1, enables *element-sensitive transmission*. This may be used to detect the amount of specific elements in an absorber, which often is referred to the host. This is the key feature of characteristic emission measurements (see Section 5.5.3). In element-sensitive transmission two sequential measurements are performed, one with energy just above the K-edge of the element in question, and one with energy just below. The attenuation properties of the host are approximately equal at these energies, whereas they are significantly different for the element in question. The ratio of these measurements is thus very sensitive to the element concentration [5]. The jump in the attenuation coefficient at the K-edge is element dependent. For the elements shown in the plots in Figure 3.7 it is about 7.9 for iron and 4 for lead. This method can of course also be applied to the other edges.

For thickness measurement of thin sheets and films, β^--particle transmission may be used instead of γ-rays. As discussed in Section 3.1.3 the shape of the transmission curve (see Figure 3.4b) may be approximated by Lambert–Beer's exponential decay law [Equation (5.36)], provided the absorber thickness x is less than the maximum electron range R_{max} by some margin. The reason for this is that β^--particles are not emitted at a single energy, but with a spectrum of energies all the way up till a maximum energy E_{max}. This is very fortunate because the transmission may be determined through simple intensity

* Characteristics X-ray emissions, which have lower energy than do γ-ray emissions, may be used for measurements on the smallest path lengths (see Table A.1).

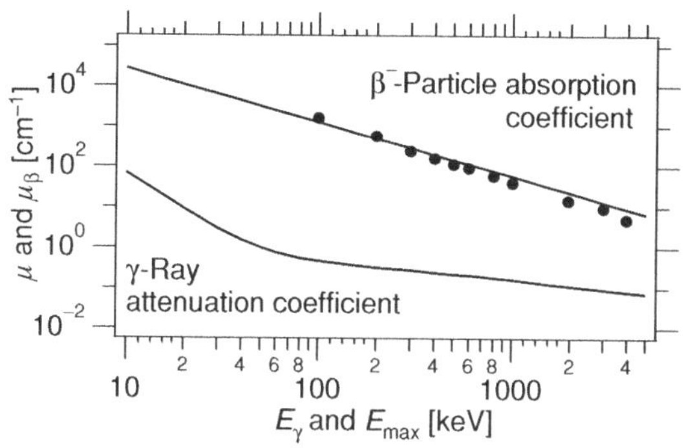

Figure 5.24 γ-Ray attenuation coefficient (μ) and β^--particle absorption coefficient (μ_β) of aluminium as functions of γ-ray energy and maximum β^--particle energy, respectively. Data for μ from [12], and for μ_β from Equation (3.6) [15] and Reference [109] (legends)

measurement similar to that used for γ-ray transmission. The absorption coefficient (μ_β) of β^--particles is, as can be seen from the plot in Figure 5.24, 2–3 orders of magnitude higher than the γ-ray attenuation coefficient. This means that maximum measurable thickness is correspondingly less, but on the other hand the measurement resolution is much higher. Depending on the material and energy (source), β^--particle transmission may be applied for thickness measurements in the range between about 200 μm and 2 cm.

Attenuation in air cannot be neglected for β^--particle transmission. However, for fixed conditions its effect may be approximated in a similar way to the γ-ray vessel wall attenuation given in Equation (5.37), and thus be compensated through calibration measurements. For measurements of even smaller thicknesses (nm range) and higher measurement resolution than is possible with β^--particle transmission, α-particle (or heavy ion) transmission may be used. This, however, is out of the question for industrial process gauges because air absorption now is so significant that operation in at least moderate vacuum is necessary. Moreover, because α-particle emissions are mono-energetic, this also requires dE/dx measurements by means of a PHA, and not intensity measurements through pulse counting.

5.5.2 Scattering

Transmission measurements are often preferred to scatter measurements because their measurement function is accurately given by Lambert–Beer's exponential decay law, or may be approximated by this. Because of the random nature of scattering processes one is restricted to the use of semi-empirical models for the measurement function of scattered radiation. However, in many situations simple calibrations at known conditions are sufficient to enable the required information to be extracted from the output signal of a gauge. Scatter measurement has an advantage in that it only requires access to one side of the process. For vessels with large diameter where transmission cannot be applied, scatter measurement may be the only option.

In this case where the source and detector are positioned close to each other we talk about *backscatter* measurements. This is the most widely used configuration; however,

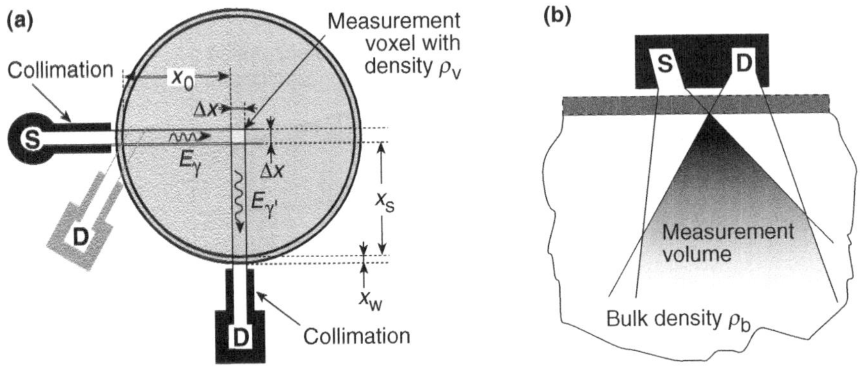

Figure 5.25 Illustration of γ-ray scatter geometries with (a) strict collimation for measurement in small volume elements (voxel) and (b) relaxed collimation for bulk measurement. For the latter the near boundaries of the measurement volume are defined by the intersection of the source beam and the detector view, whereas the far boundaries are diffuse and determined by the attenuation properties of the medium. The near side of the measurement volume contributes the most to the measurement as indicated by the gradual shading

scatter measurements may be performed at any angle as indicated in Figure 5.23. The scatter response is normally found by intensity measurements as for transmission. For γ-ray scatter PHA energy measurements may also be used because the scatter energy, according to Equation (3.15), carries information about the scattering angle. This is also the only way to discriminate scattered events from full energy transmitted ones if the detector is exposed to both. This may be the case when the detector is positioned along the dashed line indicated in the scatter illustration in Figure 5.23. This type of discrimination requires fairly good energy resolution because of the relatively low-energy transfer to forward scattered photons, particularly at low γ-ray energies (see Figure 3.10).

There are basically two approaches that may be applied for measurement of Compton scattered γ-rays. One uses strict collimation of source and detector to define a small measurement volume, as illustrated in Figure 5.25a. Ignoring multiple scatter this will be the only volume contributing to the scatter response (I_s) measured by the detector. The second approach uses more relaxed collimation of source and detector for bulk measurement of μ (Z_{eff} or ρ), as shown in Figure 5.25b.

Considering the strict collimation set-up with the right detector shown in Figure 5.25a, we can set up an ideal model for the scatter response in the detector when scatter is generated by monochromatic γ-rays with energy E_γ:

$$I_s = I_0 \overbrace{e^{-\mu_w x_w}}^{1} \overbrace{e^{-\mu x_0}}^{2} \overbrace{\frac{\mu_\sigma}{\mu}(1 - e^{-\mu \Delta x})}^{3} \overbrace{e^{-\mu'(\Delta x/2)}}^{4} \overbrace{e^{-\mu' x_s}}^{5} \overbrace{e^{-\mu'_w x_w}}^{6} \overbrace{c_g}^{7} \tag{5.39}$$

where the different terms are

1. Relative transmission of incident radiation intensity I_0 in the vessel wall, where μ_w is the linear attenuation coefficient of the wall at E_γ, and x_w its thickness.

2. Relative beam transmission over the path length x_0 before reaching the measurement volume (voxel). Here μ is the linear attenuation coefficient of the process medium at E_γ.

3. Relative generation of scatter over the path length Δx inside the measurement volume. Here μ_σ is the linear Compton attenuation coefficient at E_γ so that μ_σ/μ is the fractional number of Compton interactions to the total number of interactions [see Equation (3.12)]. The number in parentheses is the relative attenuation over Δx.

4. Average relative transmission of scatter over the path length Δx inside the measurement volume. The linear attenuation coefficient of the process medium, μ', is slightly different from μ because the energy of the scattered radiation, $E_{\gamma'}$, is less than E_γ.

5. Relative transmission of the scattered beam over the path length x_s towards the radiation detector.

6. Relative transmission of the scattered beam intensity in the exit vessel wall.

7. Coefficient accounting for incomplete stopping efficiency in the detector and geometrical effects. Only a fraction of the scatter generated inside the measurement volume is scattered towards the detector. Further, Δx is in many cases not small compared to x_0 and x_s as is assumed in Equation (5.39).

A real measurement set-up using strict collimation normally has the detector positioned at the beam entrance side as shown in Figure 5.25. By doing so $x_0 \approx x_s \approx 0$, such that the beam attenuation in the process medium [terms 2 and 5 in Equation (5.39)] outside the voxel is very low and often negligible. Or at least it may be assumed that the attenuation properties here are the same as in the voxel. This is actually a necessity because there would be no point in measuring the attenuation properties locally in a small voxel and assuming they are identical everywhere outside of that volume. With this more realistic geometry it is convenient to skip term 4 in Equation (5.39) and redefine x_s to be the average path length from the centre of the voxel to the exit wall. A further advantage of this geometry is that the scatter response will be much higher for a given source activity: partly because of less attenuation in the process medium, and partly because the (absolute) scatter generation is higher at the beam entrance side. This is important because in many cases the major drawback with this method is the relatively low response (I_s), unless very high source activities are used. Finally, the influence of multiple scatter, which is not accounted for in Equation (5.39), will be far less with this geometry.

By introducing a second scatter measurement it is possible to measure the density of a substance completely surrounded by another substance. This has been applied within medicine for density measurements on tissue such as bone [108]. The technique is based on defining a voxel somewhere in the substance of interest, very much as outlined for the centre voxel in Figure 5.25a. The second measurement now enables so-called *matrix compensation*. This means that attenuation of incident and scattered radiation in the surrounding substance is compensated for. The second measurement is either performed at a different scattering angle with an additional detector, or using two energies and energy-sensitive detectors with window counting. In both cases there are two scattered radiation energies that have different attenuation.

In most applications of γ-ray scattering measurement it is the bulk density, or the average density in a larger volume inside a vessel that is of interest. A set-up with relatively relaxed collimation of source and detector is then used as illustrated in Figure 5.25b. The scatter response (I_s) is now much higher than in the case with strict collimation, and a given measurement accuracy is achieved in shorter time, as we saw in Sections 5.3.2 and 5.3.4. In the design of scatter gauge geometry there is always a balance between increasing scatter generation and reducing attenuation of scatter back towards the detector. In transmission measurement, attenuation at all positions along the path length between the centres of the source and the detector, contributes equally to the total attenuation. On the contrary, for scatter measurements interactions in positions close to the source and detector contribute more than those further away (see the illustration in Figure 5.25b). This is because of the combination of higher scatter generation and less attenuation of scatter at these close positions. Often it is desirable to adjust the degree of collimation for bulk measurements to define a measurement volume as a layer in the plane of the vessel cross section. The number of close positions is then significantly reduced in the length dimension of the vessel, and the scatter response will be correspondingly reduced. The influence of deep interaction positions may be increased using higher radiation energy so that the depth of the measurement volume is extended. But still interactions in close positions dominate the scatter response. This analysis is, however, not complete; multiple scatter, which may give a significant contribution to the scatter response, has, for instance, not been taken into account. The design of scatter gauges is a very good example where so-called *Monte Carlo simulations* of radiation transport are very useful. We will come back to this in Chapter 8.

Needless to say the scatter response is highest in the energy range where Compton interactions are dominant: This yields optimal scatter generation and lowest possible attenuation of the scatter towards the detector. The measured parameter is thus the density. Semi-empirical models for backscatter bulk measurements of density in uniform media have been developed [14]:

$$I_s(\rho) = \rho \, e^{a+b\rho+c\tau} \tag{5.40}$$

where a, b and c are model constants and τ is the photoelectric cross section of the medium. This model assumes there is no other material, such as vessel wall, between the process and the source/detector. This is not often the case and it is possible to include walls in the model. The simplest approach though is to establish the empirical relationship between the scatter response and the process density directly through calibration measurements at several known densities. In other cases it is not necessary to measure the exact density, but rather relative changes in density with time and/or position. No matter which approach is taken, it is important to acknowledge that the scatter measurements have high sensitivity to geometry: Small variations in source/detector positions, wall thickness, etc., may cause significant changes in the scatter response because of either differences in the scatter generation, the attenuation of scatter, or both.

Since a considerable fraction of β^--particles interact with matter by elastic scattering, this phenomenon can be used for measurement using backscatter geometry. This method is primarily used to measure the thickness of coatings and sheets on a backing material.

The scatter response $I_{s\beta}$ fits a relationship of the form [8]

$$I_{s\beta} = I_{s\beta s}(1 - e^{-kx}) + I_{s\beta 0} \tag{5.41}$$

where $I_{s\beta s}$ is the saturation response for an infinite sheet thickness, x the sheet thickness, and $I_{s\beta 0}$ the scatter response with zero sheet thickness. Further, k is an empirical constant depending on the β^--particle energy spectrum and the composition of the sheet material.

Elastic neutron scattering, moderation, is frequently applied to measure the bulk density of hydrogen inside process vessels: but this is mainly employed as a process diagnostics method and there are few permanently installed neutron gauges around. This is a backscatter concept using a fast neutron radioisotope source, such as ^{252}Cf or ^{241}Am/Be (see Table 5.6), positioned next to a detector sensitive to slow neutrons only. We saw in Section 3.6 and Equation (3.26) that hydrogen is a very efficient moderator because 50% of the neutron energy in average is transferred to the hydrogen nucleus in a single collision. For oxygen ($M = 12$) this number is only about 14.2%. The presence of hydrogen in the vicinity of a fast neutron source will thus give rise to a high density of slow neutrons. To optimise the detector response it is important to keep the detector close to the source, or more precisely, the moderator, which is the origin of the slow neutrons. The effective measurement volume extends typically 100–150 mm into the vessel [6]. This is why the backscatter concept is used, although on small diameter vessels the detector could be positioned anywhere around the vessel as indicated in the scatter configuration in Figure 5.23. The vessel wall does not influence the measurement in any other way than that it separates the detector from the moderator, unless it contains elements with high absorption cross section for slow neutrons. In practice this method is seldom applied to vessels with wall thickness above 40 mm [6].

Neutron backscatter is for bulk measurement, although it is possible to use neutron collimators to define a more restricted measurement volume. Further, it is also difficult to model the scatter response to measure the hydrogen concentration in absolute terms. In practice the relative response at different process conditions is sufficient calibration. Hydrogen-rich materials in the vicinity of the source and detector but outside the process affect the scatter response, and so does the presence of elements with high absorption cross section for slow neutrons in or outside the process. Nevertheless, neutron backscatter is a powerful tool particularly for distinguishing between materials that are close in density and thus a difficult case for γ-ray methods. The required condition is that these materials have different neutron moderation and absorption properties.

5.5.3 Characteristic Emissions

We now move on to the measurement modalities based on emission of radiation by elements in the process medium, which in these cases are called *the host*. The categorisation used in Figure 5.23 is one of several possible; however, it is convenient to consider those modalities where the emission is caused by interactions with an external radiation source, as one group. As we will see these have much in common with scatter measurement. In this section we shall focus on the emission of ionising electromagnetic radiation, characteristic of the elements in the process: that is, *characteristic X-rays* (fluorescence) by atomic electrons and *prompt* γ*-rays* from the nucleus. These emissions may be regarded as fingerprints of the process elements and thus useful for some degree of *chemical* or *elemental analysis* of the process medium. The major difference between these is the type of source used for excitation or activation, and the energy range of the emissions. Characteristic X-rays have energies below about 100 keV, whereas prompt γ-ray energies are about 2 orders of magnitude higher. Further, we will focus on the use of radioisotope sources, i.e. neutron sources in the case of γ-ray emissions, and γ-ray sources in the case of characteristic X-rays.

These methods are principally associated with laboratory analysis of samples or specimens brought in from the process. High-intensity sources and high-resolution (cryogenic) radiation detectors are used. Nevertheless, there is potential for using these concepts in in-line industrial processes, even if the performance requirements have to be relaxed. In contrast to laboratory analysis, we often look for the concentration of one or a few elements in an in-line process analysis. This enables the measurement to be carried out with room-temperature detectors and window counting, instead of full PHA and spectral analysis against emission libraries in computers. The two basic properties of these methods are that their emission energies are element specific, and that the intensities are related to the concentration of the element. The intention with the following presentations is merely to give an idea of the basic physics and possibilities of these methods.

X-ray fluorescence analysis (XRF or XRA) is most often performed on K- and L-shell emissions with a high-intensity X-ray tube as the excitation source. The use of radioisotope excitation and room-temperature detectors implies a lower emission intensity and poorer energy resolution. The former is because the intensity of radioisotope sources is typically restricted to about 10^7 photons/(s·sr) compared to about 10^{12} for X-ray tubes [11]. This may to some extent be compensated for by using a geometry with the source closer to the sample.

The fluorescence intensity of a system with geometry identical to that presented in Figure 5.25a, may be modelled. The incident radiation is now the monochromatic excitation radiation (E_γ) with intensity I_0 and the output beam is the characteristic radiation of an element j in a homogeneous sample with concentration C_j. The net fluorescent intensity free from influences such as background, overlap, etc., can then be expressed as [112]:

$$I_{\mathrm{f}} = I_0 \overbrace{e^{-\mu_{\mathrm{w}} x_{\mathrm{w}}}}^{1} \overbrace{e^{-\mu x_0}}^{2} \overbrace{C_j \frac{\mu_j}{\mu_h}}^{3} \overbrace{(1 - e^{-\mu \Delta x})}^{4} \overbrace{\frac{r_{\mathrm{K}}-1}{r_{\mathrm{K}}} g_{\mathrm{K}_\alpha} \omega_{\mathrm{aK}}}^{5} \overbrace{e^{-\mu'(\Delta x/2)}}^{6} \overbrace{\frac{A_{\mathrm{det}}}{4\pi D^2}}^{7} \overbrace{e^{-\mu' x_s}}^{8} \overbrace{e^{-\mu'_{\mathrm{w}} x_{\mathrm{w}}}}^{9} \tag{5.42}$$

where the different factors are

1. Relative transmission of incident radiation intensity, I_0, in the vessel wall, where μ_{w} is the linear attenuation coefficient of the wall at E_γ, and x_{w} its thickness.

2. Relative beam transmission over the path length x_0 before reaching the measurement volume (voxel). Here μ is the linear attenuation coefficient of the process medium, the host, at E_γ.

3. Fraction of radiation absorbed by the fluorescent element of attenuation coefficient μ_j, where C_j is its concentration. Absorption of radiation by the species present in the sample is ignored, under the assumption of low concentration.

4. Fraction of the radiation that is attenuated in the volume element, where Δx is the length of the voxel in direction of the radiation source.

5. Excitation factor, which is the product of three probabilities:
   $\frac{r_K - 1}{r_K}$: Absorption jump factor, fraction of the absorbed intensity by element i, which leads to K ionisation [11].
   g_{K_α}: Probability of emission of a K_α-line in preference to other K-lines.
   ω_{aK}: K-line fluorescence yield.

6. Average relative transmission of scatter over the path length Δx inside the measurement volume. Here μ' is the linear attenuation coefficient of the process medium at the energy of the fluorescent radiation.

7. Solid angle subtended by the detector collimator at the voxel, where D is the distance from voxel to detector collimator and A_{det} the area of the detector. An approximation is used here because the distance from the voxel to the detector is much larger than any of the linear dimensions of the detector collimator.

8. Relative transmission of the fluorescent radiation over the path length x_s towards the radiation detector.

9. Relative transmission of the fluorescent radiation intensity in the exit vessel wall.

The detector is assumed to have 100% stopping efficiency. It is also possible to utilise the amount of scattered radiation for matrix compensation, as mentioned in Section 5.5.2. When the sample is irradiated with polychromatic radiation it is necessary to consider all of the primary energies in the useful range, and knowledge of the spectral intensity distribution of the radiation emanating from the radiation source is required. Experiments show that this model provides a good estimation of the fluorescent intensity [112]. As a first approach in the design of a radioisotope XRF system the following consideration is useful: The probability of producing fluorescence at any emission angle for a given concentration, is expressed by the product of the fluorescence yield and the photoelectric attenuation coefficient just above the edge, i.e. $\omega_{aK}\mu_{\tau K}$ (terms 4 and 5, see Table A.2). This product has its maximum (see Table A.2) for elements with atomic number around 30. Regrettably, the average K-line fluorescence energy is only about 10 keV in this region (see Figure 3.8 or Section A.3). This presents a disadvantage because low-energy photons are more likely to be attenuated before they reach the radiation detector. The implications of this on the application of XRF directly on industrial processes are as follows:

- The use of low attenuation radiation windows, preferably open processes or samples, is essential to achieve minimal attenuation of the fluorescence as well as the excitation radiation [terms 1 and 9 in Equation (5.42)].
- Elements with atomic number in the range of 30–50 yield the best response for K-line emissions. L-line emissions may be used for higher Z elements. Even though the L-line fluorescence yield is lower, the photoelectric attenuation is higher owing to the lower energies.
- Low attenuation process media or host materials are preferred to minimise attenuation of the excitation and fluorescence radiation, and the generation of Compton scattered

Table 5.5 Recommended balanced filter pairs for detection of K-line fluorescence from various elements [6][a]

Element	Pb	Hg	W	Ta	I	Sn	Cd	Mo	Nb	Zn	Cu	Ni	Co	Fe	Mn
Filter pair	Re Ir	W Re	Er Tm	Ho Tm	In Sn	Pd Ag	Ru Rh	Y Zr	Sr Y	Ni Cu	Co Ni	Fe Co	Mn Fe	Cr Mn	V Cr

[a] K-edge and fluorescence energies are listed in Section A.3.

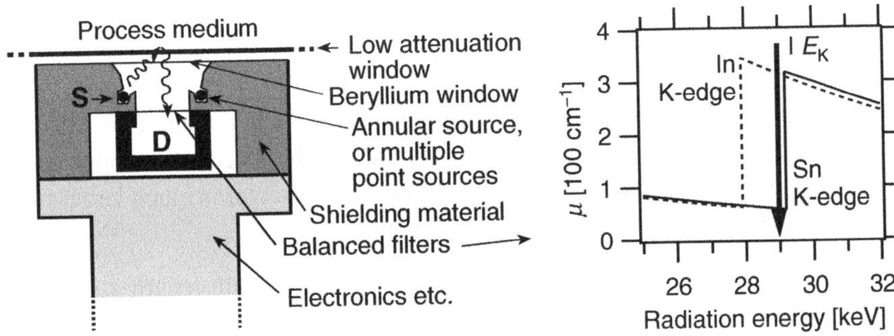

Figure 5.26 Schematic representation of the geometry of a radioisotope excitation system using annular source (left). Alternatively a number of point sources may be used [15]. A real system would also have shutter in front of the beryllium window. The plot (right) shows an example of a balanced filter pair – indium and tin – recommended for the detection of the K-line emission from iodine [6]

events, which often produce spectral background at the fluorescence emission energy. As was the case with scatter measurements, it is clear that geometries with the source and detector close to each other give the best performance. The influence of terms 2 and 8 in Equation (5.42) is then reduced.

Long counting time and temporal averaging may be used to improve the measurement resolution in cases with low fluorescence intensity, even with the presence of Compton background. This background is, according to Figure 3.11, lowest at 90° scattering angle, making this the optimal angle as defined by the source, process medium and detector geometry. A compact, annular source measurement geometry as suggested in Figure 5.26 may be used for optimal response.

Balanced filters are used to increase the sensitivity to a particular fluorescence energy line, and suppress other energies [6, 113]. A pair of materials is used: the first has its K-edge just below the emission line of interest, whereas the second has it just above this energy. This is illustrated in Figure 5.26, using iodine K-line emission as an example. In the first measurement, using the first filter material, the fluorescence line will be heavily attenuated, but not so in the second measurement with the second filter. By comparing the two spectra, high sensitivity is obtained for the fluorescence line because this is in the band-pass region between the two K-edges. Background radiation outside this region will be suppressed. One could of course envisage doing these measurements simultaneously using a parallel detector pair with different filters. Recommended filter pairs for detection of some elements are listed in Table 5.5.

One great advantage of using a radioisotope for excitation instead of a traditional X-ray source is that a single excitation energy produces far less Compton background. A source with emission energy closest possible above the edge of the element of interest should be chosen for excitation. Applicable sources are ^{241}Am, ^{93m}Nb and ^{57}Co, and characteristic X-ray emissions may be used as well as γ-ray emissions. Compound semiconductor detectors are good candidates for detection of X-ray fluorescence for industrial gauges, particularly for analysis of high-Z elements. At lower energies room-temperature silicon detectors and proportional counters may be used.

Industrial online XRF is a very good example of possible application of the Fluor'X tube presented in Figure 2.13. This is ideal for this purpose because the excitation energy can be tuned to the element of interest; the excitation intensity is much higher than that of a radioisotope source making the scattering background far less than that of a traditional tube. On the other hand a radioisotope source has the advantage of perfectly stable emission intensity.

The essence of the fluorescence measurement function is the relationship between the measured intensity in a spectral window (I_f), the concentration of an element (C_j), and the lowest detectable concentration. In practice this is not as trivial as indicated in Equation (5.42), particularly not for low-intensity applications measured by detectors with moderate energy resolution. Whenever possible, calibration against process media samples with known elemental concentrations is a reliable approach. References [8, 11, 15, 108, 114, 115] are recommended for further reading on XRF and related methods.

Prompt γ-ray neutron activated analysis (PGNAA) is based on the detection of γ-ray emissions immediately ($<10^{-12}$ s) after nuclear reactions or inelastic scattering initiated by neutrons. Nuclei involved in nuclear reactions are often unstable and will disintegrate according to their half-life. γ-Rays emitted after such disintegrations are called delayed γ-rays and are the foundation of *delayed γ-ray neutron activated analysis* (DGNAA or just NAA). PGNAA is suitable for online measurements because it is immediate during the irradiation. In contrast to DGNNA it also enables detection of elements that produce only stable isotopes, and it produces less residual activity after irradiation.

The majority of neutron activation analysis is carried out with a nuclear reactor as the neutron source. The flux of thermal neutrons is then very high. There are also neutron generators based on compact particle accelerators [9, 10]: however, radioisotope neutron sources are of particular interest for permanently installed industrial gauges because of their simplicity and ruggedness. The drawback with these is once again the much lower available fluxes: In a reactor facility thermal neutron fluxes of 10^{12} neutrons/(cm^2·s) would be fairly typical [6], and an accelerator-based generator would typically produce a flux of 10^9 neutrons/(cm^2·s) [8]. The isotropic neutron emission fluxes of the ^{241}Am/Be and ^{252}Cf isotopic sources are listed in Table 5.6 alongside other properties. In spite of the lower flux and its higher γ-ray background, the ^{241}Am/Be source is often preferred to the ^{252}Cf source because of lower cost and longer half-life. The ^{241}Am/Be type sources can also be made switchable by means of a mechanical operation so that the neutron emission can be switched off when the source is not in use [118].

These isotopic sources emit fast neutrons, making the use of a moderator necessary to produce slow neutrons suitable for the (n, γ) reaction. Efficient moderators such as heavy water are used for this purpose. Sometimes graphite is preferred to a hydrogen-rich

Table 5.6 Properties of isotopic neutron sources [116, 117]

Source	Reaction	Isotropic neutron emission intensity	Half-life	Emission energies	
				Maximum	Average
^{241}Am/Be	^{9}Be(α, n)^{12}C	2.2×10^{6} neutrons/(s·Ci) (0.3 g/Ci)	433 y	10–12 MeV	4.4 MeV
^{252}Cf	Spontaneous fission	2.3×10^{12} neutrons/(s·g)	2.7 y	6–8 MeV	2.35 MeV

moderator to avoid excessive loss of neutrons via capture by hydrogen, and through that also reducing the γ-ray background. In other cases the process medium is used as moderator, particularly when it is desirable to include prompt γ-rays from inelastic scattering. The latter is a requirement for analysis of oxygen and carbon that have a very low slow neutron absorption cross section. In this case ^{241}Am/Be is also recommended because of its higher neutron energies.

The characteristic emission energies of PGNNA are high compared to those of XRF, making its attenuation less of a problem. Nevertheless the detector should be positioned as close as possible to the measurement volume from which the prompt γ-rays are emitted, but preferably shielded from the neutron flux. Neutron interactions in the detector may shorten its life through radiation damage and also produce radiation that increases the background. The two most appropriate detectors for permanently installed gauges at these high energies are NaI(Tl) and BGO scintillation detectors [116, 119]. The NaI(Tl) detector generates intense internal radioactivity of ^{24}Na and ^{128}I even at exposure to low neutron fluxes, making spectral interpretation a lot more difficult. The BGO detector is therefore often used in spite of its higher cost and its high sensitivity to temperature changes as we saw in Section 4.6.7. Its strength is of course very good stopping efficiency, which is important at these high energies. As already mentioned shielding of the detector is important to protect it from neutron irradiation damage and to reduce the γ-ray background. Design examples of compact gauges using isotopic sources and scintillation detectors are given in References [116, 120, 121]. Anticoincidence Compton suppression is often used to reduce spectrum background in laboratory detection systems, but seldom in field systems.

The prompt γ-ray emission spectra of many elements can be very complex with hundreds of emission lines [21]. Fortunately, most of these are rare taking place only in a few percent of all neutron captures, whereas others are more or less distinct and applicable for element identification. A list of recommended emission lines (E_r) has been complied for 51 elements by Chung [15] and is included in Section A.4. For each element the thermal neutron cross section (σ_{th}) for the radiative capture (n, γ) reaction is given alongside the probability (I_r) that there will be an emission at E_r after a neutron capture. The minimum detectable amount of any element with atomic mass M can now be estimated for a given background count-rate (n_b), counting time (τ_I) and the absolute detection efficiency of the detector (ε). The dependency of the detection limit (DL) to these parameters can then be expressed as [15]

$$\mathrm{DL} = \frac{329M\sqrt{n_b \tau_I}}{N_A \varepsilon(E_r)\bar{\Phi}\bar{\sigma} I_r(E_r)} \quad [\mathrm{g}] \tag{5.43}$$

Table 5.7 PGNAA detection limit for elements analysed with $\Phi = 10^7$ neutrons/(cm^2·s), $\tau_I = 10^5$ s and conditions otherwise typical for a nuclear reactor analysis facility using a HPGe detector [15]

Detection limit range	Elements
0.1–1.0 μg	B, Gd, Cd
1.0–10.0 μg	Sm, Hg
10.0–100.0 μg	Er, Ag, Nd, Rh, In, Kr, H
100.0 μg –1.0 mg	Eu, Ti, Dy, Au, Ta, Cl, Lu, Tm, Hf
1.0–10.0 mg	Ho, Co, Xe, Os, Mn, La, Ir, Cr, Ge, Ca, Sr, V, Mo, Ni, Te, Br, I, S, Fe, K, Y, Zn, As, Cs, Cu, Li, Yb, Ar, Sc, N
10.0–100.0 mg	Se, Ga, Si, W, Ru, Na, Pr, Zr, Al, Re, Pt, Sn, Mg, P, Tb, Ba, Pb, Ce
100.0 mg –1.0 g	Nb, Be, Tl, Sb, Ne, C, F, Bi, Rb
1.0–10.0 g	O

Source: Reproduced by permission of John Wiley & Sons, Inc.

where $\bar{\Phi}$ and $\bar{\sigma}$ are the weighted averages or neutron flux and reaction cross section, respectively. Chung has calculated this for several elements and provided a very useful list sorting these by orders of magnitude after their detection limit (see Table 5.7) [15]. This is for typical conditions using a reactor as neutron source and a high purity Ge (HPGe) detector. For a radioisotope system typical detection limits may be a few orders of magnitude higher. Nevertheless, Table 5.7 gives a clear indication of the feasibility of analysing the various elements at different conditions as given by Equation (5.43). Field PGNAA is an inherently robust method that finds an increasing number of field applications. The number of counts in a spectral line may be modelled as a function of the concentration of the element in question. In practice, however, the best solution is often, as for XRF, to calibrate at known conditions against process media samples with known elemental concentrations whenever possible. For further reading on PGNAA, References [15, 116, 122, 123] are recommended.

5.5.4 Tracer Emission

Radioisotope tracing differs from the other radioisotope measurement principles presented in the preceding sections in that it is not radiation interactions with the process medium that form the basis of the measurement. The basic principle of any tracer investigation is to label a substance, an object or a phase and then study its behaviour through the system. Provided the tracer behaves exactly as the marked material, its dynamics can be studied by measuring the emission intensity of the tracer isotope as function of time and position. The basic requirements of a tracer are as follows: It should behave in the same way as the material under investigation; it should be easily detectable at low concentrations; detection should be unambiguous; injection, sampling and/or detection should be performed without disturbing the system; and last but not least, the residual tracer concentration in the process or its product should be minimal. The latter implies that short-lived isotopes must be used. This in turn also means that these isotopes require relatively quick transfer from their production facility to the measurement site. Such facilities are mainly nuclear reactors and accelerators that rarely are on-site equipment. A compromise thus has to be made and isotopes with half-lifes in the range of 2–40 h are typically used [6]. The only economical and practical on-site isotope generator is the so-called *gamma-cow* presented in Section 2.2.3.

The practical outcome of this is that radioisotope tracers by far are used as part of process diagnostic or laboratory tools. On the other hand, here they are the foundation of several powerful methods with many important applications. For process diagnostics, tracer studies are frequently used for measurement of flow rate and residence time and for leakage detection. For flow measurement for instance, a typical system uses two γ-ray detectors collimated towards a narrow pipeline cross section: one placed on the pipeline just downstream of the tracer injection point, and the second at a measured distance further downstream. Using a sharp pulse injection the average flow rate then is readily available through time difference measurement between the two detectors. Similarly, for measurement of residence time in a process vessel these detectors are placed at the vessel inlet and outlet. In this case the difference in pulse shape (intensity vs. time), the so-called *response curve*, at the two detector positions also carries information about the process conditions, such as mixing and blending. These time difference measurements are typically in the range of seconds or more, meaning the pulse shape is obtained by a large number of sequential intensity measurements each with counting times in the range of 1–100 ms. Further details on the use of radiotracers for process diagnostics are given in References [6, 124–126].

Radiotracers are also the foundation of a range of advanced laboratory methods primarily used to provide experimental process data otherwise not available. These data provide improved understanding of various processes and their dynamics and are often the key for the development of accurate process models and their validation. The most widely known method is PET where a β^+-emitter, such as ^{18}F with ~110-min half-life, is used as the tracer isotope. For each β^+-disintegration a β^+e^--annihilation will take place in the close vicinity, typically within a radius of less than 3 mm. The position and distribution of a source is found by detecting the interaction positions of several back-to-back annihilation photon pairs in two or more 2D PSDs operated in coincidence (see Figure 5.23). A minimum of two uncollimated PSDs is required, such as illustrated in Figure 5.11, but modern PET cameras use a large number of detectors in a ring surrounding the object or process. A PET facility may also be used for PEPT (positron emission particle tracking) where only a single particle in a process is labelled. Its position and velocity within the process can then be monitored with high accuracy. Another, but related method uses a γ-ray emitter as the tracer isotope. This is known as single particle emission computed tomography (SPECT) and requires the detector to be collimated so that the view of each PSD element is limited to a narrow cone through the process. The tracer positions within the process can then be determined by detecting a series of γ-rays in two or more PSDs around the object facing its centre at different angles. Further details on these techniques are given in References [97, 98, 127, 128], and references therein.

On-line gauges using radioisotope tracers are rare, but do exist as we shall see in Chapter 7. For instance, flow measurement in open channels uses the gamma-cow as an isotope generator and an automatic isotope injection system. This method has potential to be used for measurements of other of the process parameters mentioned above, such as phase residence time in process vessels. Its disadvantage is that the automatic injection system introduces mechanical movement of components in the process equipment, something that tends to lower the reliability and MTBF.

5.5.5 NORM Emissions

We saw in Section 2.2.2 that NORM γ-ray emissions are frequently used to identify stratigraphical layers in boreholes (lithology). Otherwise in industrial measurements NORM emissions generally are considered as unwanted and interfering background. The accumulation of NORM scale deposited by salts in the produced water inside process equipment in the oil industry boosts this background and makes it change with time. This may also be a safety problem requiring attention. Detector systems measuring NORM emissions at the outside of a pipeline, such as suggested in Figure 5.23, may be used to monitor the extent of scaling as well as the total radiated dose. A full spectrum count may be used for an estimate, or spectroscopy or multiple window counting could be used for measurement of particular emission lines. The latter requires detector equipment much similar to that used for PGNNA.

5.5.6 Multiple Beam, Energy and Modality Systems

The process industry increasingly requires more process information motivated by key issues such as improved process control, process utilization and process yields, ultimately brought forward by cost effectiveness, quality assurance, environmental and safety demands. This puts a challenge on the design of efficient measurement systems, particularly those applied to multiphase processes. As a consequence we see two important trends within measurement science: the extraction of more information from each measurement principle, the combination of several measurement principles or modalities, and the combination of these two.

The former is basically achieved through the availability of compact units with high computing power. This allows for the use of advanced signal processing and data analysis such as was restricted to laboratory instruments only a few years ago. The combination of several measurements is a necessity in order to determine the component fractions in multiphase systems with more than two components. This may be realised in several ways, or combination of these:

- *Dual or multiple modality systems*. Several of the modalities shown in Figure 5.23 may be used simultaneously, for instance the measurement of transmitted and scattered radiation. To a first approximation low-energy γ-ray transmission depends on photoelectric absorption and Compton scattering, whereas the generation of γ-ray scatter depends on the latter only. The clue is that any added modality yields information not available from the other modalities used. In our example photoelectric absorption is highly dependent on the atomic composition (Z), whereas Compton scattering depends basically on the density (ρ). To obtain this, non-nucleonic sensing principles are also often added: for instance one measuring the average or effective relative permittivity of the process material.
- *Dual or multiple energy systems*. Here additional information is brought forward by the measurement of two or several energies simultaneously. With reference to our previous example the influence of photoelectric absorption may be found by using one γ-ray

transmission measurement at a low energy, and in addition one at a higher energy where the influence of Compton scattering is the dominant interaction mechanism and thus can be determined (see, e.g., Figures 3.6 and 3.7). The requirement is the use of sources with multiple lines and energy-sensitive detector systems. These need not necessarily be a full PHA; in most cases counting in multiple energy windows is sufficient as discussed in Section 5.2.2.
Needless to say NORM and some tracer emission systems inherently operate at multiple energies.

- *Dual or multiple radiation beam systems.* In most cases these systems are aimed at revealing something about the distribution of process components or phases if these are not homogeneously mixed. Such measurements are most often required because the assumption of homogeneity when it is not true is a significant source of error. In many cases two or a few beams are sufficient for permanently installed gauges; more seldom full tomographic imaging is necessary.

We will discuss the latter in more detail in Section 7.7 (tomographic methods). Here we will also see that a priori process information in many cases is equally important to the final measuring result as the contribution of individual measurements. Finally, it should be noted that the use of multiple beam, energy or modality measurements often yields redundancy, which increases the reliability of a measurement system (see Section 5.3.7).

6

Safety, Standards and Calibration

We mentioned in Chapter 1 that people generally fear radioactivity, most often because of lack of knowledge about the subject. In this chapter we will see that the risks associated with using radioisotope gauges are very small, even for radiation workers who use these on a daily basis. This is because of the relatively strict recommendations and legislation for using and transporting radioisotope gauges. We will study these recommendations, but deal with the legislation on a more general basis because this varies between nation states.

6.1 CLASSIFICATION OF INDUSTRIAL RADIOISOTOPE GAUGES

International Standards Organisation ISO 7205 [129] and American National Standard N538 are the two main standards used for the classification of industrial ionising radiation devices. The two standards are similar but not identical, particularly the gauge classification codes derived from the two standards for the same gauge are different. Both standards list the desirable design features for an industrial gauge and describe methods of testing these features.

6.2 RADIOLOGICAL PROTECTION

As you will recall from Chapter 1, it was not long after the discovery of X-rays that the risks and hazards associated with electromagnetic radiation became apparent. The earliest effects that were observed were acute effects of over exposure to ionising radiation, usually localised burns to the hands. As the harmful effects of radiation became recognised so the need for protection became apparent and the damage suffered by Röntgen and his co-workers led to the first attempt at a code of practice for protection against ionising radiation. This code, known as *Rollins Code*, was suggested in 1902 and used exposure of a photographic plate to assess whether or not a source was safe to handle. We shall see that today radiological protection is addressed by several international agencies and bodies.

The intention of this chapter can be summarised by quoting the International Commission on Radiological Protection (ICRP) in their publication no. 60 on radiation protection [129]:

Radioisotope Gauges for Industrial Process Measurements. Geir Anton Johansen and Peter Jackson.
 ISBN 0-471-48999-9

The Commission emphasises that ionising radiation needs to be treated with care rather than fear and that its risks should be kept in perspective with other risks. Radiological protection cannot be conducted on the basis of scientific considerations alone. All those concerned have to make value judgements about the relative importance of different kinds of risks and about the balancing of risks and benefits.

For radioisotope gauges, this is of particular importance in the design phase. Safe operation of such equipment is ensured by the commitment to regulations and recommendations [132] that are even more restrictive than the recommendations from ICRP. The balancing of risks and benefits in the design phase is done by reducing the radiation dose to the minimum level that can be accepted by the operational demands of the instrument.

Questions often raised in this connection are as follows: What is the concept of radiation dose, what are its effects and what risks are typically involved? An attempt to answer these questions is done by comparing typical doses from a typical γ-ray gauge using a single point source, with doses from natural sources in the environment. It is convenient to use γ-ray gauges for this comparison since these, because of the high-penetration capability of γ-rays, alongside neutrons are most demanding when it comes to safety. The risks involved with gauges based on α-emitters and β-emitters are much lower as long as they are treated correctly; that is, one has to make sure that the sources are properly sealed so there is no contamination from particles of radioactive material. Both α-emitters and β-emitters are hazardous when they become internal to the body, for instance through contamination of air or food, because all the particle energy will be deposited in the body. This is because, according to Section 3.1.2, the short ranges cause all the radiated energy to be absorbed in the body.

6.2.1 Radiological Protection Agencies

The principal organisation in the world for radiological protection advice is the ICRP. This body recommends levels of exposure, which if adhered to will sufficiently protect man from harm. The ICRP has no power of enforcement and recommends maximum levels of exposure. The ICRP is independent of any government and is made up of members who are selected on the basis of their scientific reputation.

The following agencies also make recommendations on various aspects of the use of radioactive material:

- United Nations Scientific Committee on the Effect of Atomic Radiation (UNSCEAR).
- World Health Organisation (WHO).
- Nuclear Energy Agency (NEA) of the Organisation for Economic Development (OECD).
- International Atomic Energy Agency (IAEA) [228].
- European Atomic Energy Community (Euratom).

Most national governments have a radiological protection body that advises the government on the best way to implement the recommendations of the ICRP. Remember that the

ICRP recommends only maximum values, what they suggest should not be exceeded. The various national bodies throughout the world have produced a large variety of measures, which are legally binding on uses of radioactive materials in their territories.

Unfortunately the interpretation of the ICRP recommendations, or more precisely how to ensure that the recommendations are adhered to, can be different. Consequently the user or supplier of radioactive material may find that their normal practice in one country is inadequate or excessive in another country. It is important therefore that before supplying any radioactive instrument internationally the supplier checks the local legal requirements, see examples in Section 6.2.8.

6.2.2 Quantities Used in Radiological Protection

The activity, A, of a radioactive isotope is defined in Equation (2.12) as the number of disintegrations per second: The SI unit of the activity is *becquerel* (Bq), which is defined as one disintegration per second. The *absorbed dose, D*, is defined as the mean energy, $\mathrm{d}\bar{\varepsilon}$ absorbed from any type of radiation per unit mass, dm, of the absorber:

$$D = \frac{\mathrm{d}\bar{\varepsilon}}{\mathrm{d}m} \tag{6.1}$$

The SI unit of D is *Gray*, which is defined as 1 J/kg. The radiation energy is deposited through ionisation of the absorber's molecules. The biological effects of nuclear radiation exposure are, however, dependent on the type of radiation and its energy. The *equivalent dose*, H_T, in a tissue or organ has therefore been introduced as

$$H_\mathrm{T} = \sum_\mathrm{R} w_\mathrm{R} D_\mathrm{T,R} \tag{6.2}$$

where $D_\mathrm{T,R}$ is the absorbed dose averaged over a tissue or organ T because of radiation R and w_R is a dimensionless *radiation weighting factor.* For low linear energy transfer (LET)-radiation such as γ-rays, X-rays and electrons (β-particles), w_R is unity for all energies. Low LET-radiation deposits its energy over a relatively long range in the absorber. An example of high LET-radiation, which has higher ionisation density, is α-particles for which $w_\mathrm{R} = 20$. For neutrons, w_R is energy dependent and equal to 5 ($<$10 keV and $>$20 MeV), 10 (10–100 keV and 2–20 MeV) and 20 (100 keV–2 MeV). The SI unit of H_T is *Sievert* (Sv), which is defined as 1 J/kg. The biological effects of nuclear radiation are also found to vary with the organ or tissue irradiated. Because of this the *effective dose*, E, is introduced and defined as

$$E = \sum_\mathrm{T} w_\mathrm{T} H_\mathrm{T} \tag{6.3}$$

where w_T is a dimensionless *tissue weighting factor.** The values of w_T are chosen so that a uniform equivalent dose over the *whole body* gives an effective dose numerically equal to that uniform equivalent dose. This is also referred to as a *full body dose*. The effective dose has the same SI-unit as the equivalent dose (Sv).

*The w_T value for the gonads is largest (0.20), while it is lowest for tissues like bone and skin (0.01). The w_T value is identical to the formerly used *Quality factor*, except for neutrons where there are some changes.

Table 6.1 Relationships between new and old units on radiation activity and absorption

Quantity	Symbol	Definition, new (SI) unit	Old unit	Relationship
Activity	A	1 Bq = 1 disintegration/s	Ci (Curie)	1 Ci = 3.7×10^{10} Bq
Absorbed dose	D	1 Gy = 1 J/kg	rad[a]	1 rad = 10^{-2} Gy
Equivalent dose	H	1 Sv = 1 J/kg	rem[b]	1 rem = 10^{-2} Sv
Exposure	X	1 X unit = 1 C/kg in air[c]	R (Röntgen)	1 R = 2.58×10^{-4} X units

[a] Radiation absorbed dose.
[b] Röntgen equivalent man. The rem unit was used with the old *dose equivalent* quantity.
[c] No SI unit has yet been defined for the exposure unit.

Another quantity that is not that frequently encountered anymore is the *exposure unit*, X. X-ray and γ-ray fields may be specified in terms of the exposure unit, which is independent of the properties of the organism being exposed. It expresses the radiation's ability to ionise air and is defined as that quantity of electromagnetic radiation, e.g. X-rays or γ-rays, that produces ions carrying one coulomb of charge per kilogram air:

$$X = \frac{dQ}{dm} \tag{6.4}$$

No SI unit is yet defined for the exposure unit. It is, however, an important quantity since it, in contrast to the dose quantities defined in Equations (6.1)–(6.3), can be measured in a relatively simple manner. The absorbed dose at a point in the radiation field is calculated as the product of the exposure unit and a conversion factor that is dependent on the absorption properties of the tissue and the radiation energy.

Although the SI units are standard today, some of the old units are still being used. Table 6.1 gives a summary of the definitions and relationships between old and new units. The associated rates (time derivatives) of the quantities defined in Equations (6.1)–(6.4) are usually denoted with the time derivative mark, e.g. $\dot{D}$, and expressed in units of hours^{-1}.

6.2.3 Biological Effects of Ionising Radiation

Ionising radiation causes both *deterministic* and *stochastic effects* in irradiated tissue. Deterministic effects result from the killing of cells which, if the dose is large enough (typically a few Gy), causes sufficient cell loss to impair the function of the tissue. Radiological protection aims at avoiding deterministic effects by setting dose limits below their thresholds. Stochastic effects may result when an irradiated cell is modified rather than killed. There are repair and defence mechanisms, but modified somatic cells may subsequently, after a prolonged delay, develop as cancer cells. Another outcome may be severe hereditary effects. ICRP uses the term *detriment* to represent the combination of the probability of occurrence of a harmful health effect and a judgement of the severity of that effect. The detriment is adopted by ICRP to be about 7.3×10^{-2} Sv^{-1} for the whole population. Three of the largest epidemiological investigations on the effect of ionising radiation, however, do not show any significant increase in the probability of cancer for doses less than about 200 mSv. The risks adopted by ICRP at low-dose levels are based on data from high-dose levels and high-dose rates [133].

Table 6.2 Average annual risk of death in the UK from some common causes and from accidents in various industries and [2]

Cause	Risk of death per year
Smoking 10 cigarettes a day	1 in 200
Natural causes, 40 years old	1 in 700
Accidents on the road	1 in 10 000
Accidents at home	1 in 10 000
Most exposed from nuclear effluents (0.3 mSv)	1 in 70 000[a]
Sea fishing	1 in 500
Coal mining	1 in 7000[a]
Construction	1 in 10 000
Radiation workers (1.1 mSv per year in average)	1 in 27 000
Food, drink and tobacco	1 in 45 000
Timber, furniture	1 in 100 000

[a] Estimated, not observed.

6.2.4 Risk

Risk is a statistical quantity expressing the probability, for instance, of developing fatal cancer by exposure to a certain radiation dose. Risk estimation for exposure to low-radiation doses over long time is associated with some uncertainty because it is based on extrapolation of data on high doses delivered over short time. Traditionally it has been assumed that there is a linear relationship between the risk of some effect and the dose, all the way from high doses to zero dose, zero risk. There are now discussion on whether this relationship may be exponential in the low-dose range [134], at least for β-particles, X-rays and γ-rays, meaning the increase in risk with dose is much lower at low doses than at high doses. Experts are even debating whether zero risk is at some threshold dose higher than zero. This implies that low doses of ionising radiation benefit our body. We will not dwell on this, but just conclude that regardless of this uncertainty risk is a useful concept because it allows comparison of hazards involved in various situations, professions, etc.

6.2.5 Typical and Recommended Dose Levels

The human body cannot sense low doses of nuclear radiation. A radiation dose can, on the other hand, be quantified with a much better accuracy than for instance toxic chemicals. This may be one of the explanations why some find it difficult to relate to nuclear radiation in a realistic manner. To get a better quantitative understanding of what an absorbed dose is, it may be useful to take a look at some typical dose levels in various known situations. Some examples are listed in Table 6.3.

In Section 6.3.4 there is an example showing that the accumulated dose for a radiation worker carrying out a series of source transfers over a day is about 40 μSv. This is the so-called *occupational* exposure; a further understanding of the dose concept may be gained by considering the annual average doses from natural radiation sources. These are given in Table 6.4 and constitute about 85% of the total annual average dose. The remaining 15%

Table 6.3 A rough comparison of effective doses at different situations [135]

Effective dose	Radiation source	Situation/time
1 μSv	The body (natural)	24 h
10 μSv	Cosmic radiation	5 h in an aircraft
100 μSv	Dental X-ray	Examination with 2–3 exposures
1 mSv	Medical X-ray	Typical examination of, e.g., back, pelvis or kidneys
10 mSv	Medical X-ray	Complex examination of, e.g., stomach/guts, CT
100 mSv	Cosmic radiation	1 year in a space station
1 Sv	Medical gamma exposure	Cancer therapy (typically 15 Sv)

Table 6.4 Average annual effective doses from natural radiation sources in Norway [136], UK [2] and USA and Canada [134]

	Average annual effective dose [mSv]			
Radiation source	Norway	UK	USA/Canada	Depends on
Cosmic	0.35	0.25	0.27	Altitude
External gamma	0.48	0.35	0.28	Geology and materials of the environment
Internal (the body)	0.37	0.30	0.4	Smoking and food habits, sex and age
Radon	4	1.30	2.0	Geology. Largest values in areas with alunite (slate)
Total	5.2	2.2	2.95	

are man-made or artificial sources such as medical (13%), fallout from nuclear weapons (0.4%) and industrial (0.2%) [2].

ICRP recommends that all use of nuclear radiation should follow the *ALARA* (as low as reasonably achievable) philosophy. This implies a cost-benefit analysis aiming at lower dose levels in all applications. The dose should not exceed the recommended limits listed in Table 6.5. For industrial applications of nuclear radiation further restrictions are given by national bodies: The Radiation Protection Institutes in the Nordic countries recommend that the annual equivalent dose from nucleonic instrumentation should be kept below 0.1 mSv for the general public and below 5 mSv for workers operating such gauges [131]. Moreover, it should be pointed out that the risks in industries using radioisotope gauges are found to be lower than risks in other industries. This is due to effective safety precautions, and the basic philosophy of ICRP is to always keep this difference. This policy is the reason why the recommended dose limits (see Table 6.5) have been lowered in recent years. ICRP 60 changed the risk from 1 in 40 per Sv to 1 in 25 per Sv.

6.2.6 Dose Rate Estimation for γ-Ray Point Sources

The absorbed dose rate from a point γ-ray source can be estimated as

$$\dot{D}(d) = \frac{\Gamma}{10^6}\frac{A}{d^2} \tag{6.5}$$

Table 6.5 Dose limits recommended by ICRP (International Commission on Radiological Protection) [130][a]

	Effective dose limits [mSv]		Annual equivalent dose limits [mSv]		
Category	1 year	5 years	Lens of eye	Skin	Hands and feet
Occupational[b]	50	100	150	500	500
Public	1[c]	5	15	50	–

[a] The annual dose from natural radiation sources (Table 6.4) and doses due to medical diagnostics and treatment are not included.
[b] More restrictive recommendations for pregnant women.
[c] A higher annual dose could be allowed in special circumstances as long as the 5-year limit is not exceeded.

assuming that the inverse square law (see Section 3.2.1) is obeyed and that there is no γ-ray attenuation in between the source and the measuring point at a distance d. Here Γ is the specific γ-ray dose rate constant (SGRDC) for the radioisotope in question. This is normally defined as the absorbed dose rate from a point source with activity $A = 10^6$ Bq ($\approx 27\ \mu$Ci) at $d = 1$ m distance, and expressed in units of mSv/h. It may be regarded as the integrated dose rate over all X-ray and γ-ray emissions of the isotope. It is tabulated for all the isotopes included in Section A.2 with data taken from [137]. This reference also contains data for other source geometries.

The total rate of absorbed dose in the body due to a point source is found by modifying Equation (6.5) to include the effect of shielding:

$$\dot{D}(\mu, x, d) = \dot{D}(d)\, e^{-\mu x} B(\mu, x) = \frac{\Gamma A}{10^6 d^2} e^{-\mu x} B(\mu, x) \tag{6.6}$$

where μ is the linear attenuation coefficient of the shield material and x its thickness. It is assumed that x is small compared to d. The linear attenuation coefficient is energy dependent, but for radiation protection purposes the value at the principal emission energy of the sources may be used to obtain an estimate. The build-up factor, $B(\mu, x)$, is, as mentioned in Section 3.5.1, dependent on the linear attenuation coefficient and the thickness of the shield. The build-up factor for lead is plotted in Figure 3.13, and it can be seen that $B(\mu, x)$ approaches unity in the case of thin shields or low radiation energies.

In Figure 6.1 dose rates predicted by Equation (6.6) are plotted as functions of lead shield thickness. These are estimates for $A = 3.7 \times 10^5$ MBq (10 Ci) and $d = 1$ m and only intended as a guide. For each source the linear attenuation coefficient of lead at the principal energy (see Section A.2) is used. Approximate values of $B(\mu, x)$ are taken from [22].

It can be seen from the figure that the required lead thicknesses to reduce the absorbed dose rate to e.g. 1 μSv/h, are approximately 2 mm, 3 cm, 10 cm and 20 cm for ^{241}Am, ^{133}Ba, ^{137}Cs and ^{60}Co, respectively. This means, according to Table 6.3, that 1 h spent this close to these high activity radiation sources (1 m) with this amount of lead shield, gives an absorbed dose comparable to that of natural radiation from the body over 24 h. One should, however, keep in mind that the absorbed dose is proportional to d^{-2} [see Equation (6.6)] and that even operators of γ-ray gauges will not stay this close (1 m)

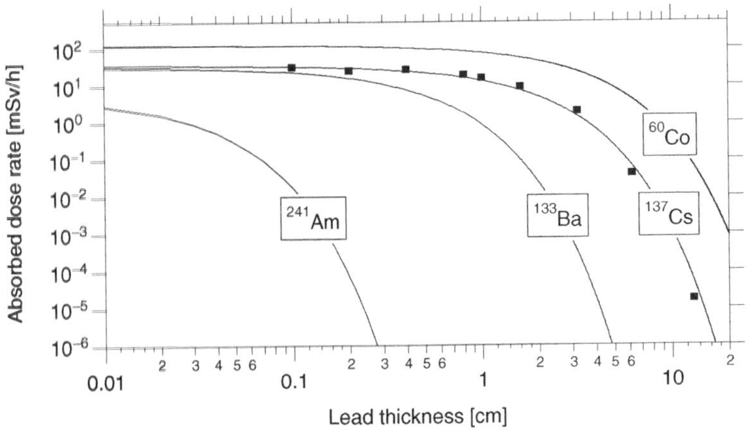

Figure 6.1 Estimated absorbed dose rates at 1 m distance as function of shielding thickness for four typical γ-ray point sources with 3.7×10^5 MBq (10 Ci) activity. The markers are data for ^{137}Cs as calculated by a semi-empirical model [12]

to the radiation source for more than very short periods. It is, due to the problems in predicting the exposure time and source-body separation, difficult to estimate the exact annual absorbed dose to an operator of this type of γ-ray gauge. However, in most cases it is far below the typical annual dose levels listed in Table 6.4 and the recommended limits listed in Table 6.5. This means that the extra risk involved with γ-ray gauge systems is negligible in comparison to other risks.

More accurate dose rate estimations are obtained using Monte Carlo simulation software (see Section 8.5). This method is widespread within radiological protection. Efficient models of the human body have been developed for a variety of radiation source geometries and radiation types. Alternatively software with semi-empirical models of the build-up may be applied [12]. Experimental benchmarking is normally carried out to validate the models used by these types of software. However, for γ-ray point sources relatively simple calculations provide a fairly good estimate of the dose rate as can be seen from the plot in Figure 6.1 where the results using a semi-empirical model has been included in the case of ^{137}Cs.

6.2.7 Dose Rate Estimation for Neutrons

Neutron dose is very dependant on energy, Table 6.6 shows the neutron flux of a given energy required to produce a dose rate of 25 μSv/h.

6.2.8 Examples on National Legislation

Let us for instance take a density gauge supplied to users in three different countries. The gauge is fitted with a 1 GBq (27 mCi) ^{137}Cs source in a spherical lead shield about 15 cm diameter. Such an arrangement would have a dose rate on the surface of the container of

Table 6.6 The neutron flux of a given energy required to produce a dose rate of 25 μSv/h[a]

Neutron energy	Neutron flux [n cm^{-2} /s] for 25 μSv/h
Thermal (0.025 eV)	670
5 keV	570
20 keV	280
100 keV	80
500 keV	30
1 MeV	18
>10 MeV	10

[a] These numbers are based on a standard man body specific gravity of 1.07 g/cm^2, and that 1 kg of tissue will have a surface area of 95.6 cm^2, and with values of w_R as given in Section 6.2.2.

about 7.5 μSv/h. National limits of radiation to be received by a non radiation worker are all based on the same recommendations from the ICRP but the way the dose to the worker is calculated differs. In one country this interpretation of the ICRP maximum dose rate for workers on the site where the gauge is to be installed, assumes that they could be exposed to the full dose rate on the surface of the sources shielded container, and furthermore that the worker could be in this dose area for the whole of his working days. This argument would result in an assumed annual dose for workers of 7.5 μSv × 40 h working week × 50 weeks worked = 15 mSv.

In a second country the regulation may differ, by assuming that for the worker to receive a whole body dose in his working life next to the gauge. Then the dose rate to be considered is that which can be measured where the center of the workers body may be when he is standing close to the gauge. This is assumed to be 30 cm from the surface of the shielded container. At this point the dose rate would be about 1 μSv/h and this would result in a calculated annual dose of 1 μSv/h × 40 h × 50 weeks = 2 mSv.

The third country in addition to allowing for the dose to be assessed at 30 cm from the surface of the container may take into account occupancy of the gauge site. If the gauge is inaccessible and seldom visited then the dose rate may be allowed to be higher to take into account the short exposure time. The consequence of these different interpretations of the way in which the dose to the worker is assessed, is that whereas in country one the maximum source allowed in this shield is 1 GBq (27 mCi) in country two is could be increased into 7 GBq (189 mCi). In the third country this could be even higher depending on occupancy of the site.

6.3 RADIATION MONITORS AND SURVEY METERS

Survey meters or radiation monitors should be available whenever radioactive material is handled. The monitors can be categorised into the following groups: Contamination monitors, Dose rate meters, Active dosimeters, and Passive dosimeters. A selection of radiation monitors and survey meters is shown in Figure 6.2.

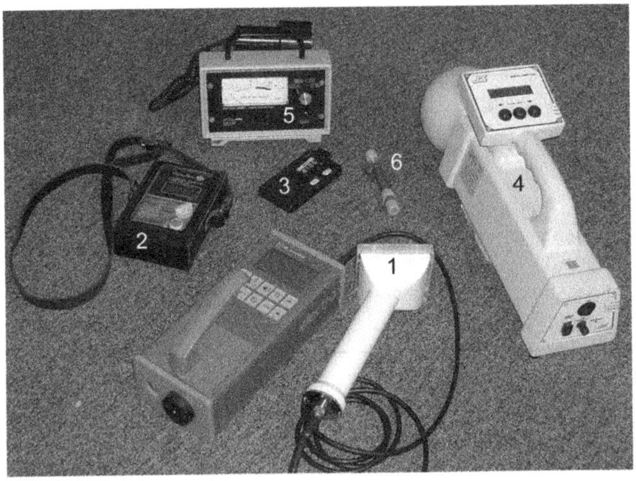

Figure 6.2 A selection of radiation monitors: (1) NE Technology electra contamination monitor (see Figure 6.3), (2) Tracerco intrinsically safe γ-ray monitor, (3) Mini Instruments 6100 personal monitor, (4) JCS neutron monitor, (5) Mini instruments 900 contamination monitor with end window GMT probe and (6) QFE dosimeter

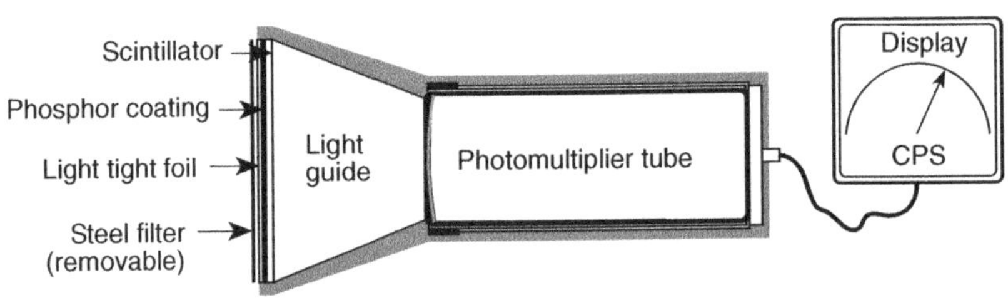

Figure 6.3 A typical contamination monitor

6.3.1 Contamination Monitors

Contamination monitors are usually scintillation devices with a high sensitivity to enable very low levels of contamination to be detected. The readout is in counts per second or minute. The choice of scintillator depends on what is to be surveyed, monitors with a thin plastic scintillator, coated on the front surface with a phosphor such as zinc sulphide are the most versatile allowing detection of α- and β-radiation, X-rays and γ-radiation (see Figure 6.3).

Pulse height selection allows the monitor to distinguish α-radiation from β-and γ-radiation. With the steel filter plate removed beta and γ-radiation are measured together whilst with the steel in place only γ-radiation can enter the monitor. Other types of contamination monitor may use Geiger–Müller tubes (GMT) with thin end windows, which are sensitive to beta and γ-radiation, or scintillation crystals with lightweight windows to allow entry of beta and low energy X-rays and γ-radiation. When checking for contamination of low energy radiation it is important to choose a monitor with a sufficiently lightweight entry window. Windows are usually categorised by their mass per square centimetre rather than their density and thickness. A window sufficiently light to allow the entry of α-radiation is very delicate, a wire mesh protects the window but care must be taken not to puncture the window when monitoring rough surfaces. A puncture in the window results in a light leak to the photomultiplier tube and a false high count-rate. Care should also be taken to protect the monitor front face from becoming contaminated from the area being monitored, the monitor should be held as close as possible without touching the dirty surface. Contamination is quantified in Becquerels per square centimetre whereas the monitor will read in counts per second. Since 1 Bq is one disintegration per second, then the number of counts per second should be divided by the sensitive area of the monitor in square centimetres and multiplied by a factor relating to the detector efficiency to give Becquerels per square centimetre. The efficiency is measured by calibration, which is achieved by comparison with a known standard plaque allowing conversion of the count-rate into Becquerels per square centimetre.

6.3.2 Dose Rate Meters

Dose rate meters are used to measure the instantaneous dose rate and readout is in Sieverts (Sv). Gamma dose rate meters are usually GMT based but scintillation counters and ion chambers are also used. The relatively low efficiency of GMTs allowing for high dose rates to be measured without saturation while the inherently low background of the GMT allows reasonable sensitivity to low dose rates. The GMT needs to be energy compensated (see Section 4.4.5) to give a broadly linear response to energy across a large range. The compensation is necessary because the tube responds best to low energy X-rays and γ-rays and is achieved by surrounding the GMT with a shield with a window that reduces the efficiency of detection of the lowest energies. In spite of energy compensation it is important with dose rate meters to ensure that the calibration energy range includes the energy being monitored. A typical energy compensated GMT monitor will be linear to within 20% over an energy range of 50 keV to 1.25 MeV.

As with contamination monitors, a thin entry window is required when monitoring α- and β-radiation, and low energy X-rays. For this application a GMT with a thin end window or a thin scintillation crystal is used. GMT based instruments operate on a relatively low count-rate and consequently have a long time constant so patience is required for accurate monitoring.

6.3.3 Neutron Dose Rate Meters

Neutron dose rate meters can use proportional counters or scintillation crystals. Proportional counters are gas filled with BF_3or He_3 gas while the scintillator crystal may be lithium iodide (see Section 4.10). In both cases the detectors only detect slow or thermalised neutrons. In order to detect fast neutrons the detector must be surrounded by a neutron moderator that has a high concentration of hydrogen which is very effective at slowing down the fast neutrons, thus allowing their detection. The moderator is usually a near sphere of polythene with the detector at the centre, this arrangement allows the detector to be multidirectional. Neutron monitors are very sensitive to the energy of the incident neutrons therefore dose rate calibration is very energy specific. Since the monitor is very sensitive to the degree of moderation that the neutrons are subject to then monitoring the surface of a neutron source shield with a monitor calibrated for the energy of the un-moderated source will be subject to error.

6.3.4 Personal Dosimetry

Personal dosimetry can be either active or passive. Active dosimeters are miniaturised versions of the field rate meters above. The simplest is the quartz fibre electroscope (see Section 1.3). The electroscope is pen sized and the user can read the accumulated dose rate at any time. The electroscope is usually scaled from zero to two thousand μSv although other ranges are available.

Small electronic GMT based units are more versatile, the simplest output being an audible alarm that triggers when a pre-set dose rate is exceeded. This device of course gives no information on accumulated dose rate but is a useful tool to reduce accumulated dose by making the wearer instantly aware of high dose-rates. More sophisticated devices have an LCD display where instantaneous dose rate is displayed alongside accumulated dose. An audible dose rate alarm can be set and a data logger can be downloaded to a computer where dose rate versus time curves can be constructed.

Figure 6.4 shows the dose rate and accumulated dose curves for a worker carrying out a series of source transfers over a day. Although the log is in 5-min periods it is easy to calculate the time of the high exposure periods. Taking the largest peak on the graph, the peak dose-rate is 9318 μSv/h or 9318/3600 = 2.6 μSv/s, the accumulated dose from that peak is 8.6 μSv it therefore follows that the peak dose rate lasted for just over 3 s. This single peak represents 22% of the total dose for the day and demonstrates the ability of this type of monitoring to reduce dose rates. If this operation were to be performed regularly, then the large peak would be a good starting point for dose rate reduction planning.

It is important for radiation workers to have their lifetime-accumulated dose measured and recorded. This is achieved by the wearing of a radiation sensitive badge at all times

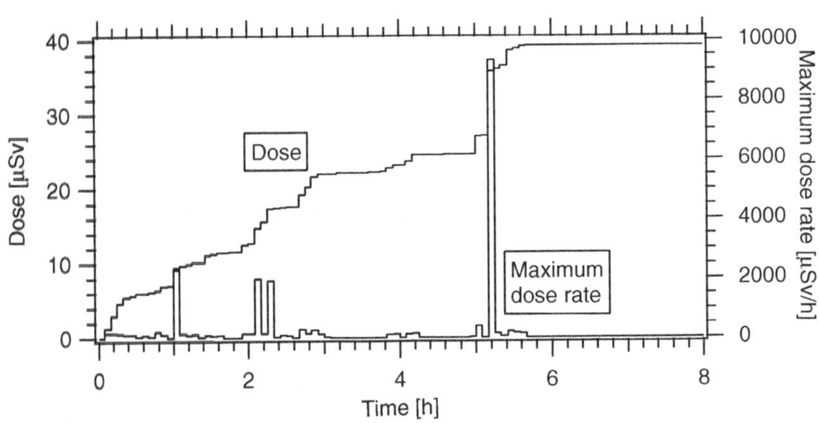

Figure 6.4 Plot of dose rate and accumulated dose as functions of time for a worker carrying out a series of source transfers over a day. The dose rate is the maximum instantaneous dose received in each 5-min period throughout the days work

Figure 6.5 Photographs of (1) Neutron badge, (2) Thermoluminescent dosimeter badge, (3) and (4) TLD finger badges

when working in a radiation area. The badge is fitted with a safety pin or clip to allow it to be worn on the outside of the clothing. Badges should be worn on the upper part of the trunk in order to best assess the whole body dose. The dose rate results are of course retrospective; record keeping is important and is usually carried out by the government appointed body responsible for processing the badge. The picture in Figure 6.5 shows a selection of passive badge style dose meters.

Two types of badge dosimeter are in common use, the *film badge* and the thermoluminescent dosimeter or *TLD*. The film badge consists of a small rectangle of photographic film, protected from light by a thin foil envelope. The film is worn on the body in a plastic carrier that incorporates a variety of filters used to categorise the absorbed radiation. The filters of plastic, lead, tin, cadmium and indium are designed to allow radiation of varying types and energies to pass through, thus allowing some energy information to be inferred from the exposure density of the different patches of the film. Although neutrons do not interact with the photographic emulsion, slow neutrons are captured by the cadmium filter, which emits γ-photons that do affect the film thus allowing estimation of neutron dose. The *thermoluminescent dose-meter* or *TLD* contains a plate of thermoluminescent material that absorbs radiation by raising electrons to forbidden bands where they are trapped until heat is applied to the material. Upon the application of a specific heating pattern the material can be made to give up the energy as light. Thus by monitoring the light output with a photomultiplier tube while heating the TLD, the absorbed energy is measured. For personal dosimetry the preferred thermoluminescent material is lithium fluoride doped with manganese because this material is most closely equivalent to body tissue. Other thermoluminescent materials are lithium borate doped with manganese for high dose dosimetry, calcium flouride doped with dysprosium and calcium sulphate doped with dysprosium are used for sensitive environmental measurements. The quantity of thermoluminescent material required for dose monitoring can be a few milligrams, which makes it ideal for use in small monitors for extremities such as finger or wrist badges.

Neutron badges contain a thin sheet of poly-allyl diglycol carbonate (PADC), which is a plastic with the ability to record the tracks of charged particles as damage to the polymeric structure. Neutrons are of course not charged but by interacting with nuclei of the plastic holder and the PADC they produce protons that leave tracks in the PADC. The minute track in the PADC is enlarged by chemical and electrochemical etching into a pit that is large enough to be counted by an automated reader. The dosimeter is not affected by X-rays or γ-rays and neutrons in the energy range from 144 keV to 15 MeV can be detected.

6.3.5 Calibration of Dose Rate Monitors

As with any measuring instrument regular calibration of radiation monitors is essential. There are ICRP recommendations, often enforced by national statute, which require that radiation monitors be calibrated every 12 months. An accredited laboratory using standard sources that are traceable to national standards must carry out the calibration (see Section 6.8).

6.4 RADIOLOGICAL PROTECTION METHODS

The exposure to external radioisotope sources is determined by the following parameters:

- Radiation energy or energies of isotope.
- Radiation intensity or activity of isotope.
- Isotope half-life.
- Distance.
- Exposure time.
- Shielding.

The first three parameters are normally determined by the application and its requirements, in conjunction with the ALARA philosophy. The latter three parameters can, however, usually be applied without disturbing the properties of the measurement principle.

We saw in Section 5.4.2 that lead normally is used as *shielding* material because it is the most cost effective. Because the intention with a shield is to maximise the energy absorption it is advantageous to use a material with high atomic number where full energy absorption of each event is most likely (see Section 4.2). This increases the fraction of photoelectric absorptions and the probability of reducing radiation leakage by Compton scattered events. A graded shield may be used to stop fluorescence radiation from lead as described in Section 5.4.2; however, this is seldom critical because its contribution to the dose-rate normally is insignificant. Contrary to intuition the more dense the shielding material is the lighter a given shield can be made. Consider a lead sphere of density about 11 g/cm^3 and 10 cm diameter with a ^{137}Cs source at the centre. The transmission from this source would be reduced by the lead to 0.5% compared to the unshielded source. The 10 cm diameter lead sphere would weigh 5.76 kg. To achieve the same dose rate reduction using a denser shield material such as one based on tungsten with a density of 17 g/cm^3 the sphere would now only need to be 6.4 cm in diameter. This sphere would weigh in at

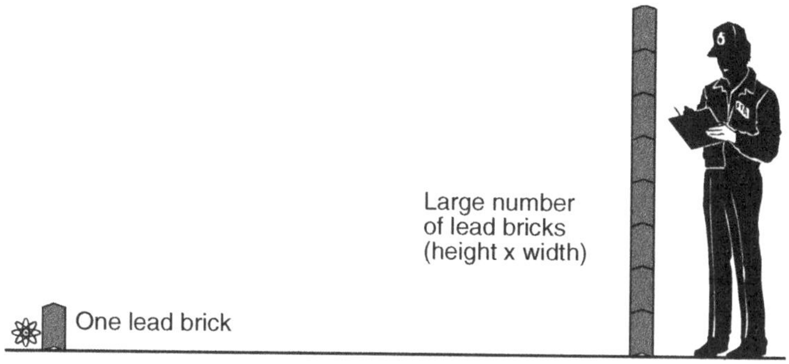

Figure 6.6 Close positioning of source shielding makes it more efficient

2.33 kg, less than half of the weight of the lead sphere. The use of tungsten and depleted uranium as shield materials is usually restricted to portable source holders, as even taking into account the considerable weight difference; the lead shield will be comparatively inexpensive.

When a large amount of shielding is required, for instance for source storage rooms etc., it is less expensive to use concrete as the shield material. The lower absorption compared to lead is then compensated for by increasing the thickness. Shielding in permanent installations should be fixed and not easily removed. We discussed shielding of neutrons in Section 5.4.2. Paraffin wax is the low cost alternative for making large, yet efficient neutron moderators.

Safe handling of radioisotope sources and the use of shielding are described in Section 6.5.6. In temporary situations such as when arming a gauge or when in the laboratory, movable shields such as lead bricks* are often used. It is worth noting here that a lot less shielding material will be required if it is placed immediately around the source (see Figure 6.6). This also reduces the amount of scatter generated by the source significantly.

Remember that internal building walls and structure are not always as substantial as they seem, in fact often they are little more than a couple of thin sheets of board and offer no shielding to people working on the other side (see Section 6.5.6).

The most effective radiological protection measure is *distance*. From Equation (6.6) we see that the dose rate reduces as the square of the distance, consequently the dose at the surface of a radioactive source (say 1 mm away) is one million times more than at 1 m away. A simple pair of pliers (10 cm long) will reduce the dose to the hands by a factor of ten thousand.

Next consider *exposure time*. For protection purposes it may be assumed that the total dose equals the product of dose rate and exposure time, although many biological effects of radiation are dependent on the dose rate. Anyway, it is important to minimise the time spent in the radiation area. Planning all operations carefully in advance ensures that the time spent near a source can be as short as possible. If anything goes wrong then retire to a safe distance and make another plan, do not stand around discussing strategy in the radiation area. When a gauge is installed on a site prominent notices should be displayed to discourage people spending unnecessary periods in the close vicinity.

*It is recommend to wear gloves for frequent handling of lead bricks because lead is poisonous. Alternatively, the lead bricks may be wrapped in an adhesive plastic tape.

Table 6.7 Excerpts from list of United Nations numbers, proper shipping names and descriptions of subsidiary risks (effective from 1 July 2001) [138]

Number	Proper shipping name
UN 2908	Radioactive material, excepted package – empty packaging
UN 2909	Radioactive material, excepted package – articles manufactured from natural uranium or depleted uranium or natural thorium
UN 2910	Radioactive material, excepted package – limited quantity of material
UN 2911	Radioactive material, excepted package – instruments or articles
UN 2912	Radioactive material, low specific activity (LSA-I)
UN 2913	Radioactive material, surface contaminated objects (SCO-I or SCO-II)
UN 2915	Radioactive material, type A package
UN 3332	Radioactive material, type A package, special form

In addition to the radiation protection techniques listed above, there are several general precautions that may be taken to ensure secure handling of radioisotope sources. Pregnant women and children should be kept away from all rooms where such equipment is used or stored. Laboratories and rooms containing nuclear radiation sources should be marked with warning signs. In addition, a survey meter should be available for periodical dose (rate) checks, and (some of) the personnel should wear dosimeters that are checked periodically.

6.5 TRANSPORT OF RADIOACTIVE MATERIALS

The United Nations classifies all materials for international transport purposes. Every substance has a UN number, which classifies it and the hazards that are likely to be encountered from it, during transport and especially in an accident situation. Radioactive materials are no exception and the UN numbers shown below are allocated to the various types of radioactive material.

Using the UN number on forms for international and national transport allows a consistent description of the goods and the related precautions, which is recognized in any country. For transport by road most countries have a government body that is responsible for the regulation of transport activities within national boundaries. The regulations differ little from country to country and stipulate the type of packaging and labelling suitable for the radioactive contents, and often require that the driver is specially qualified. A consignment note identifying the isotope, its quantity and type of packaging is normally required to accompany the shipment. Transport by air is controlled by the International Air Transport Association (IATA) Dangerous Goods Regulations and for transport by sea the International Maritime Dangerous Goods Regulations apply. All the regulations for land, sea or air are formulated with the following considerations:

- Containment of the radioactive material.
- Protection against radiation emitted by the container.
- Dissipation of the heat generated in the process of absorbing the radiation.
- Prevention of criticality when the material is fissile.

The last two only apply to reactor fuel and waste and are not relevant to transport of industrial gauging sources. Possible hazards during transport are emission of radiation from the container and the escape of container contents in an accident causing contamination of the vehicles used for transport or the storage facilities used during transit. These hazards are avoided by ensuring that the material is packed and shipped according to regulations based on the recommendations of the International Atomic Energy Agency.

During Transport containers must be properly secured and stowed to minimize the dose rate to persons and photographic materials. When shipping or rceeiving radioactive material it is important to ensure that the end user of the instrument has in place the appropriate licenses to use and hold the radioactive material. Most countries require that the site where the gauge is to be used be issued with a license to use the radioactive material. Some countries require that their own regulating authority approves the gauge type and this will require an application some months in advance of the intended installation date. When using tracers, which will be disposed of from the work site, a special disposal license will be required.

6.5.1 Source Containers

The type of packaging and labelling is regulated and the design and testing of the container is laid down in the regulations. The types range from 'strong industrial containers' for low-level solids through type 'A' packaging for intermediate level solids, liquids and gasses to type BII packaging for fissile materials. Most industrial gauge sources fall into the intermediate group and Type 'A' packaging is generally utilized. Type 'A' containers must fulfil the following design requirements:

- No dimension must be less than 10 cm
- The container must have a seal to indicate to the recipient that it has not been opened.
- All lifting or tie down eyes on the container should be so constructed as to leave the container intact should they fail in an accident.
- The container and its contents should retain its integrity at temperatures from −40°C to +70°C.

6.5.2 Testing of Type A Containers

Type A containers must either pass the tests listed in Table 6.8 or the designer must be able to prove using sound engineering arguments that the container would easily pass the tests. The pass criteria are that there must be no dispersal of the radioactive contents and the surface dose rate must not increase by more than 20%.

The severity of the penetration bar test and the free drop test are increased for the containers used for liquid or gaseous radioactive materials. Any supplier on their own behalf can carry out these tests and the designation of a container as a type 'A'. Evidence of the tests and or arguments must be held available for inspection by the authorities and often a certificate declaring that a container is as specified is requested by shippers or

Table 6.8 Tests for type 'A' containers for radioactive materials

Test	State of radioactive material (source)	
	Solid	Liquid, Gas
Water spray test	Equivalent to rainfall of 50 mm per hour for 1 h.	
Free drop test	The container must be dropped onto a flat, horizontal and unyielding target from 1.2-m height.	The container must be dropped onto a flat, horizontal and unyielding target from 9-m height.
Stacking test	Stack 6 high or load the top of the container area with 13 kPa (0.13 bar or 2 lb/in.^{2}).	
Penetration bar test	Drop a 32 mm diameter bar with a hemispherical end and 6 kg mass onto the container from 1-m height.	Drop a 32-mm diameter bar with a hemispherical end and 6-kg mass onto the container from 1.7-m height.

Table 6.9 Special form tests for sealed sources

Test	Condition
Impact test	The source is dropped onto a flat unyielding surface from a height of 9 m
Percussion test	A 1.4 kg bar is dropped onto the source from a height of 1 m
Bending test	If the source is more than 10-cm long and is slender then it is bent using a force equivalent to 1.4 kg falling through 1 m
Heat test	The source temperature is raised to 800°C for 10 min

end users. Incidentally but for the sake of completeness type 'B' containers have even more stringent design and test requirements including a fire test. Only designated (usually government) bodies can certify type 'B' containers.

6.5.3 Special Form

Special form is a designation of special types of sealed sources. If the source encapsulation does not leak when subjected to the tests listed in Table 6.9, then it is designated as special form. Special form approval is usually done by government agencies such as the Department of Transport in the UK.

The quantity of radioactive material that can be placed in a type 'A' container is dependent on the type of material and its physical form Table 6.10. lists maximum quantities laid down in the regulations for A1 and A2 levels of activity for all radioactive species. A1 quantity is the maximum quantity of activity allowed in a type 'A' container if that activity is a 'special form' and A2 quantity is the maximum if the material is not designated special form.

Sources used in industrial gauging are often special form but need not necessarily be so, when for instance an unusual source is chosen it may be built to the same standards as a special form source but small quantity production may make the designation impractical. For ^{241}Am sources special form is important because the americium when released is an

Table 6.10 A1 and A2 values for some common gauging sources and tracers

Nuclide	A1 (special form) limit		A2 (non-special form) limit	
	TBq ($T = 10^{12}$)	(Ci)	TBq	(Ci)
Americium-241	2	(54)	2×10^{-4}	(5×10^{-3})
Bromine-82	0.4	(10)	0.4	(10)
Carbon-14	40	(1000)	2	(54)
Cobalt-60	0.4	(10)	0.4	(10)
Caesium-137	2	(54)	0.5	(13)
Iodine-131	3	(81)	0.5	(13)
Iridium-192	1	(27)	0.5	(13)
Krypton-79	0.2	(5)	0.02	(0.54)
Krypton-85	20	(540)	10	(270)
Antimony-124	0.6	(16)	0.5	(13)
Scandium-46	0.5	(13)	0.5	(13)
Silver-110m	0.4	(10)	0.4	(10)
Sodium-22	0.5	(13)	0.5	(13)
Sodium-24	0.2	(5)	0.2	(5)
Tritiated Water	40	(1000)	40	(1000)
Depleted Uranium	Unlimited		Unlimited	
Xenon-133	20	(540)	20	(540)
Thorium (Natural)	Unlimited		Unlimited	
Lanthanum-140	0.4	(10)	0.45	(10)
Tantalum-182	0.8	(20)	0.5	(13)

alpha emitter and is therefore highly radiotoxic. This means that the A2 quantity listed for ^{241}Am is only 2×10^{-4} TBq, which is a bit small for many applications. The A1-value (i.e. the maximum for a type A when the source is special form) is 2 TBq or 10,000 times more.

For ^{137}Cs, which is less of a hazard if it is released, the A2 quantity is 0.5 TBq, which is usually more than any industrial gauge will require therefore, in this case, special form is less important.

6.5.4 Transport Index

Transport index is a number given to each container, which relates to the dose rate emitted by the container. The index is simply the dose rate at 1 m distance from the container in mrad or in μSv divided by 10. This enables limits to be set for certain transport vehicles or conditions, of the total number of transport indexes allowed. The transport index is written on the dangerous goods label affixed to the container exterior. Vehicles used exclusively for the radioactive shipment are permitted to carry larger quantities designated as exclusive use consignments.

6.5.5 Labelling

Three types of label may be encountered when shipping industrial gauging sources. These are shown in Figure 6.7 and their criteria are listed in Table 6.11.

Table 6.11 Labelling criteria used for shipping industrial gauging sources

Label	Condition
I-White	Transport index 0 (less than 0.05 mSv/h), and surface dose-rate less than 0.005 mSv/h
II-Yellow	Transport index more than 0 but not more than 1 and surface dose-rate more than 0.005 mSv/h and not more than 0.5 mSv/h
III-Yellow	Transport index more than 1 but not more than 10 and surface dose-rate more than 0.5 mSv/h but not more than 2 mSv/h
III-Yellow Exclusive use	Transport index more than 10 and surface dose rate more than 2 mSv/h but not more than 10 mSv/h

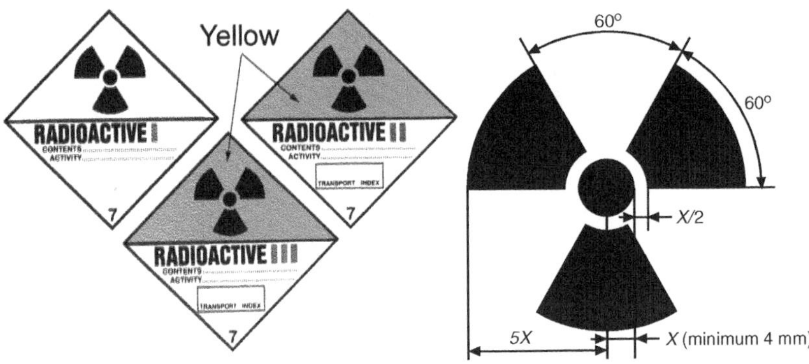

Figure 6.7 The three types of labelling that may be encountered when shipping industrial gauging sources (left). Design rules of the basic trefoil symbol (right) [138]

In addition to the above labels a label showing the UN number and a country of origin label with the countries recognised international identification letters, i.e. GB, USA, F, N, I, etc., is required, also as with any consignment the addresses of the consignor and consignee should be affixed. Finally the container must be locked or sealed so that it is opened only by the consignee upon delivery.

6.5.6 Sealed Source Handling Procedures

Before handling any radioactive material arriving in some sort of package it is important to be sure that the package contains what it is supposed to contain. This may seem a bit obvious but it is not totally unknown for sources to be incorrectly labelled. If taking delivery check that the labels agree with the other transport documents such as the carriage by road certificate or the IATA document. Is the package what was expected? No radioactive consignment should arrive without prior notification. Next check with a monitor that the dose rate measured at one meter from the surface agrees with the transport index on the dangerous goods label (see Section 6.3).

When the contents of the package are confirmed, then a plan for the transfer or unloading of the source must be made. The plan of course depends on the contents of the package and will be very different for a small ^{137}Cs and a large ^{192}Ir radiography source, but in

Figure 6.8 Examples on wrong and correct use of barriers

both cases the plan is important. Collect the right tools for the job to hand so that the transfer can be done smoothly and quickly. Ensure the workplace is clear and cover any drains, gratings or cracks where the source could be lost with plastic sheeting. Practice a dummy transfer in order to spot any unforeseen complications and to assess the time that the source will be exposed. Using data from this rehearsal (and from section 6.2.6) calculate from the exposure rate and time, the dose rate that will be received during the operation. Is the dose rate acceptable? Can it be reduced by increasing distance, reducing time or increasing the shielding? Calculate where the dose rate will be acceptable for non-radiation workers or members of the general public to have access to and place appropriate barriers and radiation warning placards at the boundary. When placing barriers remember that all walls are not good radiation shields and if working on a laboratory or workshop bench then the bench will not prevent radiation beaming downwards, see Figure 6.8. In a multi story building the same precautions apply to radiation beaming upwards through the ceiling. Concrete provides relatively efficient shielding and is often used as a low cost alternative to lead. On the other hand plasterboard walls have very low attenuation. When in doubt check the dose rate at critical locations in the building (site) with a radiation monitor (see Section 6.3).

Next consider what protective clothing is appropriate. For sealed gauging sources special protective clothing is not usually necessary but consider the possibility of the capsule being damaged. If the source is old and possibly corroded then a high level of protective clothing may be advisable. Consider the possibility of damaging the source capsule when handling, this may seem most unlikely, as the capsules of γ-emitting sources are extremely robust, but be aware of the consequences of damage. For low energy radiation sources the capsules must have a lightweight (often Beryllium) window to allow the radiation to exit the capsule. Avoid contact of sharp edges with the window; if the source needs pushing into place then use a flat-ended plastic bar, not a screwdriver. Ingestion of alpha emitters is extremely dangerous so if leakage is possible, however unlikely, then appropriate precautions such as facemasks should be used. When handling beta emitters a stout pair of safety spectacles will significantly reduce the dose to the eyes and gloves will stop all alpha dose to the hands.

Check that an appropriate radiation monitor is to hand (see Section 6.3). For most gauging or radiography sources a γ-ray monitor is appropriate and when handling old sources which may be leaking a contamination monitor is needed.

Table 6.12 Source transfer check list

Step	Action
1	Know what is in the package
2	Plan transfer
3	Collect equipment
4	Prepare the area, shielding, barriers, etc.
5	Use protective clothing
6	Use radiation monitors (Section 6.3)
7	Carry out wipe test (Section 6.6)
8	Clean up
9	Ensure safe and responsible disposal

The package can now be opened. It is wise to check the inner package for contamination before getting too far into it. Take a wipe test and check it on the contamination monitor. Follow the plan, if anything untoward is encountered do not try to retrieve the situation in a panic but retire to a safe distance and make a new plan. When the source is removed from its packing the serial number should be checked against the paperwork. The numbers are very small but do not get too close, use a magnifier, an intrascope or a closed circuit TV camera. If the source is not new then wipe test it while it is out of its container (see Section 6.6), this may be the last chance for years to perform the statutory wipe test on the actual capsule. Finally complete the transfer and monitor the area to ensure that the source is where it should be and that there is no contamination. Check the empty package with the contamination monitor and if it is clean, remove all references to radioactive material before disposal.

Handling *unsealed sources* is somewhat outside of the scope of this book but the dose minimisation philosophy remains the same as for sealed sources. Whilst reducing external dose rates is still important the problem of avoiding ingestion becomes paramount. A greatly increased level of personal protective clothing is required and depending on the volatility of the compounds, a higher standard of ventilation may be required.

6.6 LEAKAGE TESTING OF SEALED SOURCES

All sealed sources should be leak tested about every 2 years. The test is laid down in ISO 9978 Sealed Radioactive Sources Leak Test Methods. The purpose of the test is to ensure that no radioactive material is leaching from the source capsule by wiping the capsule surface and checking the wipe for radiation. When the source is installed in a shielded container as part of an installed gauge then in order to minimise the dose to personnel it is not necessary to remove and wipe the actual source capsule. The source holder may be wiped at the most likely place where any leakage from the source would exit the container. The source or holder should be thoroughly wiped over with a swab moistened with a liquid that will not attack the material of the source capsule. Water with a bit of mild detergent in it on a tissue or filter paper may be used but more convenient is a pre-packed moisturised tissue such as a medical cleansing swab or a computer screen cleaner. When wiping the actual source capsule the source must never be handled directly, use tongues or pliers and operate behind a suitable shield. The used wipe should be considered to be contaminated until the assay is completed, it should only be handled briefly and gloves must be worn. The wipe should be placed into a polythene bag and sealed for counting in order to avoid contamination of the counting equipment. Early indication of a leaking source can be achieved on site using a contamination monitor but a more precise measurement to confirm that any activity leaking from the source is below the statutory level should be carried out using a calibrated counting system. An ideal counting set up for the wipe test measurement for γ-sources is a 2 in. sodium iodide scintillation detector with a well crystal placed in a lead shield to reduce the background count thus increasing the sensitivity to low count-rates. The wipe, in its bag, is stuffed into the crystal well where counting efficiency is high. The count taken over about 100 s is compared to a background count, the difference being the count from the wipe. The detector must be calibrated using a small reference source, ideally of the same isotope as the source being tested; the source is preferably of a similar magnitude to the statutory leakage limit. The activity A on the wipe is given by:

The value calculated is assumed to be only one tenth of the actual leakage activity in order to reflect the effectiveness of the wiping process, i.e., the assumption is made that only one tenth of the activity is removed by the wipe. The limit for activity on the wipe is 185 Bq (5 nCi), which relates to 1.85 kBq (50 nCi) on the source surface.

Leaking sources are fortunately very rare but must be treated with care when they are discovered as the activity is now unsealed. When a source is leaking beyond the statutory limit it must be disposed of at the earliest opportunity. A source that is leaking significantly but is below the statutory limit should be replaced at the earliest convenient opportunity, as the leak will not repair its self.

6.7 STATUTORY REQUIREMENTS

Although the rules, regulations and laws differ from state to state there are some fairly universal regulations that we include here as a guide to the topics upon which local advice may be sought.

6.7.1 Licensing

Any premises where a radioactive source is to be kept will need to be registered with or licensed by the regulatory authority. Details of the number of sources, their location, identifying marks, isotope and activity will have to be provided with the application. This information will also need to be provided to the local emergency services in order to ensure their preparedness in the event of a site emergency.

If a source is to be imported from abroad the importer will probably be required to seek state approval by means of an import licence, the same information regarding the source will be required. The site occupants will be expected to fulfil the following requirements before the above permissions are granted.

The site where the sources are to be installed must be *secure* from easy access to the general public. If the sources are to be removed from the installed shields and retained on site (say for a vessel entry) then a secure and possibly shielded source store must be provided. The dose rate on the outside of the source store should not exceed the value laid down in state regulations. This will usually be 7.5 μSv/h but could be as low as 0.5 μSv/h. A typical store on a site with only a few sources could be a simple steel locker with shielded containers locked away inside when not in use. The store must have a prominent notice on the outside stating its purpose and displaying the internationally recognised trefoil symbol. Good records of the store contents must be maintained. The source holders must be locked to prevent unauthorised removal of the source; the key should be controlled by the radiological protection supervisor (see next paragraph). The shutter mechanism should be capable of being locked in the closed position but not in the open position.

The site occupier will be expected to appoint a responsible person or persons to oversee the operations involving the radioactive materials that may need to be carried out and to generally take ownership of the sources. This person will be the designated *radiological protection supervisor* (RPS) and will normally though not necessarily be recruited from the safety department. The RPS will be responsible for controlling any work involving the radioactive material and will issue or countersign any work permits involving the installation. The RPS will need to be sufficiently trained to be confident that all risks from the radiation are adequately controlled and sufficiently senior to ensure that instructions are obeyed. The site will also be expected to appoint a *radiological protection advisor* (RPA) who acts as a consultant to the RPS. The RPA is an accredited expert in all matters of radiological protection and need not be an employee of the user. A specialist commercial organisation or a government radiological protection department may provide the services of an RPA.

6.7.2 Labelling of Installations Shielded Containers

All containers of radioactive materials must bear the internationally recognised Trefoil symbol (see Figure 6.7) and the words 'Radioactive Material' (in an appropriate language or languages for the location). Installed shields must have a shutter mechanism which is capable of closing off the useful beam of radiation and the shutter must be clearly marked with the words 'open' and 'shut'. The container should be fitted with a label describing the isotope and its activity including the date to which the activity measurement relates. All of the above information should be indelibly engraved onto metal labels. A source container design is discussed in Section 8.3.5.

If sources are mounted on a vessel to which may be entered by means of man ways for maintenance then access to the vessel must be controlled, this is usually achieved by placing prominent notices on all entry points stating that the vessel has a radioactive source installed. The notice should direct the reader to the RPS who should be consulted before vessel entry is allowed and who will isolate or remove the radioactive source. The inside of the vessel will probably be a controlled area (see below).

6.7.3 Procedures or Local Rules

The RPA and RPS will need to produce a set of simple procedures or Local Rules, which are to be used as guidance for persons working on or in the vicinity of the installation.

The rules should be prominently displayed (usually in the control room or permit room) and should describe the processes to be followed both under normal conditions, under maintenance and in an emergency.

6.7.4 Accountancy and Training

The RPS will be expected to keep records of all radioactive sources held on site and may be expected to submit periodically to inspection by a government body. The RPS should inspect the installations routinely to confirm the presence of the radioactive source and to ensure the integrity and security of the source shield. The installation should be monitored to confirm the source is present and that the shield is intact and effective, this requires that the RPS has access to a suitable radiation monitor. A reasonable period for inspections of permanently installed sources is monthly or after maintenance, although of course most sources would be missed immediately as the gauge would stop working properly.

RPS training needs to cover only the specific hazards and procedures that are associated with the radioactive material or installations on the site. The candidate does not need to be expert in all matters of radiological protection and therefore the training required can probably be undertaken through a short, specific, training course given by the RPA. Courses are also available through commercial training organisations and may be provided by the supplier of the installation.

The course could include the following topics: The basic processes of radioactivity, radiation detection, biological effects of radiation, radiological protection methods, radiation monitoring, local radiological legislation and specific training with reference to the installation(s). The first five subjects are covered in Chapters 2, 3, 4 and 6 in this book, with particular focus on this chapter.

A radiological protection advisor needs to hold a professional qualification, which involves substantial training and experience.

6.7.5 Restricted Radiation Areas

Radiation controlled areas will be established because an employer has recognised an area where people must follow special radiological protection procedures. Typical conditions requiring the establishment of a radiation controlled area are as follows: (a) When any person working in an area is required to follow special procedures to restrict significant exposure to ionising radiation in that area. (b) If any person working in the area is likely to receive an effective dose of over 6 mSv per year. (c) If the dose rate in the area exceeds 7.5 μSv/h. (d) To restrict access to persons who may normally be in the area but are not involved in the work with ionising radiation while that work is being undertaken. (e) If there is a significant risk of spreading radioactive contamination from the area.

Controlled areas relating to instrument operations are likely to be of a temporary nature established during the short period of source insertion or removal. Once the source is in the shielded container, there should be no need for a controlled area around a well-designed installation.

A *supervised area* is any area where it is necessary to keep conditions of dose rate and access under review in order to determine whether it should become a controlled area.

A supervised area is also an area where a person is likely to receive an effective annual dose of more than 1 mSv per year. Such an area is most likely to be found at the detector of a nucleonic gauge where occupancy should be assessed in order to judge whether it should be controlled. Typical dose rates on level gauge detectors are 2 to 5 μSv/h and therefore the occupancy would have to be in the range 200–500 h per annum for a controlled area to be required. This level of occupancy of any single place on a process plant is unlikely but not impossible, hence the requirement for supervision of the area.

6.8 CALIBRATION AND TRACEABILITY

Previously we have used the term *calibration* on several occasions, for instance to establish the relationship between the measured peak content of one emission line in a XRF or PGNAA spectrum, to known concentrations of a specific element in the process or object. This relationship is then implemented in the measurement function in the data processing algorithms. But how can we make sure these 'known concentrations' are the true concentrations? Any calibration has to be relative to some other measurement or knowledge of the quantity in question. There is thus an uncertainty attached to calibration, and to minimise this uncertainty traceability is required.

6.8.1 Calibration

Generally *calibration* is defined as *determining* and *documenting* the deviation of the estimation using the measuring instrument from the conventional 'true' value of the measurand. To limit the effect of statistical fluctuations (error) this estimation usually is the average value of a series of measurements under identical conditions. Any deviation between measured (estimated) and true value is then due to *systematic errors*. (Note that when we defined the concept error in Section 5.3.1 we only referred to the statistical error.) Even though an instrument may have no systematic error upon purchase from the manufacturer, it will in most cases acquire one with time due to drift, ageing etc. The error being systematic means that the deviation between measured and true value changes in a particular direction, such as a gradually increasing system gain error.

In some cases when we reveal a systematic error in an instrument we may remove it by hardware adjustment of for instance gain and/or offset in an instrument with a linear measurement function. This is not calibration. Calibration is merely about determining and documenting the error. For critical measurements such fiscal measurements it is important to keep track of the accuracy history of the instrument, for instance for evaluation of Type B measurement uncertainty presented in Section 5.3.2 In these situations the error is accepted but corrected for through a controlled change in the measurement function by software. If an adjustment is made a new calibration has to be carried out immediately after to establish its effect and to preserve the relationship to the history.

6.8.2 Traceability

The term *traceability* means a process whereby the indication of a measuring instrument can be compared with a *national standard* for the measurand in question in one or more

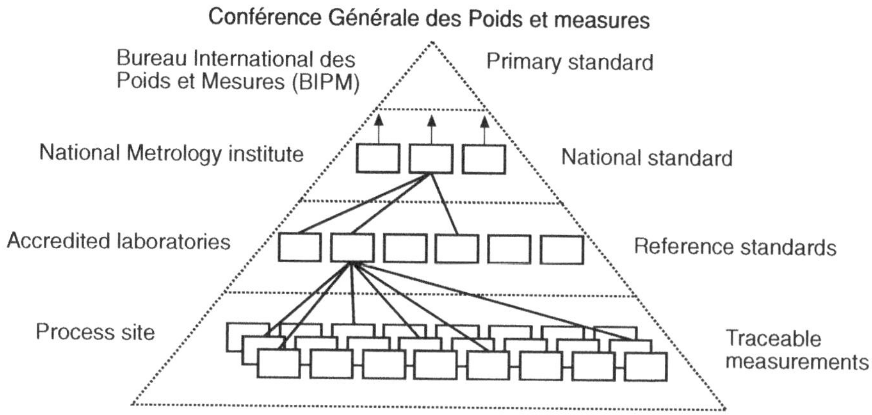

Figure 6.9 The traceability ladder ensuring traceable site measurements through calibrated instruments. The measurement uncertainty is highest for the instruments at the bottom of the hierarchy and decreasing towards the top

stages. In each of these stages a calibration has been performed with a *standard*, the metrological quality of which has already been determined by calibration with a *higher level standard*. We thus have a calibration hierarchy, as shown in Figure 6.9, and the measurement uncertainty of site instruments is traceable towards the ultimate or *primary standard* maintained by the Bureau International des Poids et Mesures (BIPM). The primary standard is designated or widely acknowledged as having the highest metrological qualities and whose value is accepted without reference to other standards of the same quality. The National Metrology Institutes are the highest authorities in metrology in almost all countries. If the Institute does not have the required facility for maintaining national standard, it has to ensure that the measurements are traceable to the primary standard maintained in another country.

Traceability is characterised by a number of essential elements:

- An unbroken chain of comparisons going back to a standard acceptable to the parties, usually a national or international standard.
- Measurement uncertainty: the measurement uncertainty for each step in the traceability chain must be calculated according to defined methods and must be stated so that an overall uncertainty for the whole chain may be calculated.
- Documentation: each step in the chain must be performed according to documented and generally acknowledged procedures; the results must equally be documented.
- Competence: the laboratories or bodies performing one or more steps in the chain must supply evidence for their technical competence (e.g. by demonstrating that they are accredited).
- Reference to SI units: the 'appropriate' standards must be primary standards for the realization of the SI units.
- Recalibrations: calibrations must be repeated at appropriate intervals; the length of these intervals depends on a number of variables, (e.g. uncertainty required, frequency of use, way of use, stability of the equipment).

For companies, traceability of measuring and test equipment to national standards by means of calibration is necessitated by the growing national and international demand that manufactured parts be interchangeable: supplier firms that make products and customers which install them with other parts must measure with the 'same measure'. But there are legal as well as technical reasons. The relevant laws and regulations have to be complied with just as much as the contractual provisions agreed with the purchaser of the product (guarantee of product quality) and the obligations to put into circulation only products whose safety is not affected by defects if they are used properly.

Traceable calibrations are carried out in controlled environments to keep the influence of variations of *environmental quantities* low. These are quantities, such as temperature, to which the sensor or the detector is sensitive in such way that it affects the output estimate.

6.8.3 Accreditation

Laboratory accreditation provides a means of determining the competence of laboratories to perform specific types of testing, measurement and calibration. It enables people who want a product, material or instrument to be checked or calibrated to find a reliable testing or calibration service able to meet their needs. It also provides feedback to laboratories as to whether they are performing their work in accordance with international criteria for technical competence. Manufacturing organisations may also use laboratory accreditation to enhance the testing of their products by their own in-house laboratories. Very importantly, laboratory accreditation provides formal recognition to competent laboratories, thus providing a ready means for customers to identify and access reliable testing and calibration services.

Many countries around the world have a formally recognised organisation responsible for the accreditation of their nation's laboratories. Most of these accreditation bodies adopt the criteria in an international standard, called ISO 17025 (previously ISO/IEC Guide 25) [139], as the basis for the accreditation of their country's testing and calibration laboratories. Furthermore, many countries have signed an international agreement so that an accreditation of a test or calibration procedure in one country is valid in all other member countries. This is one of the cornerstones of the accreditation system.

Laboratories can be audited and certified to an international management systems standard called ISO 9001. This standard is widely used in manufacturing and service organisations to evaluate their system for managing the quality of their product or service. Certification of an organisation's quality management systems against ISO 9001 aims at confirming the compliance of the management system to this standard, but does not specifically evaluate the technical competence of a laboratory. Laboratory accreditation assesses factors relevant to a laboratory's ability to produce precise, accurate test and calibration data. This includes the technical competency of staff; validity and appropriateness of test methods and measurements and calibrations to national standards; suitability, calibration and maintenance of test equipment; testing environment; sampling, handling and transportation of test items; and finally the quality assurance of test and calibration data.

Laboratory accreditation also covers the quality systems elements addressed in ISO 9001 certification. To ensure continued compliance, accredited laboratories are regularly re-examined to check that they are maintaining their standards of technical expertise. These laboratories may also be required to participate in regular proficiency testing programs as an on-going demonstration of their competence. To find out if your country has one or more laboratory accreditation bodies, try contacting your national standards body or your ministry for industry or technology.

6.8.4 Calibration of Radioisotope Gauges

Traceable calibration of nucleonic instruments almost without exception applies to laboratory instrumentation used for various types of analyses. A permanently installed radioisotope gauge seldom performs critical measurements requiring traceable calibration. On the other hand, for radiation monitors and survey meters on certain sites traceable calibration is required, see Section 6.3.4. These monitors carry a label containing an identification number, date of the last calibration, and how long this is valid. International Standards Organisation ISO 4037 describes sources and methods for X-rays and γ-rays and ISO 6980 relates to β-radiation.

7 Applications

In Section 5.5 we presented the different measurement methods or modalities applicable for industrial radioisotope gauges. In this chapter we will study examples on how these are used in various applications. Some of these represent the vast majority of installed gauges worldwide – density, level and thickness gauges, whereas others are more sophisticated or recently developed and hence not in widespread use. A large number of radioisotope measurement methods were developed over a couple of decades after the Second World War [8, 140], and many of these are still in use today [6, 141]. Traditionally, radioisotope gauges have often been preferred to other measurement principles because of the high-penetration capability allowing clamp-on installation. This is still true, particularly for process diagnostics applications where radiation specialist workers bring their equipment to process plants for the required measurements. Avoiding shutdown or any other disturbance of the processes being investigated is then very much preferred. For permanently installed γ-ray gauges, however, a quick survey of recent developments indicates that many of these use low-energy radiation and thus forsake the clamp-on possibility. This is partly because these gauges often are part of multiple modality systems requiring intrusive installation anyway, and partly because of better or different sensitivity at lower energies. Examples on recent developments of radioisotope methods are given in [142] and other references quoted in the following sections and in Chapter 8. Again, the intention of this chapter is to give some examples, as complete coverage of gauges and applications would be too extensive. There are also many excellent principles in use today, which are not reported or published for proprietary and protective reasons.

7.1 DENSITY MEASUREMENT

7.1.1 The γ-Ray Densitometer

When γ-rays travel through matter they are attenuated to an extent that depends upon the density and composition of the matter and the distance the rays travel in it. γ-ray attenuation is thus a function of both the thickness and density of the medium. So, by selecting γ-rays of the correct energy it is possible to measure the thickness of material of constant density or the density of material of constant thickness.

Radioisotope Gauges for Industrial Process Measurements. Geir Anton Johansen and Peter Jackson.
 ISBN 0-471-48999-9

At high gamma energies where Compton scattering is the dominant interaction mechanism (e.g. using ^{137}Cs or ^{60}Co sources) the mass absorption coefficient (μ_M) depends on energy (i.e. on the isotope being used) but is virtually independent of absorber composition. This is not the case at low energies, where the mass absorption coefficient depends on both gamma energy and the chemical composition of the absorber. So, low-energy gamma transmission can provide useful information on the chemical composition of the absorber.

A gamma densitometer is used to measure the density inside a medium with fixed dimensions, i.e. the thickness of the absorber is known. This gauge is often used as a clamp-on meter on pipes where the density of the flow varies with time. Typical applications of this meter are

- mining and metallurgical industries,
- pulp and paper,
- food and animal feed processing,
- chemical and petrochemical industries and
- offshore drilling fluid/mud applications.

A commonly used gamma densitometer is shown in Figure 7.1. This gauge has a built-in computer that controls the stability of the detector and compensates for the decay of the source. A NaI(Tl) scintillation counter (PMT) operated in pulse mode is most frequently used. The activity of the ^{137}Cs source is typically between 10 and 100 mCi, and the time constant, i.e. integration time, is adjustable from 1 to 1000 s. The signal output is a standard 4–20 mA current loop or fieldbus. A hand-held terminal permits remote calibration, and as for all industrial γ-ray gauges the source can be shut and locked. The source holder is shielded by lead, which is covered by stainless steel, and the dose at 1-m distance is guaranteed to be 7.5 μSv/h or less depending on the national legislation. Depending on the actual configuration, density measurement resolution of less than 1 mg/cm^3 is achievable with these gauges.

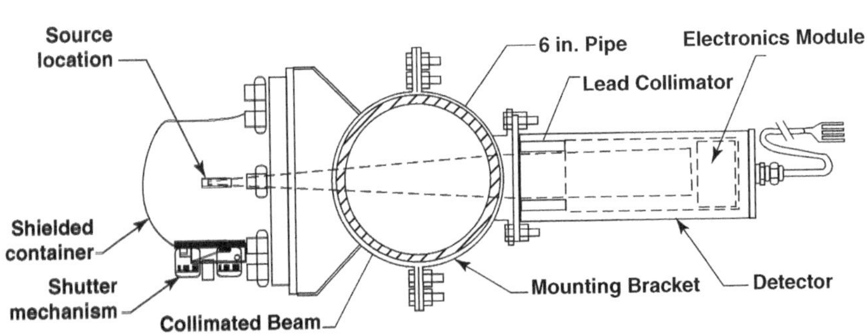

Figure 7.1 Cross section of a typical density gauge with mounting bracket for 6 in. diameter pipe. The specified sensitivity of this gauge is 0.001 g/cm^3. Courtesy of Tracerco

7.1.2 Belt Weigher

A density gauge output may be combined with another transducer to produce mass flow information. One example of this is the belt weigher, used to measure the mass of solid material being transported on a conveyor belt usually in mining or quarrying operations. The gauge is arranged on a 'C' frame (see Figure 7.13) with a detector beneath the belt and a shielded source suspended above the belt (see Figure 5.22). The attenuation of the beam is used to calculate the mass per unit length of material on the belt and this signal is combined with a belt speed measurement to indicate the mass flow rate. It is important, as discussed in Section 5.4.12, to use short counting times, typically a few tens of milliseconds, to avoid errors due to the rapidly changing amount of material between the source and the detector. The measurement accuracy of a belt weigher gauge is typically ±3% (one standard deviation).

7.1.3 Smoke Detector

The smoke detector is by far the most common radioactive instrument and because of mass production and simplicity, by far the least expensive. It is not a density gauge in the strictest sense as it does not measure the density of the individual smoke particle but it does measure the concentration of the smoke particles in the chamber. The smoke detector uses a very small ^{241}Am source of about 33 kBq. The americium is rolled into a thin gold foil about 1 μm thick, which is attached to a silver foil backing about 250 μm thick and the front is sealed with a 2 μm thick layer of palladium, which is thin enough to allow the α-particles to exit the source. The alpha detector is an ionisation chamber that simply consists of a pair of electrically charged plates with air in the space between. The α-particles ionise the air in the gap between the plates, knocking electrons off the oxygen and nitrogen atoms, leaving positively charged oxygen and nitrogen ions and free electrons. Under the applied electric field the ions drift to the cathode plate and the electrons flow to the anode, producing a very small electrical current flow in the detector. When smoke particles enter the ion chamber they capture the charged particles and reduce the current flow through the detector, thus triggering the alarm. Notice that the detection mechanism utilised here is not strictly absorption of the radiation but absorption of the result of the radioactive interaction.

7.2 COMPONENT FRACTION MEASUREMENTS

The density gauge shown in Figure 7.1 can be configured to measure component fractions of mixtures in any closed system, such as a vessel or a pipeline. This is on the basis that the densities of the components are known, either through sampling or calibration measurements, and that they do not change. Provided the mixture fills the measurement volume at all times and the sum of component fractions is unity, only one measurement is required. This principle is applied to solid/liquid, solid/gas, liquid/liquid and liquid/gas systems. In the following we use measurement of the gas volume fraction (GVF) in a homogeneous mixture of gas and liquid, the so-called *void fraction*, as an example on a two-component system. We shall also see how this can be extended to three-component measurement in liquid/liquid/gas systems. In this case two independent measurements are required, again taking into account that the sum of component fractions is unity.

7.2.1 Two-Component Fraction Measurement

The volumetric component fraction may be derived from measurements of the linear attenuation coefficient, μ_{mix}, of a homogeneous mixture of two components, provided there is sufficient difference in the density (contrast) of the components. According to Equation (3.22) we have

$$\mu_{\text{mix}} = \mu_{\text{g}}\alpha_{\text{g}} + \mu_{\text{l}}\alpha_{\text{l}} \tag{7.1}$$

where μ_{g} and μ_{l} are the linear attenuation coefficients of the gas and the liquid, respectively. Further α_{g} and α_{l} are the corresponding volume fractions, i.e., the fractions of the total volume that are occupied by the respective components. Since the sum of α_{g} and α_{l} is unity in a pipe, Equation (7.1) can be expressed as

$$\mu_{\text{mix}} = \mu_{\text{g}}\alpha_{\text{g}} + \mu_{\text{l}}\alpha_{\text{l}} = \mu_{\text{g}}\alpha_{\text{g}} + \mu_{\text{l}}(1 - \alpha_{\text{g}}) = \mu_{\text{l}} + \alpha_{\text{g}}(\mu_{\text{g}} + \mu_{\text{l}}) \tag{7.2}$$

The system has to be calibrated to determine the beam intensity I_{l} when the pipe is filled with liquid, i.e. when α_{l} is unity and α_{g} is zero, and the beam intensity I_{g} when the pipe is filled with gas, i.e. when α_{g} is unity and α_{l} is zero. By applying Equations (3.26) and (5.37)

$$I_{\text{l}} = B_{\text{l}} I_{\text{E}} e^{-\mu_{\text{l}} d} \quad \Rightarrow \quad \mu_{\text{l}} = -\frac{1}{d} \ln\left[\frac{I_{\text{l}}}{B_{\text{l}} I_{\text{E}}}\right] \tag{7.3}$$

and likewise

$$I_{\text{g}} = B_{\text{g}} I_{\text{E}} e^{-\mu_{\text{g}} d} \quad \Rightarrow \quad \mu_{\text{g}} = -\frac{1}{d} \ln\left[\frac{I_{\text{g}}}{B_{\text{g}} I_{\text{E}}}\right] \tag{7.4}$$

where d is the inner diameter of the pipe and I_{E} the beam intensity with the pipe empty. The linear attenuation coefficient of the mixture is in the same manner expressed as

$$I_{\text{mix}} = B_{\text{mix}} I_{\text{E}} e^{-\mu_{\text{mix}} d} \quad \Rightarrow \quad \mu_{\text{mix}} = -\frac{1}{d} \ln\left[\frac{I_{\text{mix}}}{B_{\text{mix}} I_{\text{E}}}\right] \tag{7.5}$$

where I_{mix} is the measured intensity when both components are present in the pipe. The gas fraction is then found by combining the Equations (7.1)–(7.5)

$$\alpha_{\text{g}} = \frac{\mu_{\text{mix}} - \mu_{\text{l}}}{\mu_{\text{g}} - \mu_{\text{l}}} = \frac{\ln\left[\frac{I_{\text{mix}}}{B_{\text{mix}} I_{\text{E}}}\right] - \ln\left[\frac{I_{\text{l}}}{B_{\text{l}} I_{\text{E}}}\right]}{\ln\left[\frac{I_{\text{g}}}{B_{\text{g}} I_{\text{E}}}\right] - \ln\left[\frac{I_{\text{l}}}{B_{\text{l}} I_{\text{E}}}\right]} \approx \frac{\ln(I_{\text{mix}}) - \ln(I_{\text{l}})}{\ln(I_{\text{g}}) - \ln(I_{\text{l}})} = \frac{\ln\left[\frac{I_{\text{mix}}}{I_{\text{l}}}\right]}{\ln\left[\frac{I_{\text{g}}}{I_{\text{l}}}\right]} \tag{7.6}$$

if it is assumed that the build-up factors B_l, B_g and B_{mix} are approximately equal. Instead of using build-up factors we could use the effective attenuation coefficients of the components as explained in Section 3.5.3. The result is thus dependent only upon knowledge of I_l and I_g, which are found through calibration measurements. To reduce the total measurement uncertainty the counting time for these calibration measurements should be say 10 times longer than the counting time of the actual measurement.

The most frequently used isotope for these measurements is ^{137}Cs with energy 662 keV and activity in the range between 10 mCi and 1 Ci. This source can be used for steel pipes with diameter up to several tens of centimetres. However, the contrast between the components is in many cases better when using low-radiation energies, such as 59.5 keV from the ^{241}Am source, because of the much higher fraction of photoelectric interactions at low energies. Depending on the wall attenuation, the ^{241}Am source may be used for vessel diameters of about 10 cm or more on the basis of the density of the fluid.

The read-out electronics for the two-component meter is similar to that of the densitometer. To avoid counting errors induced by gain drift in the detector system, a low counter threshold is normally applied as discussed in Section 5.4.6.

7.2.2 Multiple Beam Two-Component Metering

The gas volume fraction expression given in Equation (7.6) for the two-component gauge was derived on the assumption that the process components are homogenously mixed. This is very often not the case, particularly for gas/liquid flows where a variety of *flow regimes* is possible [143]. The type flow regime at any instance depends on the component velocities and fractions, and also the fluid properties and the orientation of the pipe. In vertical pipes variations of *annular flow* are most common, whereas in horizontal pipes variations of *stratified flow* are predominant. In some cases these regimes are stable and therefore predictable, in other cases we may have temporal variations between annular or stratified flow at one instance and *slug flow* at the next. The radiation beam of the densitometer does, as illustrated in Figure 7.1, usually not cover the entire cross-sectional area of the pipe. The measured GVF will then be underestimated in the case of annular flow as illustrated in the plot in Figure 7.2. The flow regime induced errors are here calculated for narrow beams in the case of annular and stratified flows as shown. As can be seen both cases exhibit a nonlinear relationship.

These geometrical induced errors pose a problem to the reliability of GVF meters because the type of flow regime may change with time and thus is unknown at a given instant unless additional information is available. Several approaches have been proposed as to how this can be solved. The first is to ensure homogeneous mixture of the flow components at the measurement cross section. This may be achieved by using an in-line mixer [144], installing the GVF meter just after a blind T-bend on the piping or by measuring across a pipe restriction where the flow is more turbulent. An example of the latter is presented in Section 7.5.2. A second solution is to use a broad beam covering the entire cross section of the pipe, the so-called *one-shot* densitometer. Design rules for the meter and the data analysis algorithms have been proposed [145, 146]. However, these indicate that there will be a compromise between the linearity on the one hand and the sensitivity on the other. In cases where the GVF meter is installed for instance on a vertical

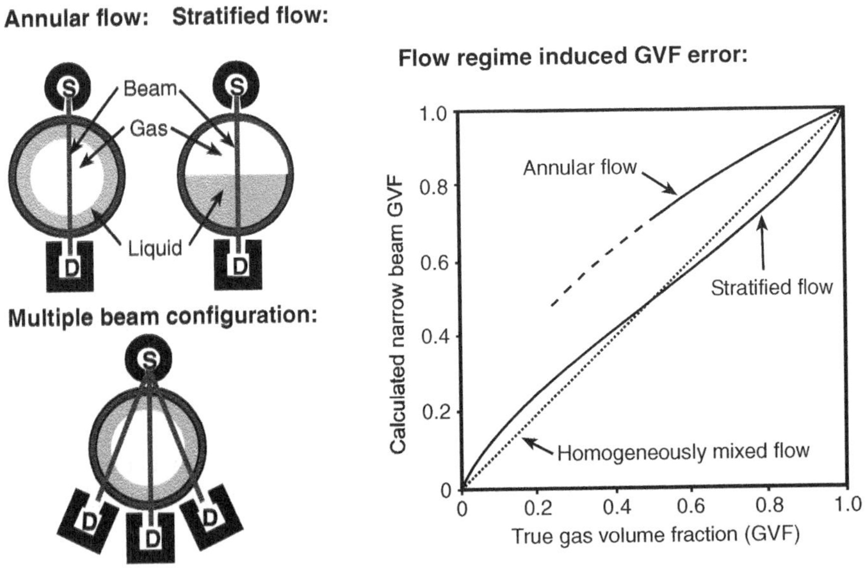

Figure 7.2 Narrow beam configuration applied to annular and stratified gas/liquid flows. The plot shows the error in the resulting GVF as a function of the true GVF for these cases and homogeneously mixed flow. The dashed line for annular flow indicates that annular flow does not occur at low GVFs. Also shown is a possible three-beam configuration

pipe with upwards directed flow, the one-shot meter and proper modelling may be applied because the flow regime to some extent is predictable. This is particularly true when the GVF-meter is used in conjunction with a flow meter because this adds information that may be used to predict the flow regime [147]. A final solution, applicable particularly when the flow regime for some reason is unpredictable, is to used a multiple beam GVF meter whereby the flow regime may be predicted so that the correct model may be used to estimate the GVF. One such application is down-hole metering where measurements on inclined flows are required [148]. A possible multiple beam GVF configuration is shown in Figure 7.2. There are several approaches as to how such a meter may be designed: In most cases a fan-beam collimated source covering the entire pipe cross section is applied. In some solutions un-collimated detectors are used [149]; however, the large extent of build-up then makes it difficult to estimate the GVF analytically. The use of artificial intelligence such as neural networks has proven to be applicable in these [150] and related cases [151]. It has also been demonstrated that the use of multiple collimated detectors (narrow beams) yields flow regime identification and accurate estimations of the GVF at varying flow conditions [143], even with as few as two or three beams [152]. Finally, a complete different measurement principle, for instance capacitance measurements, may be used to identify the flow regime.

We have now discussed errors imposed by variations in the spatial distribution of the flow components in cases where these are not mixed. It is equally important to consider temporal errors for instance because of slug flow causing rapid changes in the flow regime from what we may consider to be homogeneously mixed flow in one instance, to stratified or annular flow in the next. The counting time thus needs to be small compared to the flow velocity. Therefore the total measurement is split into several measurements made over short intervals from which the mean values are found over a longer interval. The moving average approach described in Section 5.4.13 is often applied here.

7.2.3 Three-Component Fraction Measurement

The three-component γ-ray fraction meter utilises the mixture components' relative difference in radiation attenuation at two different radiation energies [232]. This is thus a *dual-energy* measurement principle. Provided there is sufficient density contrast between the components this method may be applied to any three-component combinations. We will use gas/oil/water pipe flow as an example. The measurement method uses an isotope (or two isotopes) with two emission lines $E_{\gamma 1}$ and $E_{\gamma 2}$ as the radiation source. The corresponding linear attenuation coefficients $\mu_{1\text{mix}}$ and $\mu_{2\text{mix}}$ are derived for the two intensities $I_{1\text{mix}}$ and $I_{2\text{mix}}$, which are measured by an energy sensitive detector system counting the transmitted photons in two energy windows enclosing $E_{\gamma 1}$ and $E_{\gamma 2}$, respectively:

$$I_{1\text{mix}} = I_{1\text{E}} e^{-\mu_{\text{mix}} d} \quad \Rightarrow \quad \mu_{1\text{mix}} = -\frac{1}{d} \ln \left[\frac{I_{1\text{mix}}}{I_{1\text{E}}} \right] \tag{7.7}$$

$$I_{2\text{mix}} = I_{2\text{E}} e^{-\mu_{\text{mix}} d} \quad \Rightarrow \quad \mu_{2\text{mix}} = -\frac{1}{d} \ln \left[\frac{I_{2\text{mix}}}{I_{2\text{E}}} \right] \tag{7.8}$$

where $I_{1\text{E}}$ and $I_{2\text{E}}$ are the corresponding incident intensities at $E_{\gamma 1}$ and $E_{\gamma 2}$, and d is the effective inner pipe diameter. The attenuation coefficients may also be expressed as

$$\mu_{1\text{mix}} = \mu_{1\text{g}} \alpha_{\text{g}} + \mu_{1\text{o}} \alpha_{\text{o}} + \mu_{1\text{w}} \alpha_{\text{w}} \tag{7.9}$$

$$\mu_{2\text{mix}} = \mu_{2\text{g}} \alpha_{\text{g}} + \mu_{2\text{o}} \alpha_{\text{o}} + \mu_{2\text{w}} \alpha_{\text{w}} \tag{7.10}$$

where $\mu_{1\text{g}}, \mu_{2\text{g}}, \mu_{1\text{o}}, \mu_{2\text{o}}, \mu_{1\text{w}}$ and $\mu_{2\text{w}}$ are the linear attenuation coefficients of gas, oil and water at the two energies, respectively. Likewise, α_{g}, α_{o} and α_{w} are their volume fractions. In some literature these are referred to as α, β and γ, respectively. The attenuation coefficients of the components are derived from six calibration measurements of the following intensities:

- $I_{1\text{g}}$ at $E_{\gamma 1}$ and $I_{2\text{g}}$ at $E_{\gamma 2}$ with gas-filled pipe ($\alpha_{\text{g}} = 1, \alpha_{\text{o}} = \alpha_{\text{w}} = 0$).
- $I_{1\text{o}}$ at $E_{\gamma 1}$ and $I_{2\text{o}}$ at $E_{\gamma 2}$ with oil-filled pipe ($\alpha_{\text{o}} = 1, \alpha_{\text{g}} = \alpha_{\text{w}} = 0$).
- $I_{1\text{w}}$ at $E_{\gamma 1}$ and $I_{2\text{w}}$ at $E_{\gamma 2}$ with water-filled pipe ($\alpha_{\text{w}} = 1, \alpha_{\text{g}} = \alpha_{\text{o}} = 0$).

The last equation needed to determine α_{g}, α_{o} and α_{w}, in addition to Equations (7.9) and (7.10), is

$$\alpha_{\text{g}} + \alpha_{\text{o}} + \alpha_{\text{w}} = 1 \tag{7.11}$$

if it is assumed that the build-up factors in all the cases are approximately equal so that the effects of scattered radiation are cancelled out by the calibration. To develop expressions for α_{g}, α_{o} and α_{w} it is convenient to use the difference in attenuation coefficients between the components so that $\Delta\mu_{1o,g} = \mu_{1o} - \mu_{1g}$, $\Delta\mu_{1o,w} = \mu_{1o} - \mu_{1w}$, $\Delta\mu_{1w,g} = \mu_{1w} - \mu_{1g}$ for the low energy and likewise $\Delta\mu_{2o,g} = \mu_{2o} - \mu_{2g}$, $\Delta\mu_{2o,w} = \mu_{2o} - \mu_{2w}$, $\Delta\mu_{2w,g} = \mu_{2w} - \mu_{2g}$ for the high energy. The volume fractions of the components may then be expressed as [153]

$$\alpha_g = \frac{\left\{\frac{1}{d}\ln\left(\frac{I_{1E}}{I_{1mix}}\right) - \mu_{1w}\right\}\Delta\mu_{2o,w} - \left\{\frac{1}{d}\ln\left(\frac{I_{2E}}{I_{2mix}}\right) - \mu_{2w}\right\}\Delta\mu_{1o,w}}{\Delta\mu_{1g,w}\Delta\mu_{2o,w} - \Delta\mu_{1o,w}\Delta\mu_{2g,w}} \tag{7.12}$$

$$\alpha_o = \frac{\left\{\frac{1}{d}\ln\left(\frac{I_{1E}}{I_{1mix}}\right) - \mu_{1w}\right\}\Delta\mu_{2g,w} - \left\{\frac{1}{d}\ln\left(\frac{I_{2E}}{I_{2mix}}\right) - \mu_{2w}\right\}\Delta\mu_{1g,w}}{\Delta\mu_{1o,w}\Delta\mu_{2g,w} - \Delta\mu_{2o,w}\Delta\mu_{1g,w}} \tag{7.13}$$

$$\alpha_w = \frac{\left\{\frac{1}{d}\ln\left(\frac{I_{1E}}{I_{1mix}}\right) - \mu_{1g}\right\}\Delta\mu_{2o,g} - \left\{\frac{1}{d}\ln\left(\frac{I_{2E}}{I_{2mix}}\right) - \mu_{2g}\right\}\Delta\mu_{1o,g}}{\Delta\mu_{1w,g}\Delta\mu_{2o,g} - \Delta\mu_{2w,g}\Delta\mu_{1o,g}} \tag{7.14}$$

The measurement uncertainty in these component fractions caused by the statistical errors in the measurements is given as [153]

$$\sigma_{\alpha_o} = \frac{\frac{1}{d}\left\{\left(\frac{\sigma_{I_{1mix}}}{I_{1mix}}\Delta\mu_{2g,w}\right)^2 + \left(\frac{\sigma_{I_{2mix}}}{I_{2mix}}\Delta\mu_{1g,w}\right)^2\right\}}{\Delta\mu_{1o,w}\Delta\mu_{2g,w} - \Delta\mu_{2o,w}\Delta\mu_{1g,w}} \tag{7.15}$$

expressed for one standard deviation for the oil volume fraction. Similar types of equations can be used to predict the errors in the gas and water volume fractions. Two energy selection criteria must be applied for the three-component γ-ray fraction meter: As for single energy meters, an average attenuation of about 86% gives the minimum statistical error. For the dual energy meter, an additional criterion applies: the highest energy should be chosen where Compton scattering is the dominant attenuation mechanism in the mixture. The linear attenuation coefficients of the components are then proportional to their densities. The lowest energy should be in the range dominated by photoelectric absorption where the linear attenuation coefficients are strongly dependent on the effective atomic number or composition. The ratio between the attenuation coefficients of water and oil will then be different at the two energies since the effective atomic numbers of the water and oil are different. This is evident from Figure 7.3. This ratio will define the meter's ability to resolve the water and oil components, as discussed by Van Santen et al. [154]. The lowest energy is, however, limited downwards by the first criterion of 86% attenuation, even though low attenuation radiation windows are used in the pipe wall. Several of the dual energy meters in use today utilise a characteristic X-ray emission line for the lowest energy. Energy combinations reported are ^{241}Am (59.5 keV) and ^{133}Ba (356 keV line) [153], ^{241}Am (17.8 keV and 59.5 lines) [153, 155–157], and ^{137}Cs (32.1 and 661.6 keV) [158, 159]. Needless to say, flow regime induced errors and corrections as discussed in Section 7.2.2 also apply to dual energy or three-component fraction meters.

A dual energy gauge requires an energy sensitive detector system with window counting. This system also requires gain stabilisation (Section 5.4.7) and often also background correction (Section 5.4.8) for the low-energy peak to ensure proper operation. Different detector types have been used for these gauges: NaI(Tl) scintillation detectors, silicon PIN detectors and CdZnTe detectors. The latter two have also been applied with thermoelectric coolers (Section 4.8) to reduce noise and reduce the window counting error [156]. Those

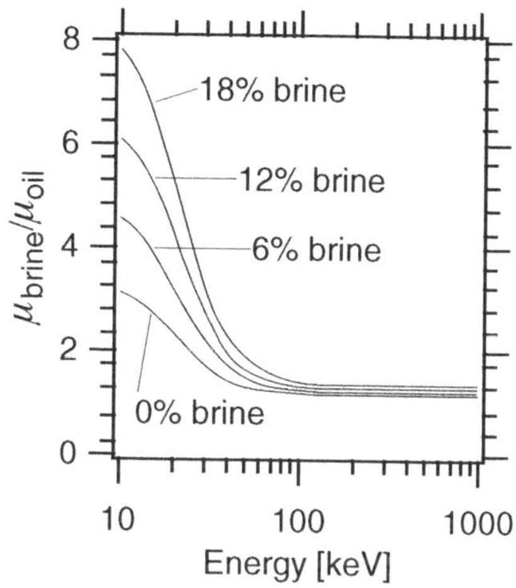

Figure 7.3 Ratio of the linear attenuation coefficients of brine and oil (Exxsol D100) as a function of radiation energy. This ratio is approximately equal to the corresponding density ratio above 100 keV where Compton scattering is the dominant attenuation mechanism

using the X-ray fluorescence peaks of ^{241}Am also require low attenuation windows, such as polyetheretherketon or carbon fibre reinforced epoxy (see Section 4.9.2). A design example of a system using a ^{137}Cs source and a NaI(Tl) scintillation detector is given Section 8.3.

The accuracy of the dual energy or three-component meters is very much dependent on the exact configuration with respect to energy selection, pipe diameter, meter installation, etc. [157]. One problem encountered is the dependency of the attenuation coefficient on the salinity of the water component (see Figure 7.3). This is particularly noticeable in the photoelectric absorption region because of the relatively high atomic number of chlorine or other high Z salt constituents. This problem has received great attention for several reasons: Increased oil recovery by water injection causes changes in the produced water salinity since the injected water and the formation water have different salinity. There may also be horizontal and/or vertical gradients in the formation water salinity across the reservoirs, and this may cause sudden changes in the salinity of the produced water in the case of 'water breakthrough' [160]. This problem is of increasing importance since new technology has made it economically feasible to produce so-called marginal wells with more than 80% water content. One approach to solve this problem is to sample the flow regularly, measure the water salinity and enter this data to the flow meter's computer to correct the output. This method will not pick up sudden changes in salinity, and is even impossible in applications where there is no access to the production line. A more elegant solution is the *TEGRA method* (triple energy gamma ray absorption) [142, 161, 162]. This is based on the same principle as the dual energy meter, but incorporates a third energy to determine the salinity. Another approach is to combine scatter and transmission measurements (see Section 7.2.4).

7.2.4 Dual Modality γ-Ray Densitometry

The dual modality principle is another approach to solve the salinity dependency of the fraction measurements in gas/oil/water pipe flow as discussed in Section 7.2.3. Here the

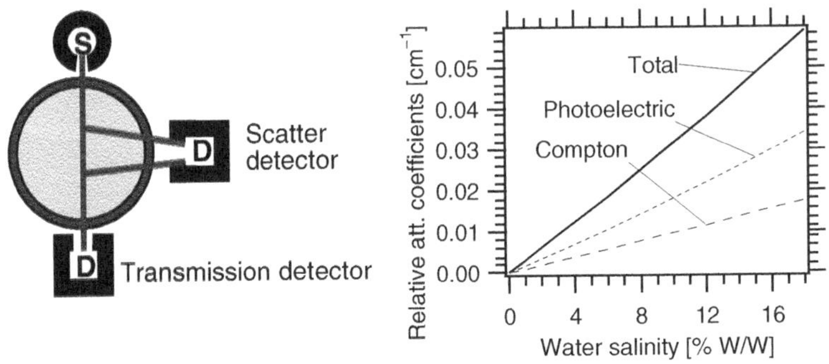

Figure 7.4 (Left) The dual modality densitometry measurement principle using one transmission detector and one scatter detector. (Right) The composition of the linear attenuation coefficient as a function of the water salinity, expressed as the difference in salinity relative to 0% WW salinity at 59.5 keV. Compton scattering is still the dominant interaction mechanism at this energy; however, photoelectric absorption accounts for the largest increase in the total coefficient with the salinity

different response in photoelectric attenuation and Compton scattering to changes in salinity is utilised. The total attenuation coefficient is found through traditional transmission measurements with a detector positioned outside the pipe wall diametrically opposite the mono-energetic source. The scatter response is measured with a second detector positioned somewhere between the source and the transmission detector (see Figure 7.4) [163]. This is thus a measurement of the Compton scatter cross section when attenuation of the scattered radiation is corrected for.

The 59.5 keV emission line of the ^{241}Am is used as the radiation source because both photoelectric absorption and Compton scattering contribute to the total attenuation in hydrocarbons at this energy (see Figure 7.4). A model has been developed for the scatter response; however, the gas fraction may be determined independent of changes in the salinity by a simple empirical relationship. This measurement principle is also dependent on the flow regime and new models are being developed to cope with this problem [164]. An additional third measurement is required to determine all three-component fractions in gas/oil/water pipe flow.

7.2.5 Component Fraction Measurements by Neutrons

Another 'dual modality' principle to measure all three-component fractions in gas/oil/water pipe flow is based on the transmission and scattering of fast neutrons [165]. Fast neutron transmission is used to determine the volume fraction of the gaseous phase, or equivalently the liquid fraction, since oil and gas are expected to be equally effective in removing fast neutrons. Neutron scattering, on the other hand, can be used to determine the ratio of water to oil. This is done by taking advantage of the fact that salt-carrying water is a stronger slow neutron absorber than oil. The geometry is similar to that presented in Figure 7.4, except that a collimated fast neutron detector is used for transmission measurement, and an un-collimated slow neutron detector is used for scatter measurements.

7.2.6 Local Void Fraction Measurements

A system utilising the γ-ray scatter method shown in Figure 5.25a has been developed for measuring local void fraction or density inside vessels [166].

7.2.7 Dual-Energy Ash in Coal Transmission Measurement

Environmental legislation has had a significant impact on coal utilisation in limiting emissions of potentially hazardous materials to the environment. This is especially true for coal combustion for power generation. For the most part, such emissions derive from the inorganic constituents (ash) in the combustible coal. There are many methods applied for analysis of the coal [167], including a variety of nuclear methods [141, 168]. Some of these are on-line methods [169], and of these the dual energy transmission (DET) gauge for measurement of ash in coal on conveyer belts is frequently used [170].

The foundation of the DET gauge principle is the same as that of the three-component fraction meter described in Section 7.2.3; the ash or mineral components have a higher atomic number ($Z \sim 12$) than the matrix (coal) ($Z \sim 6$). However, the DET gauge is different in that a conveyer as opposed to a pipe is an open system. A similar plot to that shown in Figure 7.3 could be produced demonstrating the difference in attenuation at low and high γ-ray energies for ash and coal. The DET gauge utilises a C-frame inserted under and over the conveyer belt as shown in Figure 7.13 in the case of a metal sheet. A narrow beam collimated source with two isotopes, ^{241}Am (59.5 keV) and ^{133}Ba (356 keV emission line), is positioned below the belt, whereas a collimated NaI(Tl) detector is placed above it. Alternatively, a ^{137}Cs may be used instead of ^{133}Ba. The ash content may then be approximated as [170]

$$C_{\text{ash}} \approx a_1 \frac{\ln\left(\frac{I_{1E}}{I_{1\text{mix}}}\right)}{\ln\left(\frac{I_{2E}}{I_{2\text{mix}}}\right)} + a_2 \quad \text{or} \quad C_{\text{ash}} \approx \frac{1}{N}\sum_{i=1}^{N}\left[a_1 \frac{\ln\left(\frac{n_{C1E}}{n_{Ci1\text{mix}}}\right)}{\ln\left(\frac{n_{C2E}}{n_{Ci2\text{mix}}}\right)} + a_2\right] \tag{7.16}$$

whereas the weight per unit area of coal may be expressed as [170]

$$M = a_3 \ln\left(\frac{I_{2E}}{I_{2\text{mix}}}\right) \quad \text{or} \quad M = \frac{1}{N}\sum_{i=1}^{N}\left[a_3 \ln\left(\frac{n_{C2E}}{n_{Ci2\text{mix}}}\right)\right] \tag{7.17}$$

where subscripts 1 and 2 represent the measured intensities at the low and high energies respectively; subscript E represents the empty conveyer belt intensity whereas a_1, a_2 and a_3 are constants. Now, the right-hand expressions take into account the effect discussed in Section 5.4.12; the counting time needs to be small compared to the conveyer belt transit time. Therefore, the total measurement is split into several measurements made over short intervals from which the mean values of are found over a longer interval. Here n_{Ci} denotes count-rate in the ith period of N periods, each with short duration. The moving average approach described in Section 5.4.13 may also be applied here. The value of the counting interval (Δt) depends on source activity and conveyer belt speed. A thorough presentation of the DET gauge and its properties are given in reference [170].

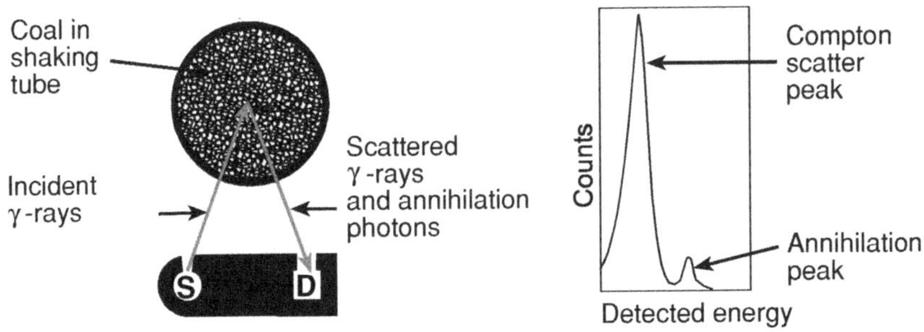

Figure 7.5 The pair production gauge used for the determination of the ash content of coal moving through a vertical shaking tube [141, 172]. The detection spectrum (right) contains the Compton scatter peak and the annihilation peak

Compared to chemical assays the gauge is accurate in determining the ash content in coal to within ±0.5% by weight, provided the composition of the ash does not change significantly after calibration. Besides the counting statistics the main source of error in the DET gauge in this application is variations in the concentration of Fe_2O_3 in the coal. [141, 170].

The DET gauge is routinely used for separator control where lumps of gangue are separated from those of coal by a pneumatic gun as they fall of the end of the conveyer belt [171].

7.2.8 Pair Production Ash in Coal Measurement

There is another technique whereby the ash content in coal can be determined. This is also a dual modality method utilising the difference in the effective atomic number of ash and coal discussed in the Section 7.2.7, but in this case it is based on the pair production response and not the photoelectric response. We saw in Section 3.3 that the pair production cross section is approximately proportional to Z^2. The measurement set-up is shown schematically in Figure 7.5. The coal is irradiated with high energy γ-rays, which interact partly by Compton scatter and partly by pair production. The latter gives rise to annihilation photons of which some interacts in the detector alongside Compton scattered photons. The probability of the latter is approximately proportional to the bulk density of the coal whereas that of annihilation photons depends on Z^2.

Window counting is used to find the content of the Compton (C) and annihilation (P) peaks illustrated in the detection spectrum shown in Figure 7.5. The ash content is then calculated using an equation of the form

$$\text{Ash} = f(P + gC) + h \tag{7.18}$$

where the constants f, g and h are determined by least square fitting the chemical laboratory ash analysis and the measured values of P and C. The advantage of this gauge over that described in Section 7.2.7 is its reduced sensitivity to changes in the ash composition [141].

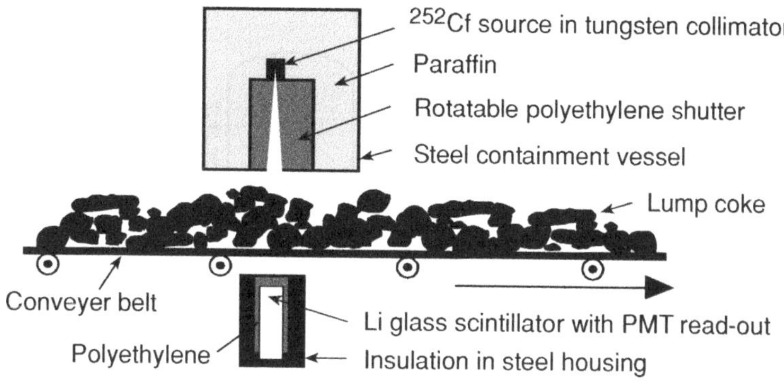

Figure 7.6 The fast neutron and γ-ray transmission gauge for determination of moisture in coke [169].

7.2.9 Coke Moisture Measurements

A technique for measurement of coke moisture in iron works has been developed based on simultaneous transmission of fast neutrons and γ-rays from a ^{252}Cf spontaneous fission source [142, 169, 173]. Fast neutron transmission depends predominantly on the hydrogen concentration per unit area, whereas γ-ray transmission depends on mass per unit area. By combining such measurements the moisture can in most cases be determined independently of the mass per unit area. The principal components of the gauge are shown in Figure 7.6. On the basis of calibrations the accuracy of this conveyer gauge is reported to be within 0.4 wt.% moisture over a 9-month period of plant trial [169].

7.3 LEVEL AND INTERFACE

7.3.1 Level Measurement and Control

The measurement and control of levels and interfaces inside process reactors and vessels is an application of great importance in assessing and improving plant performance. Such measurements can often indicate the source of process malfunction. In addition, the operation of many processes can be considerably simplified by accurate interface control. For process diagnostics interface measurements are frequently carried out by *γ-ray scanning*: A γ-ray source of appropriate activity and energy is positioned on one side of the vessel and a radiation detector is positioned on the opposite side. Source and detector are moved together up and down the vessel and the transmitted radiation is recorded. The difference in ρx between the two phases in the vessel is large in most systems encountered, and the position of the interface is thus indicated by a large change in the transmitted intensity. The technique is rapid, versatile and accurate (better than ± 2 cm in most cases). Because all equipment is external to the vessel, the measurement is applicable to any process material and is not impaired by conditions of high temperature, high pressure or by corrosive, viscous or toxic materials. The technique has been used on vessels of diameter varying from a few centimetres to 10 m or more and with wall thickness up to 20 cm of steel. Typical examples are as follows:

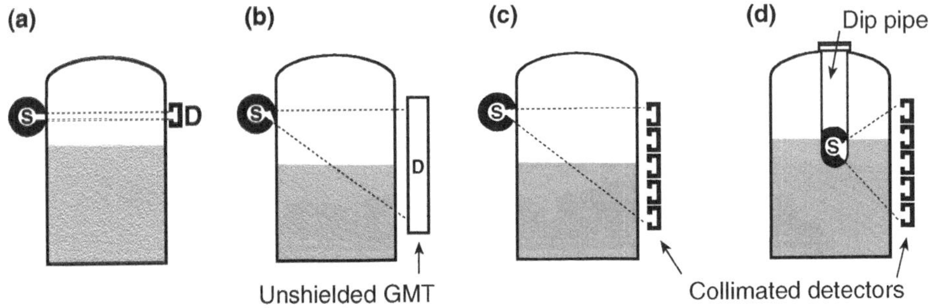

Figure 7.7 Typical gauge configurations use for level measurement and alarms. (a) High or low alarm, depending on positioning of source/detector and (b)–(d) level gauge configurations. The one in (b) uses a wedge of wall material absorber in front of the source so that the beam path length through wall material is equal in all vertical detector positions. The detector output signal is then proportional to the level (see Section 7.3.2). The dip pipe configuration (d) may also be used for alarm devices

- Measurements of liquid levels in storage vessels, still bases and reactors.
- Measurement of the level of catalyst beds in reactors.
- Measurement of packing levels in absorption towers.
- Monitoring the loading of road and rail tankers to ensure maximum utilisation of storage volume and to prevent over-filling.

γ-ray scanning is also used for trayed tower diagnostics to find damaged or collapsed trays [6, 174]. γ-ray scanning as described above is not very suitable for a permanently installed gauge because it involves mechanical motion. There are many cases where there is a demand for a permanently installed gauge, for instance for automatic control. To meet these demands, a range of nucleonic level and density gauges has been developed. These instruments are extensively used in situations where the nature of the process material gives problems with conventional level systems in which the sensor employed requires intimate contact with the material. Figure 7.7 illustrates some of the possible configurations for level systems. Arrangement (a) is the simplest type of system, widely used as a high/low-level alarm. Arrangements (b–d) are useful in giving level indication over a range of vertical height. Configuration (d) uses a closed insert tube (dip pipe) housing a source (or series of sources). A feature of this system is that source activities are low; only one wall has to be penetrated and the source/detector distance can be short. Often the source activity required is only $1/100$ of that which would be required for the external source system. As mentioned in Section 5.4.5 this configuration may also be used for measurement on vessels with thick walls and/or large diameter where transmission measurements otherwise are difficult or even impossible. The configuration with one long detector provides sufficient accuracy for many applications; it is also beneficial from a cost perspective. Its response may also be linearised with a few simple steps (see Section 7.3.2). Geiger–Müller tubes are very popular for this purpose; however, there are also systems on the market based on bundles of long scintillation fibres connected to one PMT.

Figure 7.8a shows a simple high-level alarm installed on, e.g., still boilers. The count-rate of the detector output falls dramatically as the level rises to the point where it intercepts

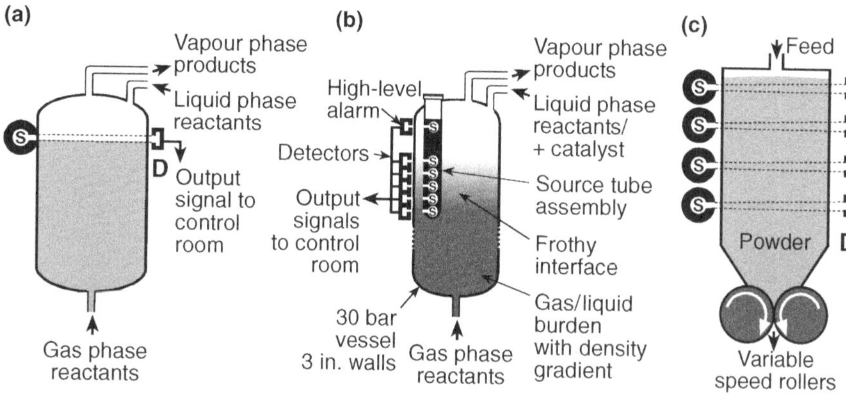

Figure 7.8 Illustrations of nuclear controls systems using (a) high-level alarm gauge installed on, e.g. still boilers, (b) proportional indicator and high-level alarm on a gas–liquid reaction vessel and (c) level gauge designed for the continuous production of tablets

the γ-ray beam. The change in detector output is sensed by an electronic control unit (mounted in the plant control room) and causes a relay to de-energise. The relay operates a valve that allows product to flow out of the vessel. Such a system is extremely reliable due to the low component count that results in a typical MTBF of 75 years (see Section 5.3.7).

Figure 7.8b illustrates proportional indication over 1.5 m and a high-level alarm on a gas–liquid reaction vessel. This is an instance of a frothy (and corrosive) interface, so that 'level' is a matter of definition. This can be made meaningful for the γ-ray system in terms of a particular mass per unit distance (ρx) taken, for example, in the case of the high-level alarm as the vapour density that does not give unacceptable carry over. Figure 7.8c shows a level control system designed for the continuous production of tablets. These examples are sufficient to indicate the generality of radioisotope methods and their relative indifference to difficult process materials or environment.

7.3.2 Linearity in Level Gauges

If a level system had a linear count-rate along the whole length of the detector and the detector response was linear then the level gauge output verses level would be linear. In practice neither of these conditions are always realised and various methods are used to linearised the output of the gauge. Firstly, the radiation field at the detector can be linearised using a shielding wedge that is placed in the beam at the front of the source shield (see Figure 7.7b). The wedge is designed so as to reduce the effect of the beam angle by reducing the output at the top of the level range where the path length is shortest and leaving it unaffected at the bottom of the level range where the path length is longest. Care is also needed where more than one source is used to ensure that there is no gap or overlap in the radiation field at the point where the two beams meet at the vessel wall, see Figure 7.9.

In addition to a linear radiation field the detector should be as linear as possible. With GMT detectors where the detector is made from a string of long Geiger tubes then with careful design the response will be linear. In the case of long plastic scintillation

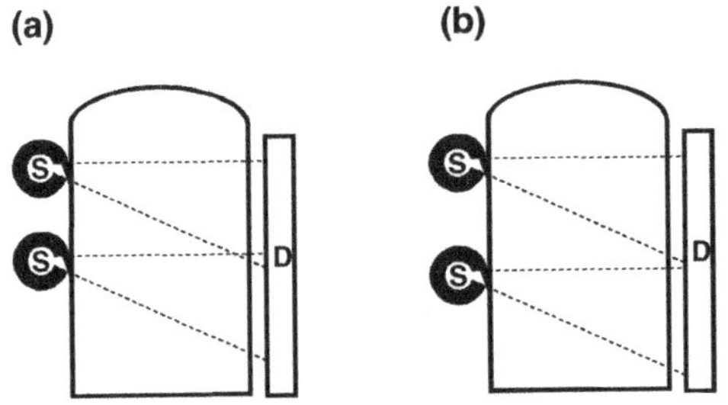

Figure 7.9 Long-level gauges may use two or more source holders. The arrangement on the left will give full level cover at the overlap whilst the arrangement on the right will have a small portion of vessel where there is no change in output as the level changes, although the dose rate along this detector length will be more linear

detectors the response to radiation is greatest near to the photomultiplier tube. Some compensation for this nonlinearity may be gained by placing the photomultiplier tube at the bottom of the level range, thus placing the most sensitive part of the detector where the radiation field is the weakest. In many level control applications the absolute accuracy of the level measurement is not as important as the reproducibility of the output but where absolute accuracy is important the gauge must be calibrated with a vessel fill and the output linearised electronically. In small vessels where the radiation is only partially absorbed by the vessel contents it is of course necessary to measure the count-rate when the vessel is full as well as empty thus giving two points of calibration.

7.3.3 Pressure Consideration in Level Systems

In high-pressure systems the gas above the liquid level can have a significant density and leads to significant attenuation of the beam. For instance in a polyethylene polymeriser the operating pressure is 250 bar and the vapour density of the ethylene at this pressure is 210 kg/m^3. In this case the empty vessel count-rate for calibration must be measured with the vessel under operating pressure or serious errors in level indication will transpire.

7.3.4 Interface Measurement

Whilst all level measurement is interface measurement, here we use the term to more specifically define the level of two non-gaseous media such as sand in water or oil on water. The requirement for such systems is particularly high in the oil production industry where incoming fluids from the oil well, which contain oil, water, sand and gas, are required to be separated. The separation is usually carried out in gravitational separators and a typical sand level monitor is shown in Figure 7.10.

The source used in such a device would normally be ^{137}Cs or ^{60}Co, both of which have a sufficiently high γ-ray energy to allow the assumption to be made that the absorption coefficients for the two fluids are independent of fluid composition. For attenuation in the two fluids the count-rate at the detector is given by

$$I = I_0 e^{-\mu_{M1}\rho_1 x_1} e^{-\mu_{M2}\rho_2 x_2} \tag{7.19}$$

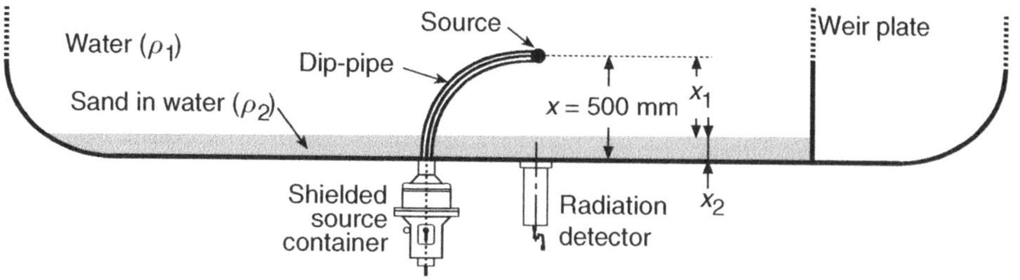

Figure 7.10 A sand level monitor for gravitational separators. The source may be retracted into the source container

which simplifies to

$$\frac{-\ln \frac{I}{I_0}}{\mu_M} = \rho_1 x_1 \cdot \rho_2 x_2 \tag{7.20}$$

Using this equation and the knowledge that the sum of the depths of the two fluids is equal to the total depth below the source, i.e. $x_1 + x_2 = x$. Then from these two simultaneous equations we can calculate the position of the interface:

$$x_2 = \frac{\ln \frac{I_0}{I}}{\mu_M} - \frac{\rho_1 x}{(\rho_1 + \rho_2)} \tag{7.21}$$

With this arrangement for interface measurement the radiation dose rate on the bottom of the vessel can be high when the vessel is empty and care should be taken with the creation of a possible controlled area below the vessel (see Section 6.7.5).

7.3.5 Installed Density Profile Gauges

The distribution of process material inside a reaction vessel can be investigated using the γ-ray attenuation technique. This is particularly useful in measuring the extent and density of foam layers above reacting process liquids. The entrainment of liquid droplets in the gas streams from reactors (carry-over) can be similarly studied. Thus, carrying out scans at different process rates can check the performance of a demister pad and steps can then be taken to reduce the carryover. Gas entrainment or bubbling in process liquids can also be quantitatively assessed by means of γ-ray density scans. A similar application is the detection of voids in catalyst beds or packed volumes. Scanning measurements are normally carried out by a process diagnostic team rather than by an automatic gauge. In some cases, however, continuous density profile is required and permanently installed gauges are required.

In these cases a permanent density profile instrument may be used [175, 176]. Such an instrument consists of a vertical array of sources and detectors mounted in a pair of adjacent dip pipes inside the vessel (see Figure 7.11). In effect the instrument is a multiplicity of density gauges, each giving an independent measure of the density at a fixed level within the vessel. It can thus measure the positions of any number of interfaces in a multiphase system simultaneously and in addition the extent of phase–phase dispersion (mixing) at every interface between dissimilar materials. In other words, the density profiler

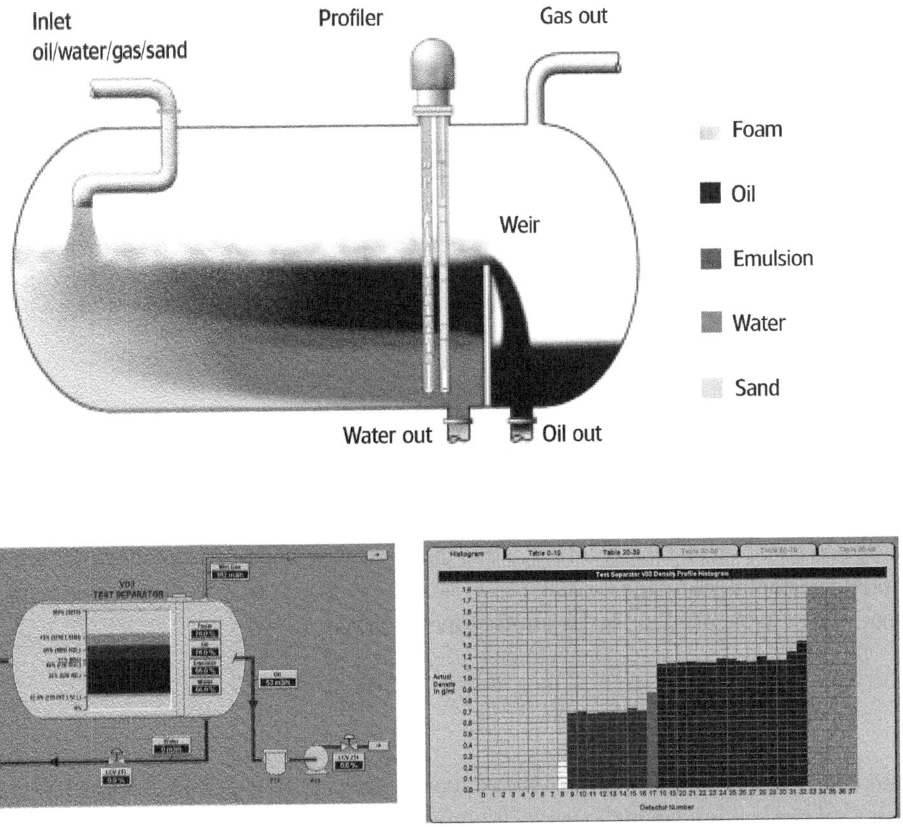

Figure 7.11 A density profiler installed in a gravitational separator where interfaces between layers of gas, foam, oil, emulsion, water and sand can be detected (top). In this case 37 detectors are used. Typical control room output screens are shown to the right. Reproduced by permission of Tracerco a trading division of Johnson Matthey PLC

measures both the elevation and the quality of each interface in a vessel containing multiple phases.

Two arrangements have been used: firstly with ^{137}Cs sources and a path length in the vessel of about 30 cm and secondly using ^{241}Am sources with a path length in the vessel of about 7 cm. Each source must be collimated in order to ensure that its radiation falls only on the associated detector, thus achieving a clear density reading at each elevation. The most common application for density profilers is in gravitational separators used in the oil industry to separate oil well fluids. Each individual density reading is transmitted to a process controller where they are combined in display and control functions.

Samples of displays available to the operator are also shown in Figure 7.11. The histogram indicates the presence of sand (sensors 33–37), water (18–32), emulsion (17), oil (9–16) and foam (8). The weir is at an elevation that corresponds to sensor 11. Presented with this information, the operator will add small quantities of anti-foam and de-emulsifier chemicals and will consider performing a sand wash. The vessel mimic relates to vessel conditions at a different time. It indicates the levels of water, emulsion, oil and foam, and expresses these measurements as a percentage of vessel height.

When ^{137}Cs (662 keV) sources are used the attenuation coefficient can be assumed to be independent of the fluid composition in the beam. This is not the case with ^{241}Am (60 keV)

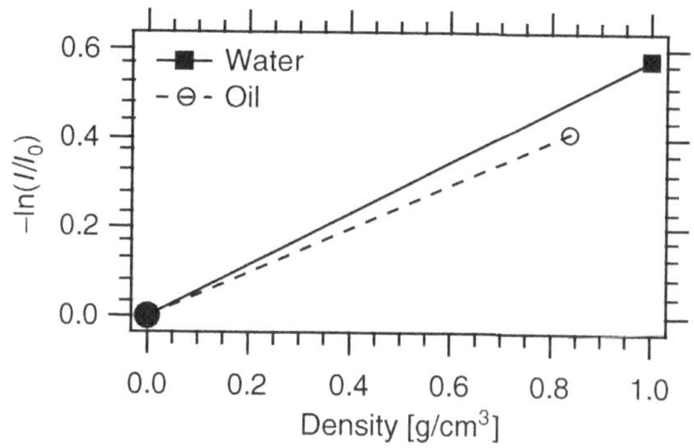

Figure 7.12 The chart shows a comparison between the true calibration using oil and water and the calibration achieved by using water and empty only (that is assuming the attenuation coefficients are equal)

because a significant fraction of the interactions is by the photoelectric effect at this low energy. The attenuation coefficients can be corrected logically if the arrangement and the coefficients of the various fluids are known but individual fluids give some interesting and useful results on correction. If a mean of the attenuation coefficients is used then the resultant density differences (which is what we are really interested in) are exaggerated, which allows for fluids of almost identical densities to be easily differentiated (see Figure 7.12).

7.4 THICKNESS MEASUREMENTS

In many industries, radioisotope gauges are used to monitor and control the thickness of sheet materials ranging from thin plastic to sheet steel. Thickness measurements are also useful in carrying out checks for corrosion and erosion of pipes, ducts and (particularly) the tube bundles of heat exchangers. Basically, a γ-ray source and a miniature radiation detector are inserted simultaneously down adjacent tubes in the bundle and the radiation transmitted through the tube walls is recorded. The transmitted signal is related to the thinning of the tube walls in the direction of measurement, following calibration. By systematically carrying out this procedure for each pair of tubes, a comprehensive picture of the position and degree of the corrosion over the entire bundle is built up. The technique is rapid compared with other inspection methods available and is capable of high accuracy (0.1 mm thinning is readily detectable). The scanning of heat-exchanger bundles in this way is often incorporated into many plant shutdowns so that the progress of corrosion can be monitored. The technique is also used in the emergency situation to identify areas of high corrosion and thus to facilitate decision-making as to whether to replace or block off badly corroded tubes.

7.4.1 γ-Ray Transmission Thickness Gauges

A γ-ray thickness gauge is used to measure the thickness of a material whose density is known. One application of this technique is in the production of flat-rolled steel and non-ferrous metals. Thin aluminium sheets can be made by hot-rolling large bars, typically 30 cm thick, down to about 4 cm. The temperatures in this process vary from 300 to 600°C.

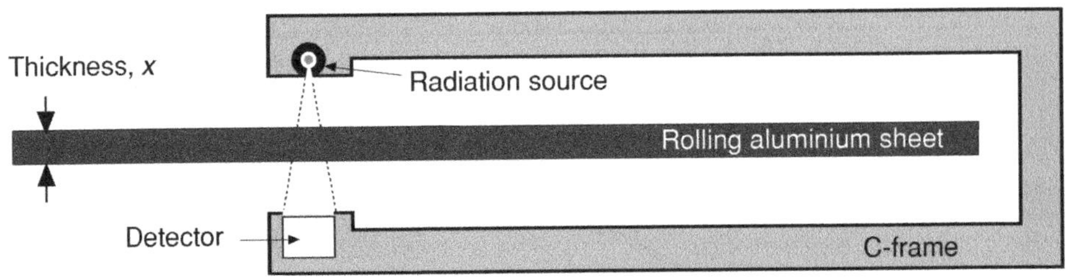

Figure 7.13 Schematic cross section of radiation gauge mounted on a C-frame for continuous thickness measurement of rolled aluminium. The rolling speed of the aluminium sheet is about 1000 m/min

The thickness may be further reduced down to 2 mm in a cold-rolling process. In both cases there is a need of a continuous measurement of the thickness with gauges. These have to withstand high ambient temperature, steam and pollution without being damaged and without introducing measurement errors. A non-contacting arrangement based on a C-frame, as shown in Figure 7.13 for measurement on aluminium, is thus ideal for this purpose.

We discussed the use of transmission techniques to measure thickness in Section 5.5.1 and the associated measurement accuracy in Section 5.3.4. By measuring the beam intensity, I, the sheet thickness can be determined according to Equation (5.14):

$$x = \frac{1}{\mu} \ln \left(\frac{I_0}{I} \right) \tag{7.22}$$

here I_0 is found by measuring the intensity without any sheet present. Accurate values of the linear attenuation coefficient μ of the sheet are best determined by calibration at known thickness. This means the effective attenuation coefficient is used in the case of build-up (see Section 3.5.3). On the other hand the build-up is very low in thin sheets. The main limitation of this technique is that the linear attenuation coefficient is quite sensitive to the composition of the aluminium alloy, particularly at low energies. The rolling speed of the aluminium sheet is about 1000 m/min and the measurement accuracy should be better than $\pm 5\%$. To achieve this sources with activity up to 30 Ci are used. Generally, fairly accurate prediction of the measurement error is obtained using Equation (5.15).

7.4.2 Thickness Measurement Using γ-Ray Scatter

Gamma-ray backscatter may also be used for thickness measurements of sheets etc. although it is most widely used for measurements of density and related parameters such as component fractions. The scattering of gamma radiation can be applied in a number of ways to the investigation of plant performance. These techniques are generally less applicable than those based upon γ-ray attenuation, but in certain circumstances they can be used to obtain information that would be difficult, or impossible, to obtain by alternative methods. Scattering of gamma radiation is quantitatively related to the properties of the scattering medium although the relationships between incident and scattered beams are more complex. Figure 7.14 shows empirical observations for 180° γ-ray backscatter of the 1.2 and 1.3 MeV γ-rays of ^{60}Co for various media. It can be seen that this phenomenon is dependent on both density and thickness.

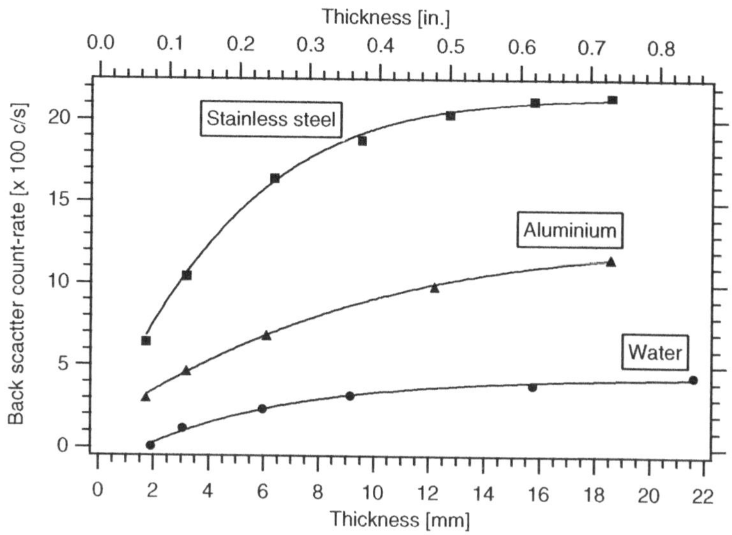

Figure 7.14 The backscatter response of ^{60}Co (1173 and 1333 keV) is dependent on both density and thickness of the material

The best accuracy for backscatter thickness measurements is obtained with relatively low radiation energies because of higher attenuation and more efficient collimation and shielding. As for β-particle backscatter discussed in Section 5.5.2 there is a saturation thickness or limit for γ-ray backscatter measurements, as can be seen from Figure 7.14. This is basically a function of the radiation energy and the material density. These determine how much scatter is generated and the penetration depth, and how much of the scatter towards the detector is absorbed. At low energies the composition (atomic number) is also important because photoelectric absorption increases the attenuation. For backscatter of 59.5 keV γ-radiation (^{241}Am) the saturation thickness in aluminium is reported to about 6 mm [177].

A practical advantage offered by gamma backscatter is that the source/detector assembly can be constructed in a single unit, capable of being used by one person when used for diagnostics purposes. In addition, since measurements are carried out from one side of a vessel only, access is much less of a problem.

Backscattered photons are reduced in energy relative to the primary radiation (e.g., for ^{60}Co the primary radiation is 1.3 MeV while the backscattered radiation is of energy 200 keV). It is thus easily possible to select the backscattered beam by electronic means to the exclusion of the primary radiation.

In chemical plant applications – such as interface detection, thickness and coating measurements or the measurement of build-up on a vessel wall – the technique is limited by the saturation of the backscatter in the vessel wall. This makes it difficult to apply the technique to vessels with wall thickness much in excess of 0.5 in.

7.4.3 β-Particle Thickness Gauges

For accurate measurement of thin films and sheets, C-frame β-gauges are used because of the higher sensitivity compared to γ-ray gauges (see Figure 5.24). Some β-gauges use

Table 7.1 Typical measurement ranges and uncertainties of the most frequently used β-particle sources for materials with density $\rho \approx 1$ g/cm^3 [a]

Source	Max. energy [keV]	Typical range [μm]	Typical uncertainty [μm]
^{147}Pm	225	Up to 275	±0.3
^{85}Kr	672	150–1500	±1
^{90}Sr	2274	1000–8000	±4

[a] The measurement uncertainty is expressed in terms of two standard deviations (2σ, i.e. $k = 2$) when using low activity sources [179].

current read-out, but, the best accuracy for a given source activity is obtained with pulse mode read-out. For transmission measurements the thickness is derived from Lambert–Beer's law as for γ-ray transmission, or alternatively semi-empirical models may be used [178]. The attenuation in air cannot be neglected when β-radiation is used; however, it is to some degree cancelled out by using calibration measurements to determine what may be regarded as an effective absorption coefficient. However, variations in environmental parameters, for instance the air humidity, will influence the measurement accuracy. Such errors may be corrected for by using a reference measurement in air only. Table 7.1 summarises the measurement range and uncertainties obtainable with the most common β-particle sources.

Thickness measurement by β-particle transmission has many applications: plastic and metal film and sheet, textiles, non-wovens, coated abrasives, food packing solutions, pharmaceuticals, metal foils, book binding, adhesives, coatings, laminates, packing materials, composite materials, blown film, rubber and vinyl, synthetic and natural fibres and battery coatings.

Thickness measurements by β-particle back scattering are also frequently applied, particularly for measurement of coating thickness on a backing material or in cases where there is access to only one side of the object. The former will only work as long as an infinite thickness of the backing material delivers a signal substantially different than an infinite thickness of the coating material [180]. This signal is a function of the atomic number and the density of the materials as discussed in Section 3.1. The effective atomic number of composite materials, such as various coatings, can be calculated using Equation (3.31) ($m = 1$) provided the composition of the material is known. One consequence of this is that the thickness of plastic coatings can be determined with higher accuracy on a heavy metal backing than a light one such as aluminium. Note that the scatter intensity for a coating on a backing material, as given by Equation (5.41), saturates for at a certain material dependent coating thickness. The measurement uncertainty increases as this thickness limit is approached.

It is also possible to use β-particle back scattering on thin sheets without backing material, or effectively with air as backing material. This is because an infinite thickness of air gives a much lower scatter signal than any solid material as a result of the lower density [the atomic density N in Equation (3.1)]. Using a ^{90}Sr source for scatter measurements on aluminium sheets the maximum measurable thickness is about 600 μm [177]. As rule of thumb estimation this saturation thickness is about one fifth of the maximum range (R_{max}) of β-particles in the material (see Section 3.1.2).

7.4.4 Monitoring of Wall Thickness and Defects

For the sake of completeness we include a few examples on techniques applicable to monitoring changes in thickness. The classical examples are thinning of pipe and vessel walls by corrosion and wear, and the other way around increasing thickness due to deposits and scale (e.g. coke, solid catalyst). Both cases may be critical and cause process malfunction if not discovered in time. For example, long-term studies of this type have been made to establish the rate of build-up of catalyst in an exit line from a vaporiser and to correlate this build-up rate with operating conditions. The problem is important in that high pressure drop at the vaporiser exit was the principle factor causing plant shutdowns.

This type of monitoring is carried out on a regular basis by a process diagnostic team because in most cases the rate of change in thickness is slow. On the other hand, there are examples of pipe bends, etc. being completely destroyed within hours by sand blasting of produced sand from oil wells. There are thus situations where critical parts of a vessel or pipeline need permanent and continuous monitoring. The methods described in this section are thus aimed at investigating the process equipment, rather than the process it self. This is consequently to some extent NDT (non-destructive testing), a field we will not cover in this book, other than it will be mentioned in the context of radiography (see Section 7.7.1). For on-line NDT on process equipment various scatter methods are very applicable [181].

Traditional γ-ray transmission methods can seldom be applied to monitor wall thickness because it is impossible to tell changes in wall thickness from variations in attenuation in the process material. Gamma-ray backscatter, however, is very applicable because proper collimation of radiation source and detector may be used to define a small measurement volume near the inner wall of the vessel, see the strict collimation backscatter example in Figure 5.25. Alternatively several detectors can be used in a ring around the source, each collimated at different positions on the incident beam. The thickness resolution is then determined by the collimation and size of the measurement volume. For this method to work there of course needs to be a measurable difference in density between the process medium and the pipe material in the case of corrosion or wear, and the scale and deposit in the case of blockage building. Radioisotope sources have been applied for this method for pipe corrosion monitoring [182, 183]; however, an X-ray tube produces a scatter signal with significantly higher intensity [184, 185]. There is of course a trade-off between the thickness measurements resolution (size of the measurement volume) and the time required to obtain a certain accuracy because of counting statistics.

A more exotic method sometimes applied for wall thickness monitoring is *thin layer irradiation*: If a steel component is placed into a beam of protons a thin layer of irradiated material is produced, the thickness of which is related to the beam energy. The thickness can be from about 10 μm to about 2 mm and will contain small but detectable quantities of ^{56}Co produced from ^{56}Fe by the reaction ^{56}Fe(p, n)^{56}Co. The half-life of ^{56}Co is 77.3 days and it emits a range of useful γ-ray energies from 847 keV to 2.6 MeV. The radiation can be detected using a sensitive scintillation detector from the outside of a machine or vessel and therefore enables the detection of minute quantities of wear, erosion or corrosion. Only a small proportion of the target material atoms are converted so no significant difference is made to the physical or chemical behaviour of the component. Examples of the use of this technique are monitoring the wear rate in the bore of an internal combustion engine and measuring the corrosion rates in chemical reactors. In the case of machine wear studies the loss of activity can also be assessed by sampling the activity level in the lubricating oil.

7.5 FLOW MEASUREMENT TECHNIQUES

7.5.1 Density Cross Correlation

By placing two density gauges a small distance apart (<0.5 m) the velocity of slugs in a multi phase flow can be monitored. The signals from the two gauges are identical in magnitude and vary only in respect of arrival time. The time difference between the two correlated signals divided by the distance between the two detectors yields the slug velocity. Each individual density gauge gives the slug density and its length. From this data and the pipe dimensions the liquid mass flow-rate can be computed. This method has been applied to multiphase flow meters [186, 187] and for studying bubble phenomena in gas–solid fluidised beds [188].

7.5.2 Mass Flow Measurement

The combination of a *Venturi* differential pressure flow meter and a density gauge also provides the mass flow rate of liquid in a two-phase (liquid/gas) pipe flow (see Figure 7.15). The Venturi meter uses differential pressure measurements (Δp) over a contraction of the pipe cross-section (from diameter D to diameter d) to provide the volumetric flow rate (q). A γ-ray densitometer is positioned to measure the gas volume fraction (α_g) across the Venturi throat where the flow is more turbulent and thus more homogeneously mixed.

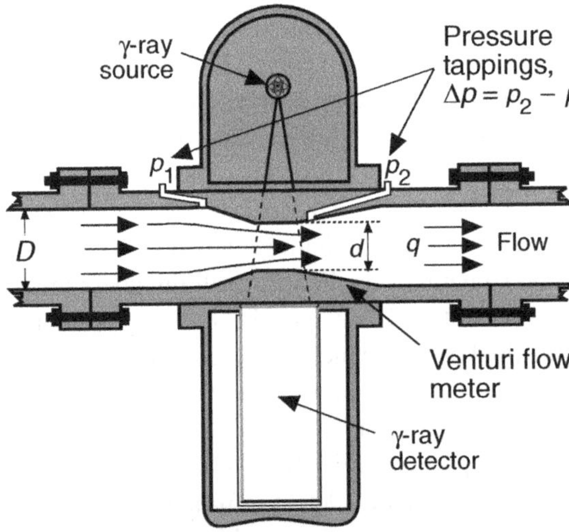

Figure 7.15 Schematic view of a combined Venturi meter for differential pressure (Δp) measurement of volumetric flow (q) and a γ-ray gauge for gas volume fraction (α_g) measurement

It can then be shown that for gas pressures less than about 100 bar where the density of the gas is negligible compared to the liquid density, the mass flow rate of the liquid component is given as [189]

$$\begin{aligned} \dot{M}_l &= \sqrt{2}\varepsilon C E A_2^M \sqrt{\rho_l (1-\alpha)\,\Delta p} \\ E &= \frac{1}{\sqrt{1-\left(\frac{d}{D}\right)^4}} \\ A_2^M &= \frac{\pi d^2}{4} \end{aligned} \tag{7.23}$$

where C is the *discharge coefficient* for the Venturi meter, ε is the *expansibility coefficient* of the fluid, E is *velocity approach factor*, A_2^M is the throat cross-section area and ρ_l liquid density. This is an important relationship used in several multiphase (gas/oil/water) flow meters.

7.5.3 Multi-Phase Flow Metering

Multiphase flows are common in the petroleum industry, and yet their measurement nearly always presents difficulties to the process engineer. The traditional solution to the problem of metering multiphase gas, oil and water flows is to separate the components first, and then measure the flowrate of each using conventional single phase instruments. While there is currently no alternative to separation and metering for fiscal purposes, there are other parts of the oil production process where this solution is both inconvenient and expensive: In applications such as allocation measurement and well monitoring and testing, there is a need for a three-phase flowmeter. The primary information required from the user of a three-phase flowmeter is the mass flowrate of the oil, water and gas components in the flow. These cannot be measured directly and independently: they are calculated from instantaneous measurements of velocity and cross-sectional fraction of each component [190, 191], such as is illustrated in Section 7.5.2.

Multiphase gas, oil and water flow measurement is an application where there has been a substantial development in radioisotope methods over the past decades. We saw several examples of this in Sections 7.2.2–7.2.5 and references therein concerning component fraction measurements. This application is also a very good example of how multiple measurement modalities are applied: Although the volume fractions of all components can be determined by γ-ray methods only [234], it is also common to combine radioisotope methods with for instance electrical sensing principles. Very often a γ-ray densitometer is used to differentiate between the gas and liquid components whereas electrical conductivity or permittivity is used to distinguish between the oil and water components [190, 191].

7.5.4 Tracer Dilution Method

The dilution flow method or total count method involves sudden injection of a radioisotope as a single pulse into the flow. The total activity A of the injected material is measured prior to injection and the total counts in the stream after a suitable mixing reach are measured using a calibrated detector or by assaying samples. The volumetric flow rate is given by

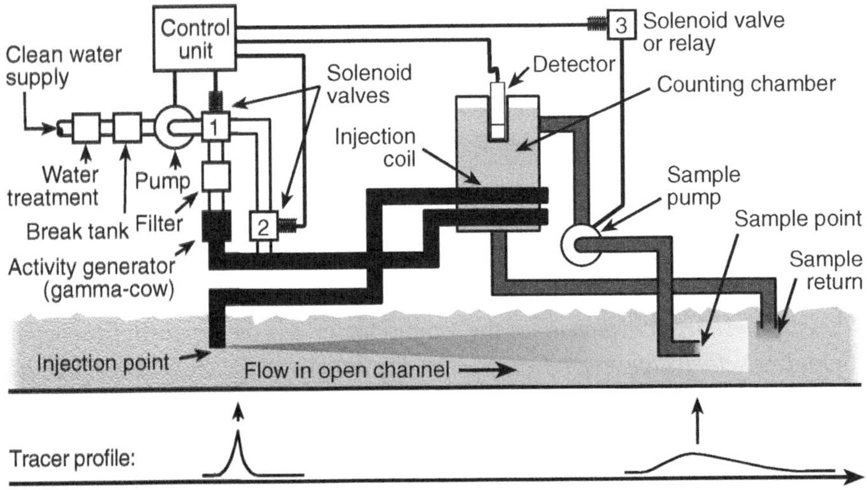

Figure 7.16 General arrangement of an isotope dilution flowmeter using a gamma-cow. Under computer control the injection pump produces a pulse of ^{137m}Ba, which is measured by the detector and then flushed into the stream with water. The sample pump delivers a constant sample from the stream back to the same counter where the count-rate in the stream sample is assayed

the following equation:

$$q = \frac{A}{\int (c - c_0)\, dt} \tag{7.24}$$

where q is the flow rate, A is the injected activity, c is the concentration of activity in the stream and c_0 is the natural activity in the stream or the background activity concentration. This method of flow measurement lends itself readily to automation using a ^{137m}Ba isotope generator as the activity source for an open channel flowmeter.

The flowmeter runs a batch process, which is repeated about every 10 min. This is just long enough for a reasonable level of activity to build up in the generator ready for the next injection. The controlled sequence starts with a background measurement on the stream sample then water is pumped through the activity generator for a few seconds to generate a pulse of active ^{137m}Ba. While the water pump is still running the solenoid valves divert the flow around the generator in order to move the discreet pulse of activity towards the injection coil that is placed under the counting chamber. When the detector, which is a sodium iodide scintillation counter, detects the approaching activity the water pump is stopped momentarily while the decay rate of the activity is measured. This is a safety check in order to stop the flowmeter from injecting ^{137}Cs. When the decay rate confirms that all is well the active pulse count-rate is recorded (A) and the activity is injected into the stream. After a suitable mixing reach in the stream a sample is continuously withdrawn and is pumped into the counting chamber that has a volume of about 10 l so as to achieve a high detection sensitivity. The sample is streamed through the counting chamber until all activity has passed the sample point and the counts from the stream sample are integrated ($\int (c - c_0)\, dt$). Now by knowing the relative counting efficiencies of sample and injection we can calculate the dilution in the steam using Equation (7.24). The half-life of the injection isotope is only 2.5 min, which is ideal for environmental measurements such as effluent flow rate measurements as often the tracer has decayed to undetectable levels even before the effluent stream crosses the site boundary. Flow rates up to 5000 m^3/h have been measured with this instrument. One 30 mCi (111 MBq) generator operated for over 5 years injecting every half hour 24 h/day. This adds up to a total activity generated of about 2.6 MCi or 97 TBq which is a vast amount of activity but remember the total activity in the system never exceeds the generator activity of 30 mCi and it all decayed before it crossed the site boundary. Such short-lived isotopes pose a not unpleasant dilemma for the regulatory authorities, is there any radioactive disposal if all the activity decays before leaving the site? One local ruling on this question decided that the disposal takes place where the radioisotope stops being useful and becomes waste, i.e., after the stream sample point.

7.6 ELEMENTAL ANALYSIS

Measurement of the elemental composition of process streams is often of paramount importance in assessing and/or controlling plant performance. The efficient operation of a process is for instance often critically dependent upon the concentration of a catalyst present as a trace element. It is thus vital to ensure that this concentration is maintained at a constant level. In some cases it is necessary to test the materials used for plant construction to ensure that they are suitable for their proposed duty. For example, the steel used in vessels carrying hydrogen at elevated pressures and temperatures must contain specified proportions of Cr and Mo. Further, to ensure that specifications on finished products are met stringent quality control is required. A rapid monitoring technique is desirable so that any off-spec material can be diverted and the plant restored to correct operation as quickly as possible.

We studied the principles behind elemental analysis using characteristic X-rays (fluorescence, XRF) and prompt γ-rays (PGNAA) in Section 5.5.3. These methods have been applied for laboratory analysis for many years; however, there is still limited use of these as permanently installed gauges. Such use of XRF is mainly restricted to metal identification, see [15, 113, 192–194] and references therein. Detection of low-Z elements is restricted by the energy resolution available when using room temperature detectors. Several of the proposed gauges are designed as outlined in Figure 5.26. In some cases the probes are immersed in the process medium, such as slurry streams, to optimise the measurement conditions. In other cases the gauges are installed at sample bypass lines to allow for longer measurement times.

PGNAA is an emerging technology for elemental analysis not having the same restrictions as XRF as discussed in Section 5.5.3. Even so it still has limited use as a permanently installed measurement principle [15, 116, 122, 123, 195]. Prompt gamma measurements are used on a regular basis to identify concealed explosives and drugs in transport containers and luggage [196].

Finally, it is also possible to use NORM for elemental analysis as discussed in Section 5.5.5, for instance for detection of shale in sedimentary iron ores [197].

7.7 IMAGING

The potentials of imaging with ionising radiation were discovered almost at the same time as the ionising radiation itself, see Section 1.3. At that time and for many years thereafter photographic film was used because efficient position sensitive detectors not were developed until more recently. The imaging modality first applied was *radiography*, and this is still important for NDT applications (and of course in medicine). This is so-called *projective imaging* and may be considered as a large number of separate transmission measurements onto one plane. For industrial processes another imaging modality is of higher interest; namely *tomographic imaging* which is cross-sectional imaging. Most people associate tomography with medical X-ray tomography, so-called CT (computerised tomography). With tomographic imaging multiple measurements are carried out at different directions through the object being investigated, followed by an image reconstruction where a 2D distribution of parameters is produced. Several sequential 2D images may also be stacked together to form a 3D image. With X-ray and γ-ray tomography the measured parameter is either the density or the effective atomic number in the Compton and photoelectric dominant regions, respectively.

7.7.1 Transmission Radiography

Fabrication of high integrity metal components, which involves casting or welding, usually requires radiography to confirm that the component is sound. Industrial radiography using a sealed source and photographic film is the most common inspection method that involves placing a source on one side of the component and a photographic film on the other side. The sealed sources most commonly used for radiography are ^{192}Ir, ^{137}Cs and ^{60}Co. X-ray generating tubes may also be used. Whilst film radiography is a little outside of the scope of this book there are obvious parallels with transmission instruments. Radiological protection considerations are identical and with the radiography sources generally larger and portable the rigorous application of safety and accounting systems is most important.

γ-ray and *X-ray cameras* also use the transmission principle but instead of a film the instruments use an array or matrix of detectors (see Section 5.2.4 and Figure 5.11). For airport security baggage scanners the X-ray energy is about 140 kV and the beam is arranged as a fan that is directed through the inspection chamber onto a linear or strip (1D) detector. The picture is built up one line at a time as the luggage passes through the beam on a conveyor belt. Gamma cameras used for imaging the fate of radioactive tracers in patients use a 2D array of detectors arranged behind an array of collimators, which ensures accurate positional information on the tracer.

7.7.2 Industrial Tomography

The application of tomographic imaging techniques to industrial problem solving has been around ever since the first commercial X-ray scanner was introduced by EMI in 1972, and even before. Since then other sensing principles, such as magnetic resonance imaging (MRI) and ultrasound, have been used. Over the past decade there has been an explosive development of electrical sensing principles, particularly electrical capacitance tomography (ECT) [198]. All sensing principles have their pros and cons. The advantage of γ-ray and X-ray transmission tomography is that photons travel in straight lines between interactions so that fairly high spatial resolution is achievable when proper collimation is used. The drawback is relatively high cost, particularly for high-speed systems.

Initially industrial X-ray tomography was mainly used in laboratories for analysis of various objects and processes, and for development and validation of process models. This is probably still the most important application of industrial tomography [97, 98, 127, 128, 199–201], also for the other measurement modalities that have been developed. Field or plant applications of tomography seem to be limited to diagnostic methods [202, 203]. Ionising radiation methods then have an advantage in that the measurements can be performed through the walls of steel vessels, etc. For permanently installed gauges in industrial processes one of the design criteria is to reduce complexity. Therefore the use of multiple measurements without image reconstruction very often provides the required data, such as discussed in Section 7.2.2. Such measurements may be combined with so-called *a priori* knowledge about the process. This may, for instance, be symmetry considerations, component distributions such as horizontal layers with the most dense component in the bottom of gravity separators. This type of cross sectional measurement without image reconstruction may be referred to as *tomometry* [175].

Finally, industrial tomography may also be realised as dual modality systems in order to image the distribution of more than two components in multiphase systems. The combination of γ-ray or X-ray tomography and capacitance tomography, which is sensitive to the dielectric constant (permittivity) of the medium, has been proposed and investigated [204–206].

7.7.3 General Design of an Industrial Tomograph

Figure 7.17 shows the design of a typical tomography system. The *sensor head* constitutes a number of sources and detectors in the case of an ionising radiation transmission system, see Figure 7.19, a number of electrodes in the case of an electrical sensing principle, etc. This may either be embedded in the vessel wall or designed as a clamp-on system. Each detector (or electrode etc.) normally has dedicated read-out electronics such as outlined

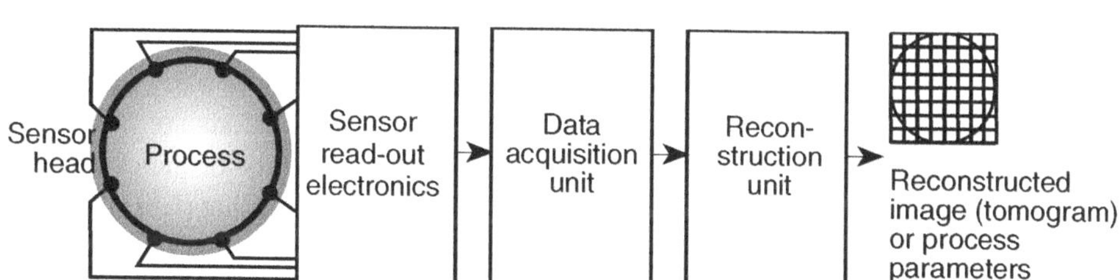

Figure 7.17 General arrangement of an industrial process tomography system with sensor head, sensor (detector) read-out electronics, data acquisition and reconstruction units. The latter also typically incorporates some image processing

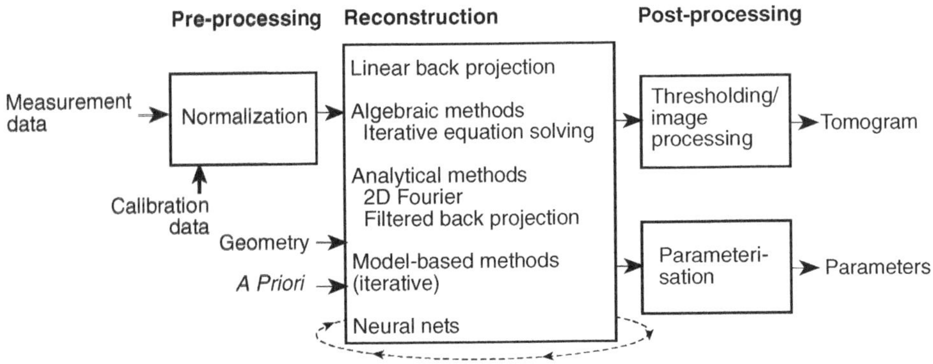

Figure 7.18 Typical layout of the image reconstruction algorithm

in Figure 5.9, to enable parallel signal processing. This is particularly important for high-speed imaging systems where the data acquisition time must be kept as short as possible.

Once data on the output of the sensor read-out electronics are captured by the data acquisition unit, the sensor read-out electronics starts acquiring data for the next frame. At the same time the previous data set is transferred to the image reconstruction unit for real time reconstruction. This pipelining structure is required for high-speed systems. Depending on the requirement defined by the application and the type of reconstruction algorithm and the computing power of the reconstruction unit, data may either be reconstructed on a $n \times n$ image grid as illustrated in Figure 7.17, or streamed to memory for off-line reconstruction. A state of the art personal computer is often used for image reconstruction, however, for high-speed real time reconstruction, parallel computing systems may be applied. In some applications there is no need for a reconstructed image (tomogram), but rather some parameters describing the process.

Figure 7.18 illustrates the different components of the image reconstruction or processing algorithms. The measurement data is first normalised to the calibration measurement data. The normalised data is then fed to the actual reconstruction or inversion process, which also has the system geometry and eventual a priori information as input. The tomographic inversion process is the most critical part. The choice of the algorithm is a trade-off between noise in measurements, computation time, number of measurements and *a priori* knowledge. The objective of the reconstruction algorithm is to inverse a set of equations relating the measurements to the image. The estimated parameters are linear attenuation coefficients. A well-suited class of inversion processes for industrial tomography is algebraic methods. The main assumption of these algorithms is the 'pixelisation' of the object. Two steps are required: definition of the objective function to minimize and definition of the algorithm that minimizes the objective function. Objective functions are usually the quadratic distance between measurements and the re-projected image or the conditional probability to get the image from the projection. A very efficient algorithm (EM) takes into account the Poisson statistical nature of γ-ray emission [207]. In addition, this algorithm can take into account *a priori* knowledge on the reconstructed image (MAP-EM) [208]. In such a case, an iterative algorithm with faster convergence properties should be used (iterative least square technique algorithm) [209, 210].

7.7.4 Industrial High-Speed Transmission Tomography

Fourth generation medical CT scanners utilise a rotating X-ray source and circular detector arrays as sensor head (see Figure 7.19). This concept has also been applied for imaging of industrial processes [211] such as pipe flows [212], however, a mechanically rotating radiation source restricts the speed of response. If the rotation is too slow compared to the process dynamics, the reconstructed image will become motion blurred because process features enter and leave the image plane before the data acquisition is completed. In order to avoid this type of inconsistent measurement, all measurements should be carried out simultaneously. This also improves the measurement accuracy at a given speed of response. For these reasons an approach incorporating several fixed radiation sources each of which faces an array of detectors on the opposite side of the process, see the instant configuration in Figure 7.19. The number of so-called *views* in such a system hence equals the number of sources, and the number of *ray-sums* in each view equals the number of detectors in each array. Each ray-sum is thus one transmission measurement. The requirement for high speed of response does not apply when the required information is temporal averaging of the process dynamics.

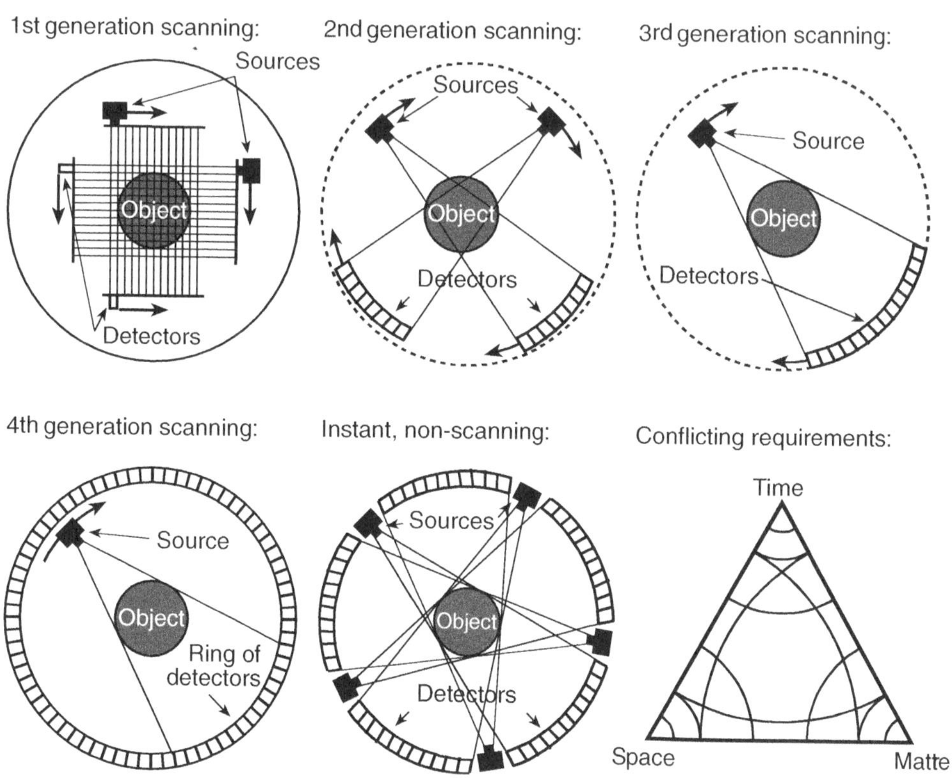

Figure 7.19 The development of measurement geometry for medical X-ray CT scanners (CT) from the first to the fourth generation. Also shown is the instant configuration in which all ray-sum measurements are carried out simultaneously, and the triangle representing the conflicting requirements of simultaneous high measurement resolution in time, space and matter

Imaging of industrial processes and their dynamics implies a trade-off between three conflicting requirements; the measurement resolution of time, space and matter. In contrast to medical tomography where the patient is kept still for the time it takes to acquire the data for one image, in process tomography the data acquisition time has to be short to avoid inconsistency as discussed above. With reference to Section 5.3.4 this means there is a compromise between statistical error and measurement time. Likewise, the spatial resolution for a given system is basically improved by reducing the size of the detectors. This in turn means fewer counts in each detector and thus a higher statistical error in each ray-sum measurement. In process tomography the solution is often, but not always, to relax the spatial resolution requirement compared to medical tomography, in order to obtain faster speed of response. The consequence is a trade-off between these three as illustrated by the triangle in Figure 7.19.

The image reconstruction of tomographic imaging implies solving an *inverse problem*: The measurement geometry and result are known so that the assignment is to determine which spatial distribution of the imaged parameter gave these results. It has been shown that a circular object can be reconstructed exactly on an $n \times n$ image grid, when $2\pi n$ equispaced views each consisting of $2n + 1$ error free ray-sums, are available [213]. In addition the reconstruction grid has to be sufficiently fine to accurately represent the true features of the object. This is a rather disconcerting requirement for a multi-source system since even an 8×8 grid requires 50 sources and at least 850 detectors in total. However, CT theory also predicts that objects exhibiting some symmetry and homogeneity properties can be reconstructed successfully from very few views [213]. The use of 5 sources is a good compromise [209, 214, 215]. Successful flow imaging systems have been built using five views each with 17 ray-sums [204, 214], that is, a total of 85 ray-sums. This is designed for an 80 mm inner pipe diameter and the spatial resolution of it is about 5 mm, whereas gas/liquid imaging is possible with a temporal resolution of a few ms [204] (see Figure 7.20). The predicted temporal resolution for a 5 source and 160 detector system is 10 ms whereas 10 mm spatial resolution is possible on a 40 cm diameter fluidised bed [215].

There are also fast X-ray imaging solutions using electronic scanning of the beam rather than mechanical scanning. These use multiple filament tubes around the object and a switchable grid voltage just in front of each filament. This enables rapid switching of the electron beam from one filament to the next, and thus rapid scanning. Reported scanning

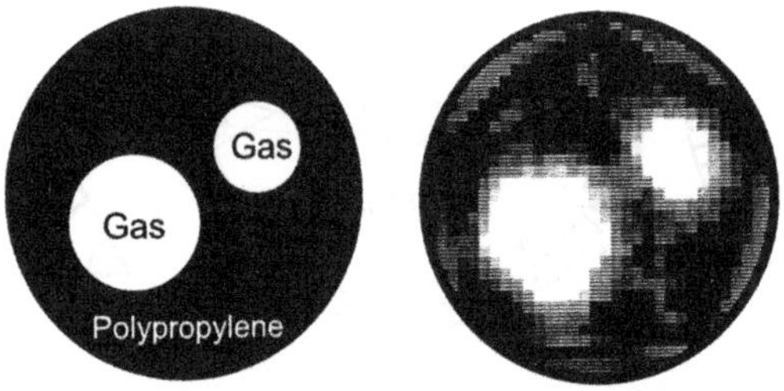

Figure 7.20 Reconstructed results using a polypropylene/air phantom in a γ-ray tomograph with 5 sources and 85 detectors [204, 216]. The pipe diameter is 82 mm. The integration or counting time is 10 ms, and seven iterations are used in the ILST reconstruction. No compensation is made for scattered radiation

times are 0.5 ms [217, 218] and 20 ms [219]. The use of polychromatic X-rays adds some complexity to the image reconstruction since the attenuation coefficient depends on the energy and is higher for the lowest energies in the X-ray emission spectrum. This is solved by corrections in the image reconstruction to remove artificial effects, which are otherwise produced. This is known as *beam hardening*.

8
Engineering

The first step in assessing a measurement problem is actually to determine which measurement principle or what combination of measurement principles are best suited to provide the required information. This is not always straight forward because people are often biased towards certain principles and have less knowledge about the capability of others. This means that very often the particular problem is solved with a non-nucleonic principle. On the other hand we know from experience that there are situations where the performance of radioisotope systems are superior.

Our design rule number one is Albert Einstein's statement: 'Everything should be made as simple as possible, but not simpler'. With this in mind we can start assessing the measurement problem. In this process we need information about the process in question and also the performance of different sensing principles that are applicable. Once it has been decided that a radioisotope technique is the answer, we need to consider what type of measurement is best suited: is it for instance transmission or scattering? We also need to decide which energy (source), activity and radiation detector. It is difficult to generalise the design of radioisotope gauges. In this chapter we will give some examples on how this can be done.

8.1 ELECTRONIC DATA

We have included a limited amount of data in the appendices of this book simply because there is a large amount of nuclear and atomic data available on the Internet. These data may be accessed on-line or by downloading software for installation on personal computers. At the Web site of this book (www.wileyeurope.com/go/radioisotope) there are some links to sites with useful information. The kind of information required is typically stopping cross sections, nuclide data like that listed in Appendix A.2. The latter is user friendly, which can be accessed either by searching for particular nuclides or through hyperlinks with tables, the map of nuclides (as in Figure 2.1), or the periodic system as user interface. All the information presented in Appendix A.2 is available, also in greater detail, and nuclear decay schemes as in Figure 2.2 may be readily looked up. One may also sort data in a variety of ways, for instance by radiation energy.

Furthermore, we may find detailed information on most materials such as process constituents, vessel walls and radiation windows, radiation detectors, radiation shields

Radioisotope Gauges for Industrial Process Measurements. Geir Anton Johansen and Peter Jackson.
 ISBN 0-471-48999-9

and collimators. Some manufacturers have excellent web sites with technical papers, conference presentations, application notes, etc. So do various organisations and bodies. One has to take the usual precautions whenever taking information from the Internet: It is not necessarily quality assured, and it may be biased in favour of the Web site owner.

8.2 RATIONALE FOR USING RADIOISOTOPE SOURCES

When deciding whether or not to use ionising radiation there are three steps involved: justification, optimisation and constraint. Justification requires that a risk assessment demonstrates that the risk is low, optimisation requires that the design complies with the ALARA (as low as reasonably achievable) principle and constraint is the requirement to operate within the regulations.

8.2.1 Justification

Justification for the use of radioactive material depends on the balance of risk and benefit. A risk–benefit analysis must be carried out for any new application; this involves predicting the outcome of all foreseeable possible circumstances, however unlikely, and assessing their probability. As with any human activity, high risk can be tolerated only if it is associated with extremely low probability and on the other hand only extremely low risk can be tolerated with a high probability. A few examples of justifiable and unjustifiable use of radioactive materials follow.

1. Good justification
   - Smoke alarm: very low risk, very high benefit.
   - Medical: moderate risk, high benefit.
   - Industrial gauges: low risk, high benefit.
   - Industrial radiography: low risk, high benefit.
   - Nuclear electricity generating plant (high hazard, low probability): very high benefit.
2. Debatable justification
   - Depleted uranium armour piercing shells
   - Luminous instrument dials
   - Routine chest X-rays
3. Worst justification (all low benefit and discontinued)
   - Use of beta lights with tritium to illuminate telephone dials.
   - Use of beta lights on fishing floats.
   - Use of alpha emitters to enhance the performance of lightning conductors.

The outcome of a risk–benefit analysis may very well be that a non-nucleonic measurement principle should be used. In general, if such a principle may be applied with similar performance with respect to measurement results and accuracy, it should be used. This is actually the first step of the ALARA principle.

8.2.2 ALARA

As low as reasonably achievable. Every application should be designed to minimise the quantities of radioactive material and to minimise the risk. Dose rates should be reduced to the lowest reasonably achievable firstly by design of shielding and secondly by good systems of work.

8.2.3 Constraint

Even when the risk assessment shows that the risk is acceptable and the exposures are as low as possible we must still operate within the relevant legislation.

8.3 DENSITY GAUGE DESIGN

In this section we will go through the different components and design considerations of a single energy density gauge step-by-step. The gauge and its components are shown in Figure 7.1.

8.3.1 Background Information

The design example is at a gas terminal where a 36 in. pipe comes ashore from an offshore gas field. The pipe carries natural gas and light condensate under normal circumstances but is flushed with water and dried with glycol periodically. The pipeline is not perfectly horizontal and the section under examination is just downstream of a hump that acts as the entrance to a liquid slug catcher. The liquids in the pipeline build up upstream of the hump until the pipe is full bore liquid whereupon the liquid slug is forced over the hump and into a long, slightly falling section of line where liquid and gas separate horizontally. The density gauge is required to measure the level of condensate in the separation length in order to control condensate pumps and to distinguish between water and glycol during dewatering routines.

The pipeline is a 36 in. (914.4 mm OD) pipe, which has a bore of 35 in. (889 mm) and a wall thickness of 0.5 in. (12.7 mm) of steel. The surface temperature of the line is ambient so the pipe is not insulated. A hot pipeline with insulation may require the detector to be mounted outside of the insulation to keep the detector below about 60°C and would effectively increase the source–detector separation. From a cross-sectional drawing the source–detector separation can be measured and in this case is 112 cm. Table 8.1 summarises the information required when customising a density gauge installation.

8.3.2 Choice of Isotope

For maximum sensitivity we require the energy to be as low as possible thus providing maximum absorption for minimum material. This applies both to the useful beam used for

Table 8.1 Information required when customising a density gauge installation

Pipe diameter	914.4 mm
Wall thickness	12.7 mm
Wall material	Steel
Insulation requirements	None
Wall temperature	Ambient
Pipe contents	Gas (methane), condensate, glycol and water
Flow regime (is pipe full?)	Horizontal stratification with condensate and gas. Full bore with water and glycol
Pipe orientation	Horizontal
Density range of interest	0–1.13 g/cm^3
Required response time for control purposes	1 and 10 s
Output required, 4–20 mA, RS485, Modbus, Hart, etc	4–20 mA
Ambient temperature range	−10 to 40°C
Site location (for licensing and local rules)	England
On site area location (for type of electrical safety certification)	Zone 0

Table 8.2 Isotope information

Radioisotope	^{241}Am	^{137}Cs	^{60}Co
Mass attenuation coefficient of water [cm^2/g]	0.227	0.089	0.063
Mass attenuation coefficient of steel [cm^2/g]	1.93	0.075	0.053
Specific γ-ray constant at 1 m and 1 MBq [μSv/h]	0.013	0.096	0.331

Table 8.3 Process information

Pipe inside diameter	88.9 cm
Total steel thickness	2.5 cm
Density of condensate	0.71 g/cm^3
Density of water	1 g/cm^3
Density of 50% glycol/water	1.07 g/cm^3
Density of glycol	1.13 g/cm^3
Density of steel	7.8 g/cm^3

the measurement and the scattered or extraneous radiation, which is more easily reduced at low energies. For greatest statistical accuracy we need the attenuation to be around the 86% level as described in Section 5.3.4, but as we will see this ideal situation is not always achievable. The practical choice of isotope is limited by the half-life and commercial availability to ^{241}Am, ^{137}Cs and ^{60}Co, see Table 8.2.

We can now use Equation 3.11 to calculate I/I_0, the transmission through the pipe for the range of conditions we expect to measure, see Table 8.3. Because both the source and detector are well collimated build up is close to unity and is assumed to be unity in the transmission calculations. The results are tabulated in Table 8.4.

Table 8.4 Transmission through the pipe at different conditions

Source	Empty pipe	Full of water	Full of 50% glycol/water	Full of glycol	Full of condensate	Water/glycol change [%]	Empty/con change [%]
^{241}Am	4.5×10^{-15}						
^{137}Cs	23	0.0085	0.0049	0.0034	0.3633	60	98.4
^{60}Co	36	0.1312	0.0887	0.0634	1.8751	51.6	94.7

First for ^{241}Am, considering the walls only i.e. the empty pipe condition, we can calculate the transmission to be 4.5×10^{-15}, which means not much gets through at all, and we can eliminate ^{241}Am from any further consideration. This does of course demonstrate how sensitive a gauge can be when ^{241}Am can be used, say on a small bore thin walled (about 3 mm steel maximum) pipe. Then for ^{137}Cs, considering the pipe walls only, we calculate that transmission is about 23%. That's better, with 23% transmission through the empty pipe; the gauge should be ideal for measuring a small condensate level in the bottom of the pipe. Now consider the pipe full of water using ^{137}Cs we get 0.0085% transmission. And finally for ^{60}Co through a pipe full of water we calculate that we will get 0.13% transmission.

The absorption for ^{137}Cs is acceptable if the gauge were required only to detect the onset of the liquid and measure a low level in the gas filled pipe. ^{60}Co is ideal when the pipe is full and will be excellent for measuring the glycol/water interface. Because we are intending the gauge to be dual purpose we must decide which of the two measurements is the most critical and then see if the gauge will still perform adequately in the other role. The process operator requires that the gauge detects the condensate level with a maximum time constant of a second because the condensate slugs arrive fast, whilst the glycol water interface will be moving more slowly and furthermore the interface will be quite diffuse so a few seconds would be an acceptable response time.

To decide whether to use ^{137}Cs or ^{60}Co we need to calculate the change we will see for both with a change from a pipe full of water to a pipe full of glycol. For ^{137}Cs the change from a pipe full of water to one full of glycol is from 0.0085% transmission to 0.0034% transmission, a signal change of 60%. For ^{60}Co the change from a pipe full of water to a pipe full of glycol is from 0.13% transmission to 0.063% transmission, a signal change of 52%. As expected the ^{137}Cs source gives a better signal change than ^{60}Co but the difference is not enough to give a significant performance enhancement. The most significant difference between the two is the actual magnitude of the signal through a full pipe.

8.3.3 Source Activity Consideration

Since it is still not obvious whether to use ^{137}Cs or ^{60}Co, we will calculate the source strength for both. Using the inverse square law (Section 3.2.1) and the absorption from the steel of the pipe wall and the pipe contents we can calculate the source strength required. The dose rate at the detector should ideally be within the limit laid down in the local legislation without requiring extra barriers to prevent access, which in this case is

7.5 μSv/h. The dose rate at 1 m from 1 MBq is given by the specific γ-ray constant (see Appendix A.2 and Section 6.2.6). For ^{137}Cs this is 0.096 μSv/MBqh, so to get 7.5 μSv/h at 1 M we require 7.5/0.09 = 83 MBq, this becomes 105 MBq to give 7.5 μSv/h at 112 cm (inversely proportional to the square of the distance). Now if we put the pipe in the beam we reduce the transmission to 23% so the source needs to increase to 105/0.23 = 454 MBq to maintain the dose at the detector at 7.5 μSv/h. Repeating this procedure for ^{60}Co we calculate that we need 73 MBq to give 7.5 μSv/h at the detector when the pipe is empty. This will give a count-rate on the detector, which is a 2 in. NaI(Tl) scintillation counter, of about 3000 c/s. The dose rate when the pipe is full of glycol and water can be calculated the same way using the factors from the table above and gives the following dose rates on the detector.

For ^{137}Cs, 0.00276 μSv/h full of water and 0.00098 μSv/h for a pipe full of glycol. For ^{60}Co, 0.02775 μSv/h full of water and 0.01345 μSv/h full of glycol. This dose rate translates to only 0.4 c/s minimum for ^{137}Cs and 5.4 c/s with ^{60}Co. Neither of these is high enough to measure the density of the full pipe with a high accuracy but the requirement is not for high accuracy but rather to detect the interface at some preset density (say 10% glycol water mix) and then divert the stream to the glycol recovery plant. We can increase the source strength somewhat to improve the performance a little but we soon come up against the statutory constraint in Section 8.2.3. The largest practical source holder, which can sensibly be mounted on top of the 36 in. pipe, will hold a maximum of 111 GBq (3 Ci) of ^{137}Cs and 7.4 GBq (200 mCi) of ^{60}Co before exceeding the allowable dose on the source container. So we can increase the counts on the detector through a full pipe to 98 c/s with ^{137}Cs and 547 counts/s with cobalt by using the largest source possible within the engineering restraints. Unfortunately, here we hit another obstacle: the detector will only operate up to a count-rate of about 100 thousand c/s before the dead time becomes too high. If we use the 111 Gbq ^{137}Cs source the detector count-rate on the empty pipe will be in excess of 700 thousand c/s, which is too high. So now we have a new source activity limit of about 15.8 GBq (427 mCi). The source manufacturers catalogue soon reveals another limitation that makes all our careful calculations seem a little too precise, the sources come in a range of fixed sizes; 11.1 GBq (300 mCi), 18.5 GBq (500 mCi) and 37 GBq (1000 mCi) are available in the region we are interested in, the 18.5 GBq being the closest to our requirement but the ALARA philosophy demands that we try the smaller 11.1 GBq source. It is now becoming clear that on a pipe of this size there is not much to choose between the performance that can be achieved with either caesium or cobalt. The main reason we would choose ^{60}Co would be if we needed more counts on the empty pipe than could be achieved with the maximum ^{137}Cs source, which is patently not the case here. In this instance ^{137}Cs will be the best choice for the primary measurement and will be adequate for the interface detection, it also has the advantage of a longer half-life.

8.3.4 Accuracy

Now that we have decided ^{137}Cs is marginally preferable we can calculate the accuracy of the gauge for the primary measurement, which is the measurement of the level of condensate with a time constant of 1 s and a source strength of 18.5 GBq. And then for

Table 8.5 The standard deviation in measured density for various conditions

Condition	Standard deviation [g/cm^3]
Empty pipe 1-s time constant	±0.00047
Pipe full of condensate 1-s time constant	±0.0037
Pipe full of water 1-s time constant	±0.0243
Pipe full of glycol 1-s time constant	±0.0384
Pipe full of water 10-s time constant	±0.0077
Pipe full of glycol 10-s time constant	±0.0121

the secondary measurement of the water–glycol interface with a 10 s time constant. The error in density is given by Equation (8.1) (see Section 5.3.4).

$$\sigma_\delta = \frac{1}{\mu_M x \sqrt{I \tau_I}} \tag{8.1}$$

This gives the standard deviations in density, calculated for the various components, which are in Table 8.5.

The gauge can now be configured to suit the process operation requirements. Either the output can be arranged with a software variable time constant that is lengthened as the count-rate falls (see Section 5.4.13) or alternatively the gauge can have two separate outputs ranged and indicating independently with two different time constants appearing to the process operator as two distinct instruments. Choosing the second option, the first gauge, which is to trip the condensate pumps, can have density ranging from 0 to 0.71 g/cm^3 and is used as a condensate level gauge for the downstream slug catcher with a trip point set at any convenient point in its range. The second output is ranged from 1 to 1.13 g/cm^3 and can be displayed as the concentration of glycol in water; this output can be set to reliably trip diversion to the glycol recovery unit when the concentration of glycol in water at the gauge reaches 50% glycol in water. Since the diversion to the glycol recovery unit is some distance downstream from the gauge, diversion can be completed before even a low concentration of glycol in water reaches it.

8.3.5 The Shielded Source Holder

The size of the source holder will of course depend on the isotope, the activity and the statutory constraints on dose rates, but all source holders should have the following common features: *Shielding* that is sufficient to ensure that the statutory dose rates are not exceeded (see Section 6.2.6). In our design example with an 11.1 GBq ^{137}Cs source 10 cm of lead would be required for shielding to give a surface dose rate on the source holder of 7.5 μSv/h. The shield has a slot or hole called a *collimator* that directs the useful beam to the detector and prevents the source beaming where it is not wanted. A *shutter* that can interrupt the useful beam; this is sometimes considered unnecessary when no possibility exists for any part of any person to access the beam. Installations without shutters are allowed with a suitably convincing risk assessment. The standards recommend the inclusion of a shutter in all cases. The shutter should be clearly labelled 'open' and 'shut'. A *locking mechanism* or mechanisms that allow the shutter to be locked in the closed position (but not in the open position) and prevents the source from being removed by unauthorised personnel. An *arming rod* that is a removable component containing the source capsule; this allows the source to be removed easily for storage during installation of the source holder or during maintenance. The holder should be *permanently labelled* with the trefoil symbol, the words radioactive or radioactive material and a fireproof notice with the source details, isotope, activity and date measured, and a serial number. The source container should survive a *fire* with the shielding intact. Ideally the beam should be shut off as a result of the fire. In the holder shown here (Figure 7.1) the molten lead shuts off the collimator slot.

8.3.6 The Detector

For ^{137}Cs a 2 in. scintillation counter with a 2 in. × 2 in. NaI(Tl) crystal is a good detector choice. The detector runs on a 12 V DC supply, the high voltage (approx. 800 V) for the photomultiplier tube being generated on board. The detector case is certified explosion proof for use in hazardous areas and incorporates a 20-mm diameter collimator in front of the crystal and a shield around the crystal. The output from the photomultiplier tube is amplified and pulses over a preset threshold are shaped into 12 V square pulses that are output to a computing unit where density or concentration is calculated. In the computing unit a range of calculation algorithms are available and various alarms and output formats can be set.

8.3.7 Radiological Considerations

A common problem with large systems that may be full of liquid or empty is exposed in this application; in order to detect in 10 s, with reasonable accuracy, the difference between a pipe full of water and a pipe full of glycol the dose rate on the detector when the pipe is empty will have to be higher than that which the maximum legal dose rate allows. In fact if we use the maximum practical source we will have a dose rate on the detector when the pipe is empty of about 250 times that which is allowed under the state regulations, whilst using the 11.1 GBq source we will have about 25 times the recommended dose. This problem is partly overcome in this example as the system is to beam vertically through the pipe and the source will be placed on the top of the pipe so the radiation beam is directed into the ground directly below the pipe where there is no possibility of a person having sufficient access to receive a whole body dose in excess of the statutory limit (see Figure 8.1). Even given this fortunate arrangement it is possible that scatter from the ground below the pipe would produce more than 7.5 μSv/h in the vicinity of the pipe that is accessible and it would therefore be wise and within the ALARA philosophy to reduce the source strength to the level where the secondary measurement requirement is just achieved, i.e., there is no justification in exceeding the specification. In other circumstances where the arrangement is not so convenient extra shielding at the detector side or barriers would be required to restrict access to the high-dose area around the detector.

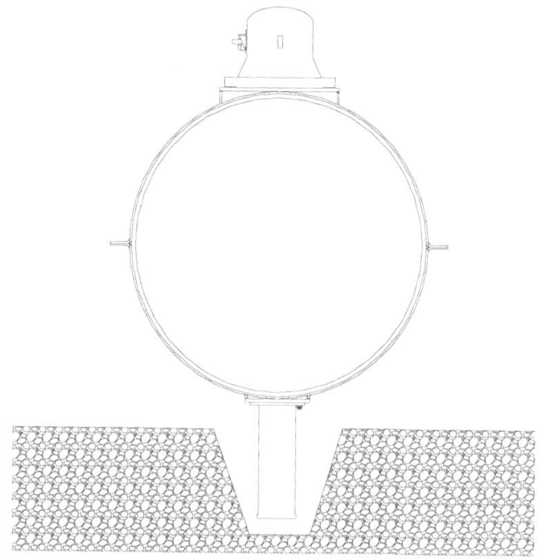

Figure 8.1 Arrangement of density gauge on the 36 in. gas pipeline

8.3.8 Installation and Handover to the Operator

Before the source is installed the supplier must confirm that the site licence to use the source is issued. It is a good idea to perform a wipe test on the source before installation because this may be the last opportunity for a long time to perform the leak test on the actual source capsule rather than on the outside of the source holder. Upon completion of the installation a source transfer receipt should be completed confirming that the new owner of the source has accepted responsibility for the installation. The supplier should demonstrate that the dose rates around the installation are as they should be and ideally should supply an isodose curve for the installation.

Note that this example is deliberately chosen for its complexity caused by the various conflicts produced by the desire to use the gauge for two dissimilar measurements and the limitations imposed by the equipment and the regulations.

8.4 DUAL ENERGY DENSITY GAUGE

Our next example is a dual energy density gauge to be used for three-component fraction measurement on a pipeline as described in Section 7.2.3.

8.4.1 The Dual Energy Shielded Source Holder

The dual energy density gauge may use one source to emit both energies from a single source holder or may use two separate sources in either the same source holder or in separate source holders. The source holder for a dual energy density gauge must fulfil the basic criteria of Section 8.3.5, but in addition the source holder must be capable of allowing the very low energy to radiate. This means that careful consideration must be given to the materials of construction that intrude into the source beam. It is not acceptable to place the source in the holder with only the thin window (see Section 2.2.4 on source construction) to protect the source from mechanical damage, the source must be protected yet the material used must be transparent to the low energy, and radiation damage resistant materials such as ceramics or polyetheretherketon are used for this purpose. Plastic materials containing chlorine, such as PVC, should not be used in close proximity to the source as radiolysis (chemical decomposition caused by the radiation) of the plastic can produce corrosive free radicals, which may damage the source. The need to produce a fireproof source container whilst not compulsory is still desirable. With source holders designed to emit very low energies this requirement is difficult to realize but can be achieved using ceramic windows.

8.4.2 Dual Energy Detector

The detector used in a dual energy density gauge must create a compromise between the detection of the two energies. The detector should detect the two energies with similar efficiencies to provide similar random errors for both measurements. This can be achieved of course by using two separate detectors, one for the low energy and one for the high energy or by using a thin scintillation crystal, the thickness of which is chosen to provide a balance between the detection of the two energies. Where a single detector is used, the output from the two regions of interest of the spectrum is used to calculate the absorption for both energies.

8.4.3 Dual Energy Design Considerations

The requirement to use a low-energy radiation for the measurement dominates the design of the dual energy density gauge. The pipe walls (or windows in the walls) must be made from some radiation transparent material such as a ceramic or a polymeric material, which if they are small can be engineered to withstand very high process pressures. The path length in the fluid must be restricted to ensure that the low energy is not completely absorbed when the pipe is full. This tends to restrict the path length to about 10 cm of liquid for radiation whose absorption is dominated by the photoelectric effect. The path length can be reduced for very low energies such as the 26 keV from ^{241}Am by placing the source in the centre of the pipe or by inserting the source or detector into the pipe in wells, but this detracts from the non-invasive nature of radioactive measurements. Another requirement is that the fluids must be homogenously mixed as they pass through the beam, again this produces flow interruption with the resultant pressure drop caused by the mixing device, which should be minimised.

8.4.4 Calibration

For an accurate calibration of the single energy density gauge, measurements should be made on an empty pipe and a pipe full of known absorber, preferably the material or

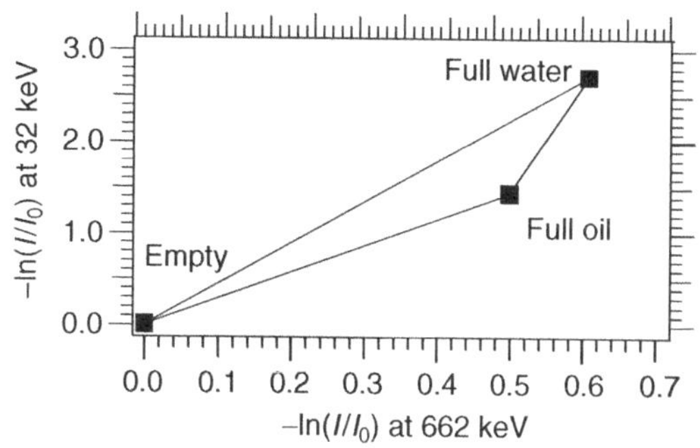

Figure 8.2 All points within the triangle represent possible oil, water and gas fractions. The area of the triangle is indicative of the precision with which the phase fractions can be measured. If for instance a higher low energy radiation were used, such as a 59.5 keV from ^{241}Am instead of the 32 keV X-Ray from ^{137}Cs, then the triangle would be narrower and would encompass a smaller area

materials that are to be measured. Measurements should take account of the background count-rate, which should be subtracted prior to density calculation. It may be difficult to calibrate with oil, water, gas, etc. and substitutes can be used such as polycarbonate cylinders in pre-installation calibrations. If the collimation is sufficient to produce near narrow beam conditions it is a safe assumption that the gauge will give a linear relationship between the logarithm of I/I_0 and density. As a last and non-ideal option the minimum requirement is an empty pipe count and if the density and absorption coefficient is known the pipe full end point of the calibration can be calculated. If the absorption coefficient is not known, it too can be reliably calculated from the chemical composition of the absorber (see Section 3.4.2).

For multi-component calibration, the requirements are the same as for single-phase measurement although the scope for accumulation of errors is greater and calibration using the correct fluids is even more desirable. The graph shows the operating envelope of a three-phase densitometer calibrated using actual oil well fluids.

8.5 MONTE CARLO SIMULATION

Monte Carlo (MC) simulation is a powerful method frequently used in nuclear engineering, radiation dosimetry and protection, radiotheraphy physics and design of radiation measurement systems such as radioisotope gauges. Because of its random nature the transport of radiation through absorbers is a complex process usually impossible to solve analytically. MC methods simulate the random trajectories of individual particles (photons) by using computer generated pseudo-random numbers to sample from the probability distributions governing the physical processes involved. By simulation of a large number of histories, information can be obtained about average values of macroscopic quantities such as energy deposition in predefined volumes, for instance in a radiation detector. Moreover, since one follows individual particle histories, the method can be used to obtain information about the statistical fluctuations of particular kinds of events. It is also possible to use MC simulations to answer questions that cannot be addressed by experimental investigation: The composition of a simulated detection spectrum may for instance be sorted after the origin of the radiation; the effect of build-up is easily determined by removing all events entering the detector with full energy, etc. The availability of high computer power at a relatively low cost also makes MC simulation an attractive alternative to more expensive experimental trials, for instance in the design phase of radioisotope gauges.

There are a variety of MC simulation codes available, all of which have four major components: The geometry definition interface, the cross-section data for all the processes being considered in the simulation, the algorithms used for the radiation transport, and finally the interface for analysis of the information obtained during simulation. The first step in using an MC simulation code is to define the three-dimensional geometry of the system being considered including the composition and density of all materials and components. Some MC software packages have user-friendly graphical interface where the geometry is defined by drawing. Some are all-round codes applicable for any geometry whereas others are restricted to special geometries. The next step basically is to define the radiation type and energy. Some codes are designed for only one type of radiation: γ-rays, neutrons or β-particles, whereas others cover several of these. The input to the simulation code is thus the geometry, radiation type and energy, material composition, and in addition specification of the type of output desired. In some MC simulation packages this is solved by defining all these parameters in a batch file, possibly with a link to a separate geometry file.

The third step is the actual simulation that is executed in a loop where each iteration covers the full history of one event from its emission from the source to the point when all its energy is deposited or carried out of the defined volume of interest by secondary radiation. Each initial photon produces a number of secondary photons or electrons, for instance fluorescence, scatter and annihilation photons, photoelectrons and recoil electrons. The simulation code processes the history of these sequentially, one by one. This is done by placing the parameters of all particles to be processed on a stack. The history of the initial photon is then not terminated until this stack is empty. The flow chart of MC simulation of photon transport is shown in Figure 8.3. The simulation of the transport of secondary electrons is not shown here. A large number of photon histories are used to reduce the statistical error; however, the random number generator period in the computer determines the upper limit of histories. If this is exceeded, the succeeding number series will have some correlation to the previous series, meaning it will not be one long sequence of true random numbers.

The last step in the simulation procedure is normally to produce some output file to be analysed in another software package. Alternatively, some MC simulation packages have built in some analysis features, in some cases also the possibility to produce a graphical plot of the particle trajectories (see Figure 8.4). The latter is very useful for educational purposes; however, in design of measurement systems, dosimetry, etc. the energy dissipation in one or several volumes within the predefined geometry is the principal information required. A typical example is the energy distribution spectrum in the active volume of a simulated radiation detector. This may be used to determine the detector's full energy or full spectrum stopping efficiency or the influence of scattered radiation, etc. This will be a noise-free spectrum, which simplifies some types of analysis. Moreover, noise can be added to such a detection spectrum to simulate a real detection spectrum.

MC simulation can be very time consuming because for each event there is a large number of interactions to be simulated before the total energy of all secondary particles is deposited and the history of the primary event can be terminated. Very often it is possible

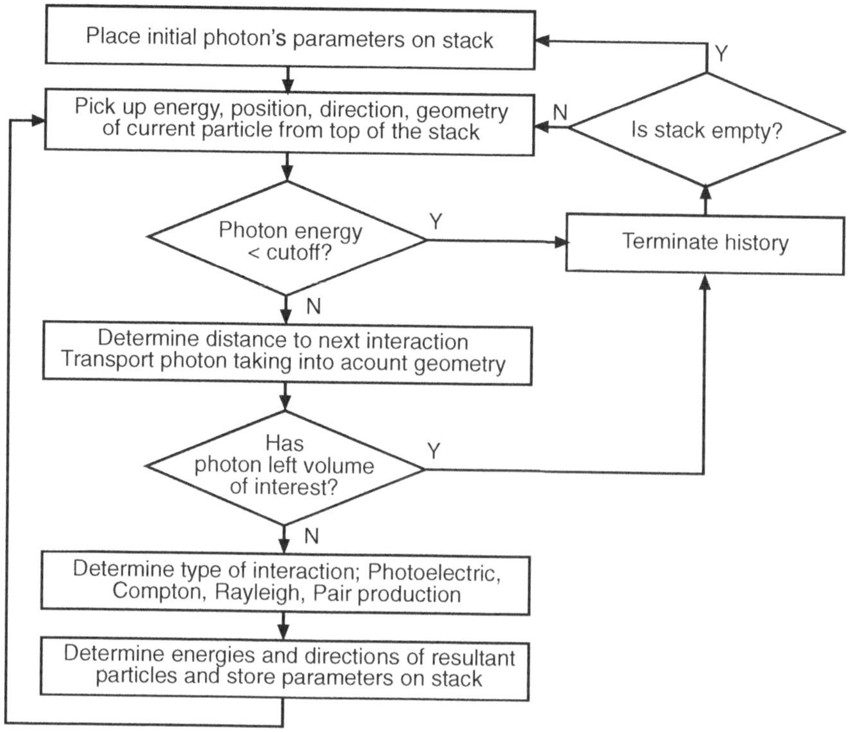

Figure 8.3 Simplified logic flow of a Monte Carlo simulation of photon transport without the associated secondary electron transport [220]

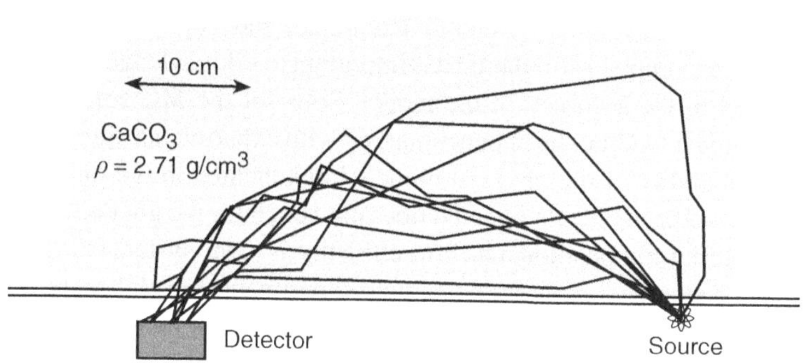

Figure 8.4 Example of 2D trajectory plot of 10 photons from a ^{137}Cs source (661.6 keV) [221]. This is a scatter gauge for density measurement, and only photons arriving in the far detector are included in the plot

to reduce the simulation time dramatically be defining a cut-off energy below which the energy of the particle or photon is deposited on the spot of interaction. To reduce the error introduced by this the cut-off energy has to be chosen carefully as a compromise between the typical range of the actual particle and the dimensions of the defined geometry. This approach is particularly efficient for electrons below a few tens of keV because these have short range, but often are generated in large numbers and undergoes a large number of scatter interactions taking up a lot of computing time. The simulation time may be further reduced by using a variety of *variance reduction* techniques.

The variance is the square of the standard deviation expressing the error caused by the random nature of the radiation emission (see Section 5.3.2). Variance reduction thus means that the statistical measurement (simulation) uncertainty for a particular problem is reduced for a given number of simulated events. The simulation time required to obtain a certain measurement uncertainty is consequently reduced, potentially by several orders of magnitude in some cases. Variance reduction is thus a powerful technique whereby design costs may be further reduced; however, applying it requires some insight to the simulation process so that errors are not introduced. Again we can quote Albert Einstein: 'Everything should be made as simple as possible, but not simpler'. Basically, variance reduction is achieved by avoiding spending simulation time on processes not relevant for the system in question. We, therefore, need to be able to judge which techniques can be applied and to what extent. For example, in many cases it is not necessary to simulate all interactions taking place in a radiation detector: The principal issue often is to know the energy distribution of the radiation entering the detector. If so, the simulation of events may be terminated once they enter the detector surface. Based on this information it is then also possible to apply *detector response functions* to produce a detection spectrum for the actual radiation detector [222, 223]. A review and discussion of MC simulation for industrial radiation and radioisotope measurement applications are given in reference [224]. This also includes variance reduction techniques and examples on applications.

As with any other computer simulation it is important to acknowledge that the simulation result is restricted to the accuracy of the model. Some of the MC simulation packages have to some degree or another been benchmarked with experiments. Such benchmarking guarantees the reliability of the physical models implemented in the radiation transport algorithms being used. On the other hand it does not vouch for the goodness of the defined geometry and material composition, i.e., how close this is to the system being simulated. A commonly applied strategy is to carry out simple experiments to validate the model. Then extensive simulations can be performed with different values of input parameters, such as one or a few dimensions in the geometry. This is very often less expensive than doing it by experiments at the same time as it also provides more information and possibilities. Once the desired design or geometry is reached another experimental validation is carried out.

Appendix A: Data

A.1 CONSTANTS

Avogadro's number: the number of molecules in the molecular weight, expressed in grams: $N_A = 6.022045 \times 10^{23}$ mol^{-1}. This value is the number of atoms in 1 mole of any element. For any element the mass of 1 mole (6.022045×10^{23} atoms) given in grams, equals the value of A for that particular element. For lead, for instance, $A = 207.19u$ so that the mass of 1 mole of lead atoms is 207.19 g. Avogadro's number is thus the reciprocal of the unified atomic mass constant.
Unified atomic mass constant: $u = 1.66053873 \times 10^{-27}$ kg, which previously was called the atomic mass unit (a.m.u.)
Speed of light in vacuum: $c = 2.9979992458 \times 10^8$ m/s
Electron (elementary) charge: $e = 1.602176462 \times 10^{-19}$ C
Electron rest mass: $m_e = 9.1093818872 \times 10^{-31}$ kg
Electron rest mass energy: $m_e c^2 = 510.998902$ keV
Classical electron radius: $r_e = e^2/m_e c^2 = 2.818 \times 10^{-13}$ cm
Proton mass: $m_p = 1.6726231 \times 10^{-27}$ kg
Neutron mass: $m_n = 1.6749286 \times 10^{-27}$ kg
Faraday's constant: $F = 96485.309$ C/mol
Planck's constant: $h = 6.62606876 \times 10^{-34}$ Js
Boltzmann constant: $k = 1.3806503 \times 10^{-23}$ J/K $= 0.8617 \times 10^{-4}$ eV/K
Permittivity of vacuum: $\varepsilon_0 = 8.854187818 \times 10^{-12}$ F/m
Standard acceleration of gravity: $g = 9.80665$ m/s^2

A.2 NUCLIDE INDEX

A selection of commonly used radioisotopes is included in the nuclide index in Table A.1. For electromagnetic emissions the average number of photons emitted relative to one disintegration is given. Some isotopes have competitive modes of disintegration not included here. γ-rays and characteristic X-rays originate from the daughter nuclides, which are shown in parentheses in the leftmost column, but are normally accessed from the parents or listed as properties of these (see Section 2.1.9). The decay time code is y = years, d = days, h = hours and min = minutes. The listed β-emission energies are the maximum energies (see Section 2.1.2). The right most column lists the specific γ-ray dose rate constant (SGRDC) Γ, defined as the absorbed dose rate from a point source with activity $A = 10^6$ Bq (≈27 μCi) at distance $d = 1$ m (see Section 6.2.6) [137].

Radioisotope Gauges for Industrial Process Measurements. Geir Anton Johansen and Peter Jackson.

Table A.1 Nuclide index of commonly used radioisotope sources [4, 225]

Nuclide	Decay type/ Half-life $T_{1/2}$	Particle emissions energy [keV]	Transition probability [%]	Electromagnetic emissions		SGRDC Γ at 1 m and 1 MBq [mSv/h]
				Energy [keV]	Photons emitted [%]	
^{3}H (^{3}He)	β^-/12.33 y	18.591	100			
^{14}C (^{14}N)	β^-/5730 y	156.48	100			
^{18}F (^{18}O)	β^+/109.77 min			511 (annihilation)	200[a]	
^{22}Na (^{22}Ne)	$\beta^+$and EC/ 2.6019 y	545.66	99.962	511 (annihilation)	181.424[a]	3.28×10^{-4}
		1820.2	0.038	1274.53	99.945	
^{24}Na (^{24}Mg)	β^-/14.959 h	280.58	0.053	1368.633	100	4.54×10^{-4}
		1392.906	99.944	2754.028	99.944	
				3866.19	0.052	
				Others	<0.002 each	
^{57}Co (^{57}Fe)	EC/271.79 d		100	6–7 (K X-rays)	55	4.97×10^{-5}
				14.41	9.16	
				122.061	85.60	
				136.474	10.68	
				692.03	0.2	
				706.40	0.025	
				Others	<0.02 each	
^{60}Co (^{60}Ni)	β^-/5.2714 y	318.13	99.925	1173.237	99.974	3.31×10^{-4}
		1491.38	0.057	1332.501	99.986	
				Others	<0.002 each	
^{79}Kr (^{79}Br)	$\beta^+$and EC/35.04 h			136.09	0.851	
				208.48	0.775	
				217.07	2.375	
				261.35	12.7	
				299.53	1.537	
				306.47	2.604	
				388.97	1.511	
				397.54	9.335	
				606.09	8.115	

				831.97	1.257	
				1115.1	0.372	
				1332.21	0.431	
				Others	<0.3 each	
^{82}Br (^{82}Kr)	β^- / 35.30 h	264.50	1.300	221.480	2.263	4.00×10^{-4}
		444.24	98.5	554.348	70.725	
				606.34	1.211	
				619.106	43.42	
				698.374	28.474	
				776.517	83.5	
				827.828	24.023	
				1007.59	1.271	
				1044.002	27.221	
				1317.472	26.4695	
				1474.88	16.30755	
				1650.37	0.742315	
				Others	<1 each	
^{85}Kr (^{85}Rb)	β^- / 10.756 y	173.09	0.434	514.007	0.434	3.68×10^{-7}
		687.1	99.563			
^{90}Sr	β^- / 28.78 y 64.1 h	546.0	100			
(^{90}Y)		519.4	0.0115			
(^{90}Zr)		2280.1	99.9885			
^{113m}In (^{113}In)	IT/1.6582 h			391.690	64.2	5.00×10^{-5}
^{133}Ba (^{133}Cs)	EC/10.51 y		100	30 (K X-rays)	} 120	8.74×10^{-5}
				36 (K X-rays)		
				53.161	2.199	
				79.6139	2.62	
				81.9971	34.06	
				160.613	0.645	
				223.234	0.450	
				276.398	7.164	
				302.853	18.33	
				356.017	62.05	
				383.851	8.94	

(*cont.*)

Table A.1 Nuclide index of commonly used radioisotope sources [4, 226]

Nuclide	Decay type/ Half-life $T_{1/2}$	Particle emissions energy [keV]	Transition probability [%]	Electromagnetic emissions		SGRDC Γ at 1m and 1MBq [mSv/h]
				Energy [keV]	Photons emitted [%]	
^{137}Cs (^{137m}Ba)	β^- / 30.07 y	513.97	94.4	32.1 (K_α X-rays)	6.3	9.63×10^{-5}
	2.552 min	1175.63	5.8	36.4 ($K_{\beta 1}$ X-rays)	1.2	
				37.3 ($K_{\beta 2}$ X-rays)	0.3	
				661.66 (via ^{137m}Ba)	85.1	
^{140}La (^{140}Ce)	β^- / 40.2744 h	1240.47	10.9	131.117	0.468	3.18×10^{-4}
		1246.14	5.68	173.543	0.127	
		1280.98	1.07	241.933	0.414	
		1297.82	5.45	266.543	0.466	
		1349.89	44	328.762	20.320	
		1414.02	4.93	432.493	2.900	
		1678.65	19.2	487.021	45.506	
		2165.67	4.8	751.637	4.331	
		Others	Low	815.772	23.278	
				867.846	5.505	
				919.550	2.662	
				925.189	6.897	
				950.987	0.519	
				1596.210	95.4	
				2347.88	0.849	
				2521.40	3.463	
				Others	<0.1 each	
^{147}Pm (^{147}Sm)	β^- / 2.6234 y	102.88	0.0057	121.22	0.0028	6.79×10^{-10}
		224.1	99.994			
^{192}Ir (^{192}Pt)	β^- / 73.831 d	258.65	5.605	136.344	0.183	1.39×10^{-4}
		538.78	41.76	295.958	28.669	
		675.12	48.03	308.457	30.002	
				316.508	82.81	
				416.471	0.664	

				468.072	47.83	
				588.585	5.148	
				604.415	8.231	
				612.466	5.309	
				Others	<0.1 each	
^{204}Tl (^{204}Pb)	β^- and EC/ 3.78 y	763.72	97.1			
^{241}Am (^{237}Np)	α/ 432.2 y	5388.23	1.6	11.9 (L_L X-rays)	0.85	1.34×10^{-5}
		5442.80	13.0	13.9 (L_α X-rays)	13.3	
		5485.86	84.5	17.8 (L_β X-rays)	19.3	
		5511.47	0.22	20.8 (L_γ X-rays)	4.9	
		5544.5	0.34	26.345	2.40	
		Others	<0.04 each	33.196	0.126	
				43.423	0.073	
				59.541	35.9	
				98.97	0.02	
				Others	<0.02 each	

[a] Exceeds 100% because each annihilation yields two 511-keV photons.

A.3 X-RAY FLUORESCENCE DATA

Table A.2 The K-shell binding energy (E_{bK}), the average energy of the K-shell fluorescence X-ray photons (E_K), the K-shell fluorescence yield (ω_{aK}), the density (ρ), the photoelectric linear attenuation coefficient on the top side of E_{bK} ($\mu_{\tau EK}$) for elements between 1 and 100 [17]

Element		Z	E_{bK}[keV]	E_K[keV]	ω_{aK}	ρ [g/cm^3]	$\mu_{\tau EK}$ [cm^{-1}]	$\mu_{\tau EK}\omega_{aK}$ [cm^{-1}]
H	Hydrogen	1	0.014			0.0899		
He	Helium	2	0.025			0.1787		
Li	Lithium	3	0.055	0.054		0.53		
Be	Beryllium	4	0.111	0.109		1.85		
B	Boron	5	0.188	0.184		2.34		
C	Carbon	6	0.284	0.279		2.62		
N	Nitrogen	7	0.4	0.393	0.001	1.251		
O	Oxygen	8	0.533	0.524	0.003	1.429		
F	Fluorine	9	0.687	0.675	0.007	1.696		
Ne	Neon	10	0.867	0.849	0.010	0.901		
Na	Sodium	11	1.073	1.041	0.013	0.97	8487.50	110.34
Mg	Magnesium	12	1.305	1.255	0.020	1.74	11553.60	231.07
Al	Aluminum	13	1.56	1.487	0.030	2.70	13203.00	396.09
Si	Silicon	14	1.839	1.742	0.041	2.33	9087.00	372.57
P	Phosphorus	15	2.144	2.020	0.051	1.82	5405.40	275.68
S	Sulfur	16	2.472	2.317	0.069	2.07	5092.20	351.36
Cl	Chlorine	17	2.824	2.636	0.086	3.17	6095.91	524.25
Ar	Argon	18	3.203	2.977	0.105	1.78	2640.32	277.23
K	Potassium	19	3.607	3.337	0.129	0.86	1135.20	146.44
Ca	Calcium	20	4.037	3.719	0.151	1.55	1767.00	266.82
Sc	Scandium	21	4.491	4.124	0.174	3.00	2709.00	471.37
Ti	Titanium	22	4.966	4.550	0.200	4.50	3415.50	683.10
V	Vanadium	23	5.465	4.998	0.227	5.80	3729.40	846.57
Cr	Chromium	24	5.989	5.467	0.253	7.19	4098.30	1036.87
Mn	Manganese	25	6.539	5.959	0.282	7.43	3640.70	1026.68
Fe	Iron	26	7.112	6.472	0.312	7.86	3458.40	1079.02
Co	Cobalt	27	7.709	7.006	0.341	8.90	3399.80	1159.33
Ni	Nickel	28	8.332	7.563	0.371	8.90	3141.70	1165.57
Cu	Copper	29	8.981	8.142	0.404	8.96	2688.00	1085.95
Zn	Zinc	30	9.659	8.774	0.435	7.14	1927.80	838.59
Ga	Gallium	31	10.367	9.367	0.466	5.91	1382.94	644.45
Ge	Germanium	32	11.104	10.015	0.499	5.32	1106.56	552.17
As	Arsenic	33	11.867	10.687	0.531	5.72	1075.36	571.02
Se	Selenium	34	12.658	11.380	0.567	4.80	796.80	451.79
Br	Bromine	35	13.474	12.096	0.595	3.12	477.36	284.03
Kr	Krypton	36	14.323	12.837	0.624	3.74	512.38	319.73
Rb	Rubidium	37	15.2	13.602	0.649	1.53	191.25	124.12
Sr	Strontium	38	16.105	14.390	0.672	2.60	299.00	200.93
Y	Yttrium	39	17.038	15.204	0.693	4.50	477.00	330.56
Zr	Zirconium	40	17.998	16.040	0.711	6.49	634.07	450.83
Nb	Niobium	41	18.986	16.900	0.730	8.55	770.36	562.36
Mo	Molybdenum	42	20	17.787	0.747	10.20	845.58	631.65
Tc	Technetium	43	21.044	18.696	0.761	11.50	874.00	665.11
Ru	Ruthenium	44	22.117	19.632	0.775	12.20	857.66	664.69
Rh	Rhodium	45	23.22	20.593	0.789	12.40	813.44	641.80

Table A.2 (*cont.*)

Element		Z	E_{bK}[keV]	E_K[keV]	ω_{aK}	ρ [g/cm^3]	$\mu_{\tau EK}$ [cm^{-1}]	$\mu_{\tau EK}\omega_{aK}$ [cm^{-1}]
Pd	Palladium	46	24.35	21.581	0.801	12.00	720.00	576.72
Ag	Silver	47	25.514	22.592	0.812	10.50	592.20	480.87
Cd	Cadmium	48	26.711	23.630	0.822	8.65	443.75	364.76
In	Indium	49	27.94	24.750	0.834	7.31	349.42	291.41
Sn	Tin	50	29.2	25.843	0.843	7.30	321.20	270.77
Sb	Antimony	51	30.491	26.965	0.852	6.68	273.88	233.35
Te	Tellurium	52	31.814	28.116	0.860	6.24	232.75	200.17
I	Iodine	53	33.17	29.291	0.868	4.92	176.14	152.89
Xe	Xenon	54	34.561	30.491	0.875	5.89	194.37	170.07
Cs	Cesium	55	35.985	31.726	0.882	1.87	58.34	51.46
Ba	Barium	56	37.441	32.988	0.889	3.50	101.15	89.92
La	Lanthanum	57	38.925	34.275	0.894	6.70	183.58	164.12
Ce	Cerium	58	40.443	35.593	0.899	6.78	176.28	158.48
Pr	Praseodymium	59	41.991	36.940	0.903	6.77	167.90	151.61
Nd	Neodymium	60	43.569	38.315	0.909	7.00	163.10	148.26
Pm	Promethium	61	45.184	39.721	0.912	6.48	143.75	131.10
Sm	Samarium	62	46.834	41.161	0.916	7.54	155.32	142.28
Eu	Europium	63	48.519	42.633	0.920	5.26	103.10	94.85
Gd	Gadolinium	64	50.239	44.132	0.922	7.89	143.60	132.40
Tb	Terbium	65	51.996	45.665	0.926	8.27	143.07	132.48
Dy	Dysprosium	66	53.788	47.226	0.929	8.54	139.20	129.32
Ho	Holmium	67	55.618	48.819	0.931	8.80	136.40	126.99
Er	Erbium	68	57.486	50.448	0.934	9.05	133.04	124.25
Tm	Thulium	69	59.39	52.108	0.937	9.33	130.62	122.39
Yb	Ytterbium	70	61.332	53.802	0.939	6.98	92.14	86.52
Lu	Lutetium	71	63.316	55.534	0.941	9.84	123.98	116.67
Hf	Hafnium	72	65.345	57.296	0.942	13.10	155.89	146.85
Ta	Tantalum	73	67.416	59.099	0.944	16.60	187.58	177.08
W	Tungsten	74	69.525	60.936	0.946	19.30	208.44	197.18
Re	Rhenium	75	71.676	62.807	0.948	21.00	216.30	205.05
Os	Osmium	76	73.871	64.718	0.949	22.40	217.73	206.62
Ir	Iridium	77	76.111	66.668	0.950	22.50	209.48	199.00
Pt	Platinum	78	78.395	68.655	0.951	21.40	189.60	180.31
Au	Gold	79	80.725	70.685	0.952	19.30	163.47	155.62
Hg	Mercury	80	83.102	72.747	0.953	13.53	108.92	103.80
Tl	Thallium	81	85.53	74.856	0.954	11.85	90.42	86.26
Pb	Lead	82	88.004	77.002	0.955	11.40	83.22	79.48
Bi	Bismuth	83	90.526	79.194	0.956	9.80	68.60	65.58
Po	Polonium	84	93.105	81.426	0.957	9.40	63.36	60.63
At	Astatine	85	95.73	83.701	0.958			
Rn	Radon	86	98.404	86.024	0.958	9.91	59.16	56.68
Fr	Francium	87	101.137	88.396	0.959			
Ra	Radium	88	103.922	90.814	0.960	5.00	27.45	26.35
Ac	Actinium	89	106.759	93.279	0.960	10.07	53.47	51.33
Th	Thorium	90	109.651	95.791	0.960	11.70	58.85	56.50
Pa	Protactinium	91	112.601	98.347	0.960	15.40	75.46	72.44
U	Uranium	92	115.606	100.964	0.961	18.90	86.94	83.55
Np	Neptunium	93	118.67	103.631	0.962	20.40	91.19	87.72

(*cont.*)

Table A.2 (*cont.*)

Element		Z	E_{bK}[keV]	E_K[keV]	ω_{aK}	ρ [g/cm^3]	$\mu_{\tau EK}$ [cm^{-1}]	$\mu_{\tau EK}\omega_{aK}$ [cm^{-1}]
Pu	Plutonium	94	121.797	106.354	0.962	19.80	84.35	81.14
Am	Americium	95	124.99	109.136	0.963	13.60	55.90	53.83
Cm	Curium	96	128.253	111.980	0.963	13.51	53.10	51.13
Bk	Berkelium	97	131.59	114.891	0.964			
Cf	Californium	98	135.005	117.867	0.964			
Es	Einsteinium	99	138.502	120.922	0.964			
Fm	Fermium	100	142.085	124.054	0.965			

Note: Also shown is the product of the latter and the fluorescence yield. This product expresses the probability of achieving fluorescence.

A.4 PGNNA DATA

Table A.3 Recommended emission lines (E_r) for PGNNA (prompt γ-ray neutron activation analysis) complied by Chung [15] with data from [21]

Element	Z	σ_{th} [barn]	E_r [MeV]	I_r [%]
H	1	0.332	2.2232	100
Li	3	0.036	2.0325	89.33
B[a]	5	755	0.4776	93
N	7	0.0747	10.8293	14.12
S	16	0.520	5.4205	59.08
Cl	17	33.2	6.1109	20.00
Ar	18	0.678	4.7450	55.00
K	19	2.10	0.7703	51.48
Ca	20	0.43	1.9420	72.55
Sc	21	2.65	8.1747	11.83
Ti	22	6.1	1.3815	69.08
V	23	5.04	6.5172	17.83
Cr	24	3.1	8.8841	26.97
Mn	25	13.3	7.2438	12.13
Fe	26	2.55	7.6311	28.51
Co	27	37.2	6.8768	8.21
Ni	28	4.43	8.5334	16.98
Cu	29	3.79	7.6366	15.71
Zn	30	1.10	1.0774	18.93
Ge	32	2.30	0.5964	33.10
As	33	4.30	1.5343	7.18
Br	35	6.8	0.7769	4.13
Kr	36	25	0.8818	84.0
Sr	38	1.21	1.8361	57.79
Y	39	1.28	6.0798	77.49
Mo	42	2.65	0.8488	14.52
Rh	45	150	0.2172	8.71
Ag	47	63.6	0.1984	23.88
Ag	47	63.6	0.1984	23.88
Cd	48	2450	0.5586	72.73
In	49	194	1.2934	17.59
Te	52	4.70	0.6031	9.34
I	53	6.20	0.4429	4.38
Xe	54	24.50	2.2252	12.56
Cs	55	29.0	1.6189	1.50
La	57	9.14	1.5962	15.36
Nd	60	50.5	0.6973	73.29
Sm	62	5800	0.7376	11.04
Eu	63	4600	1.5640	0.40
Gd	64	49000	1.1865	10.83
Dy	66	930	2.7034	2.48
Ho	67	66.5	0.5435	2.97

(*cont.*)

Table A.3 *(cont.)*

Element	Z	σ_{th} [barn]	E_r [MeV]	I_r [%]
Er	68	162	0.8160	40.96
Tm	69	103	0.6874	3.77
Yb	70	36.6	5.2657	5.78
Lu	71	77.3	0.4577	6.95
Hf	72	102	1.2064	4.60
Ta	73	21.1	0.4029	27.18
Os	76	15.3	2.2233	25.02
Ir	77	426	5.6670	1.15
Au	79	98.8	1.2018	12.17
Hg	80	376	0.3681	81.35

Note: For each element the thermal neutron cross section (σ_{th}) for the radiative capture (n, γ) reaction is given alongside the probability (I_r) that there will be an emission at E_r after a neutron capture.

[a] For B, data for the B(n, α) reaction are given.

Appendix B: Formulae Derivation and Examples

B.1 PHOTON ATTENUATION

Lambert–Beer's exponential decay law [Equation (B.3)] of photon attenuation can be derived by considering a very thin layer in a single element absorber at depth x and with thickness dx. Let n be the number of mono-energetic photons entering this layer at depth x, and dn the number of photons removed from the beam in this layer. This loss of photons is proportional to the number of photons available, n; the number of atoms per volume, N; and finally the thickness of the layer:

$$-\mathrm{d}n = \sigma_{\mathrm{TOT}} n N \,\mathrm{d}x \tag{B.1}$$

where σ_{TOT} is the proportionality factor – the *cross section*. Now, since σ_{TOT} is constant and independent of x, integration of this differential equation yields

$$-\int_{n_0}^{n} \frac{\mathrm{d}n}{n} = \int_0^x \sigma_{\mathrm{TOT}} N \,\mathrm{d}x \Rightarrow n = n_0\, e^{-\sigma_{\mathrm{TOT}} N x} \tag{B.2}$$

where n_0 and n are the number of photons before and after attenuation, respectively. These numbers are more conveniently expressed by their corresponding intensities: I_0 and I, the number of photons per second. Moreover, the product $\sigma_{\mathrm{TOT}} N$ may be substituted with μ, *the linear attenuation coefficient*, which expresses the interaction probability per unit length. Consequently

$$n = n_0\, e^{-\sigma_{\mathrm{TOT}} N x} \Rightarrow I = I_0\, e^{-\mu x} \tag{B.3}$$

B.2 COMPTON SCATTERING

B.2.1 Energy Sustained by the Scattered Photon

Equation (3.15) stating the energy of the scattered photon is based on conservation of energy and momentum. Using relativistic kinematics and the assumption that the electron's

Radioisotope Gauges for Industrial Process Measurements. Geir Anton Johansen and Peter Jackson.
 ISBN 0-471-48999-9

binding energy is negligible, energy conservation yields

$$E_{\text{kin}} = E_{\gamma} - E_{\gamma'} = E_{\text{e}} - m_{\text{e}}c^2 \tag{B.4}$$

where E_{γ} and $E_{\gamma'}$ are the energies of the incident and scattered photons, respectively, while E_{kin} and E_{e} are the kinetic and total energies of the electron, respectively. The latter includes the electron's rest mass energy $m_{\text{e}}c^2$, where m_{e} is the electron mass and c is the speed of light in vacuum.

Even though a photon is massless, it does exhibit a momentum equal to E_{γ}/c. The relativistic momentum of the electron, p_{e}, is in this context most conveniently expressed as

$$E_{\text{e}}^2 = p_{\text{e}}^2c^2 + (m_{\text{e}}c^2)^2 \tag{B.5}$$

which, by eliminating E_{e} from Equation (B.4) and some rearranging, may be written as

$$p_{\text{e}}^2c^2 = (E_{\gamma} - E_{\gamma'} + m_{\text{e}}c^2)^2 - (m_{\text{e}}c^2)^2 \tag{B.6}$$

Momentum conservation requires that the momentums $p_{\gamma} = E_{\gamma}/c$, $p_{\gamma'} = E_{\gamma'}/c$ and p_{e} for the electron must be added vectorially to form a closed triangle (see Figure 3.9). Using the cosine rule this means that

$$p_{\text{e}}^2 = p_{\gamma}^2 + p_{\gamma'}^2 - p_{\gamma}p_{\gamma'}2\cos\theta = \frac{1}{c^2}\left(E_{\gamma}^2 + E_{\gamma'}^2 - E_{\gamma}E_{\gamma'}2\cos\theta\right) \tag{B.7}$$

Now, by substituting for the electron momentum given in Equation (B.6), Equation (B.7) takes the form

$$(E_{\gamma} - E_{\gamma'} + m_{\text{e}}c^2)^2 - (m_{\text{e}}c^2)^2 = \left(E_{\gamma}^2 + E_{\gamma'}^2 - E_{\gamma}E_{\gamma'}2\cos\theta\right) \tag{B.8}$$

This can be further developed to

$$E_{\gamma'} = \frac{E_{\gamma}}{1 + (E_{\gamma}/m_{\text{e}}c^2)(1 - \cos\theta)} \tag{B.9}$$

which is Equation (3.15).

B.2.2 The Differential Klein–Nishina Formula

The *Klein–Nishina formula* predicts the Compton interaction cross section using quantum electrodynamics. The differential form of this formula gives the angular distribution of the scattered photons as plotted in Figure 3.11 [16]:

$$\frac{\mathrm{d}\sigma}{\mathrm{d}\Omega} = Z\frac{1}{2}r_{\text{e}}^2\left\{\frac{1}{[1+\alpha(1-\cos\theta)]^2}\left[1+\cos^2\theta + \frac{\alpha^2(1-\cos\theta)^2}{[1+\alpha(1-\cos\theta)]}\right]\right\} \tag{B.10}$$

This expresses the probability of a Compton photon to be scattered into a unit solid angle Ω at scattering angle θ. Here r_{e} is the *classical electron radius* and $\alpha = E_{\gamma}/m_{\text{e}}c^2$, that is,

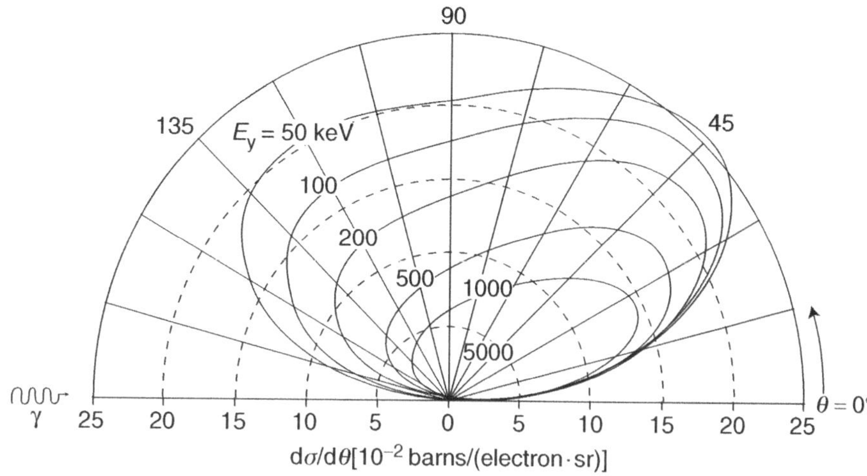

Figure B.1 The angular distribution of Compton scattered photons with γ-photons incident from left, for six incident photon energies ranging from 50 to 5000 keV, as predicted by Equation (B.11) ($Z = 1$). The radius expresses the probability of Compton interaction to take place, and for the photon to be scattered at an angle θ

the ratio between the incident γ-ray energy and the electron rest mass energy. The Klein–Nishina formula is inaccurate at low values of E_γ because of the assumption that the electron's binding energy is negligible. Sometimes it is preferred to view the combined probability of a Compton interaction and the probability of the photon being scattered at an angle θ [226]:

$$\frac{d\sigma}{d\theta} = Z\pi r_e^2 \sin\theta \left\{ \frac{1}{[1+\alpha(1-\cos\theta)]^2} \left[1 + \cos^2\theta + \frac{\alpha^2(1-\cos\theta)^2}{[1+\alpha(1-\cos\theta)]} \right] \right\} \tag{B.11}$$

This is plotted in Figure B.1. The total cross section would be the integral of this function over all angles.

B.2.3 Compton Scattering and Absorption Cross Sections

Two things happen in a Compton interaction: The incident photon is scattered and may escape the absorber without depositing its energy. Secondly, some energy is transferred to the recoiled electron, which because of its relatively short range is assumed to loose all its energy in the absorber. To express the sharing of energy between the scattered photon and the recoil electron, one may regard the Compton cross section to be made up additively by the absorption cross section, σ_a, and the scatter cross section, σ_s, so that $\sigma = \sigma_a + \sigma_s$. These may be expressed using Klein–Nishina theory [16]; however, they may also be defined as

$$\begin{aligned} \sigma_a &= \frac{\overline{E_{kin}}}{E_\gamma}\sigma \\ \sigma_s &= \frac{\overline{E_{\gamma'}}}{E_\gamma}\sigma = \left(1 - \frac{\overline{E_{kin}}}{E_\gamma}\right)\sigma \end{aligned} \tag{B.12}$$

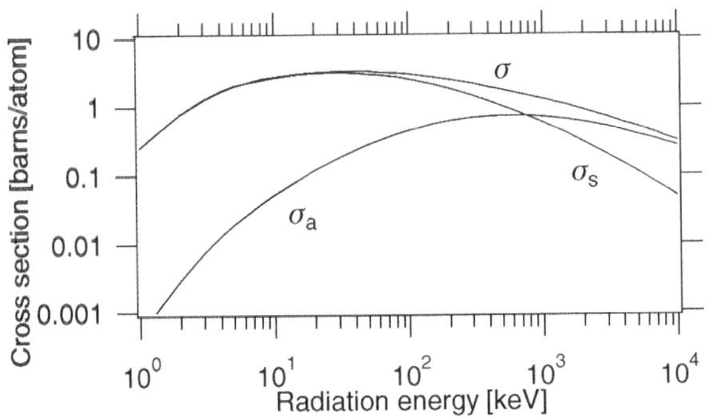

Figure B.2 Compton cross sections for carbon as a function of incident photon energy. The absorption cross section (σ_a) represents energy transferred to the recoil electron, whereas the scatter cross section (σ_s) represents energy retained by the scattered photon. Cross-section data are generated by Reference [12]

where $\overline{E_{kin}}$ and $\overline{E_{\gamma'}}$ are average energies over all scattering angles. In the plot in Figure B.2, which shows the composition of the Compton cross section for carbon, these are found by integrating Equation (B.11).

This confirms what can be seen from Figure 3.10: An increasing amount of energy is transferred to the recoil electron as the incident radiation energy increases. Yet, the scatter cross section is dominant for most energies of interest in an industrial measurement system. Finally, note that σ represents either the fractional number of photons lost from a beam or the fractional amount of energy lost from it, whereas σ_a and σ_s represent only the fractional amount of energy that is lost.

B.3 PHOTOMULTIPLIER TUBE LIFETIME ESTIMATION

The lifetime of a photomultiplier tube in (PMT) a scintillation detector may be estimated with respect to wear of the last dynode. The largest uncertainty in such estimation is the total charge throughput required to reduce the PMT gain by 50%. This value is rarely specified in PMT data sheets, but $q_{life} \approx 200$ C is generally accepted as a realistic estimate. Using Equation (4.5) the charge throughput from one full energy γ-ray interaction in a crystal may be expressed as

$$q_\gamma = \frac{E_\gamma}{w} Ge \quad \text{where } w = \frac{hc/\lambda_{max}}{Q_C Q_E (1 - L)} \quad \text{[Equation (4.5)]} \tag{B.13}$$

where E_γ is the γ-ray energy and G is the gain of the PMT. The lifetime of the tube, here defined as the time it takes to reduce the PMT gain by 50%, may then be expressed as

$$t_{life} = \frac{q_{life}}{n q_\gamma} = \frac{q_{life} w}{n E_\gamma \overline{G} e} \tag{B.14}$$

where n is the count-rate or number of γ-ray photon interactions per second. Considering a standard NaI(Tl) scintillation detector with bialkali photocathode read-out ($w \approx 100$ eV), an average PMT gain of $\overline{G} = 10^6$ and exposure to ^{137}Cs 662-keV γ-photons, the lifetime is about 6 years, for continuous operation at a count-rate of 1 kc/s. This estimation assumes full energy γ-ray interactions only; in reality some of these will be Compton interactions depositing less energy in the crystal, as discussed is Section 4.2. On the other hand there will be some contribution from dark counts, etc.

It is evident from Equation (B.14) that for high count-rate applications (high n), where lifetime often is most critical, it is advantageous to reduce the PMT gain. Likewise, low-quantum-efficiency scintillators, such as plastic scintillators where $w \approx 500$ eV, are ideal. This will of course be on the cost of other parameters such as energy resolution and signal-to-noise ratio (SNR). The conclusion is thus that there is a trade-off between SNR and lifetime.

B.4 STATISTICAL ERRORS IN MEASUREMENT

B.4.1 The Linear Attenuation Coefficient

The linear attenuation coefficient is found from Lambert–Beer's law:

$$I = I_0 \, e^{-\mu x} \tag{B.15}$$

$$\ln\left(\frac{I}{I_0}\right) = \ln(e^{-\mu x}) \tag{B.16}$$

$$\ln\left(\frac{I}{I_0}\right) = -\mu x \tag{B.17}$$

$$\mu = \frac{1}{x} \ln\left(\frac{I_0}{I}\right) \tag{B.18}$$

There are two parameters that influence the error of the linear attenuation coefficient measurement. Firstly, there will be an optimal attenuation value that yields a minimum error: If either a very small fraction of the beam is attenuated, or on the other extreme, if most of the beam is attenuated, the measurement sensitivity will be very low, and the error consequently large. The average attenuation in the measurement volume is determined by the μx-product. Secondly, it is the number of detected events that ultimately is determined by the incident intensity I_0 and the counting time τ_I (this is the time the number of detected events is counted before the intensity is calculated).

The measurement error is caused by the random emission of γ-photons or -particles from the disintegrations in the isotopic source. For large numbers of detected events, $n_C >$ 100, we saw in Section 5.3.4 that the emission is adequately described by the Gaussian distribution, and that its standard deviation is

$$\sigma_{n_C} = \sqrt{n_C} \tag{B.19}$$

In the attenuation measurement, I_0 is measured by counting the number of detected events, n_0, over a time period τ_0, so that $I_0 = n_0/\tau_0$. And likewise, $I = n_C/\tau_I$. Thus

$$\mu = \frac{1}{x} \ln\left(\frac{n_0/\tau_0}{n_C/\tau_I}\right) = \frac{1}{x} \ln\left(\frac{n_0\tau_I}{n_C\tau_0}\right) \tag{B.20}$$

The standard deviations of both n_0 and n_C will hence contribute to the standard deviation of the measured linear attenuation coefficient as

$$\sigma_\mu = \sqrt{\left(\frac{\delta\mu}{\delta n_C}\right)^2 \sigma_{n_C}^2 + \left(\frac{\delta\mu}{\delta n_0}\right)^2 \sigma_{n_0}^2} \tag{B.21}$$

since these may by regarded as uncorrelated noise source. Now

$$\frac{\delta\mu}{\delta n_C} = \frac{\delta\left[\frac{1}{x}\ln\left(\frac{n_0\tau_I}{n\tau_0}\right)\right]}{\delta n_C} = \frac{1}{x}\left(\frac{n_C\tau_0}{n_0\tau_I}\right)\left(\frac{n_0\tau_I}{n_C^2\tau_0}\right) = \frac{1}{xn_C} \tag{B.22}$$

and

$$\frac{\delta\mu}{\delta n_0} = \frac{\delta\left[\frac{1}{x}\ln\left(\frac{n_0\tau_I}{n_C\tau_0}\right)\right]}{\delta n_0} = \frac{1}{x}\left(\frac{n_C\tau_0}{n_0\tau}\right)\left(\frac{\tau_I}{n_C\tau_0}\right) = \frac{1}{xn_0} \tag{B.23}$$

so that

$$\sigma_\mu = \sqrt{\left(\frac{1}{xn_C}\right)^2 n_C + \left(\frac{1}{xn_0}\right)^2 n_0} = \frac{1}{x}\sqrt{\frac{1}{n_C} + \frac{1}{n_0}} \tag{B.24}$$

Now, the standard deviation is more conveniently expressed as a function of the intensity $I_0 = n_0/\tau_0$:

$$\begin{aligned}\sigma_\mu &= \frac{1}{x}\sqrt{\frac{1}{I\tau_I} + \frac{1}{I_0\tau_0}} = \frac{1}{x}\sqrt{\frac{1}{I_0\,e^{-\mu x}\tau_I} + \frac{1}{I_0\tau_0}} \\ &= \frac{1}{x}\sqrt{\frac{1}{I_0}\left(\frac{1}{e^{-\mu x}\tau_I} + \frac{1}{\tau_0}\right)} = \frac{1}{x}\sqrt{\frac{1}{I_0}\left(\frac{e^{\mu x}}{\tau_I} + \frac{1}{\tau_0}\right)}\end{aligned} \tag{B.25}$$

when the relationship $I = n_C/\tau_I = I_0\,e^{-\mu x}$ is also used.

The incident intensity I_0 is determined from a calibration measurement where the counting time τ_0 is usually long compared to the actual counting time τ_I, which normally has to be kept short to meet the speed of response requirements. This means that the following approximation can be used:

$$\sigma_\mu = \frac{1}{x}\sqrt{\frac{1}{I_0}\left(\frac{e^{\mu x}}{\tau} + \frac{1}{\tau_0}\right)} \approx \frac{1}{x}\sqrt{\frac{e^{\mu x}}{I_0\tau_I}} \tag{B.26}$$

Finally, it is also convenient to express the relative standard deviation, or the relative error:

$$\frac{\sigma_\mu}{\mu} \approx \frac{1}{\mu x}\sqrt{\frac{e^{\mu x}}{I_0\tau_I}} \tag{B.27}$$

From this expression it is evident that the relative error decreases with the number of detected events, $I_0\tau_I$. The relative error's dependence on the μx-product is less obvious; however, this can be investigated by derivation of the relative error with respect to μx:

$$\frac{\mathrm{d}\left(\frac{\sigma_\mu}{\mu}\right)}{\mathrm{d}\mu x} = \frac{\mathrm{d}\left(\frac{1}{\mu x}\sqrt{\frac{e^{\mu x}}{I_0 \tau_I}}\right)}{\mathrm{d}\mu x} = \frac{1}{\sqrt{I_0 \tau_I}} \frac{\mathrm{d}((\mu x)^{-1} e^{\mu x/2})}{\mathrm{d}\mu x}$$
$$= \frac{e^{\mu x/2}}{2\mu x} - \frac{e^{\mu x/2}}{(\mu x)^2} = \frac{e^{\mu x/2}}{\mu x}\left(\frac{1}{2} - \frac{1}{\mu x}\right) \tag{B.28}$$

This expression equals zero at $\mu x = 2$, meaning that the relative error has a minimum value when the beam intensity is attenuated to about 14% of its initial value, i.e. 86% attenuation.

B.4.2 The Density

In the energy region where Compton scattering is dominant, say between 100 keV and 1.5 MeV, the mass attenuation (μ_M) may be regarded as constant (see Section 3.3) for a given energy, and the attenuation is a function of the density (ρ). Equation (B.18) may then be expressed as

$$\rho = \frac{1}{\mu_M x} \ln\left(\frac{I_0}{I}\right) \tag{B.29}$$

and Equation (B.20) correspondingly as

$$\rho = \frac{1}{\mu_M x} \ln\left(\frac{n_0/\tau_0}{n_C/\tau_I}\right) = \frac{1}{\mu_M x} \ln\left(\frac{n_0 \tau_I}{n_C \tau_0}\right) \tag{B.30}$$

Equation (B.24) then takes the form

$$\sigma_\rho = \sqrt{\left(\frac{1}{\mu_M x n_C}\right)^2 n_C + \left(\frac{1}{\mu_M x n_0}\right)^2 n_0} = \frac{1}{\mu_M x}\sqrt{\frac{1}{n_C} + \frac{1}{n_0}} \tag{B.31}$$

and expresses the standard deviation in the measured density. Again the standard deviation is more conveniently expressed in terms of intensities rather than number of counts. That is, $n_0 = I_0 \tau_0$ and $n_C = I \tau_I$. Further, we use the relationship $I = I_0\, e^{-\mu x}$ so that Equation (B.31) may be further derived to

$$\sigma_\rho = \frac{1}{\mu_M x}\sqrt{\frac{1}{I\tau_I} + \frac{1}{I_0 \tau_0}} = \frac{1}{\mu_M x}\sqrt{\frac{1}{I\tau_I} + \frac{1}{I \tau_0\, e^{\mu x}}}$$
$$= \frac{1}{\mu_M x}\sqrt{\frac{1}{I}\left(\frac{1}{\tau_I} + \frac{1}{\tau_0\, e^{\mu x}}\right)} \tag{B.32}$$

If we now assume that the incident intensity I_0 is determined from a calibration measurement with counting time τ_0, which is long compared to the actual counting time τ_I, we can approximate the standard deviation in ρ due to statistical errors as

$$\sigma_\rho \approx \frac{1}{\mu_M x \sqrt{I \tau_I}} \tag{B.33}$$

B.5 READ-OUT ELECTRONICS

In this section we will take a closer look at methods and issues that are important when designing read-out electronics for detectors where the SNR is critical.

B.5.1 Experimental Noise Characterisation

When optimising the SNR for radiation detection systems it is useful to know the relative influence of different noise sources to the total system noise. In cases where, for instance, one noise source is dominant we do not want to spend time and money on reducing noise from other sources. The influence of different noise sources may often be estimated using information on datasheets, but it may also be characterised experimentally. We shall, as in Section 5.1.4, use a silicon PIN (p-type–intrinsic–n-type) detector connected to a charge-sensitive preamplifier as a case study because this detector has an additional level of complexity compared to other detectors; the detector capacitance is not constant but varies with the applied bias.

The first approach may be to measure the capacitance and leakage current of the diode as function of applied reverse bias. We then produce a plot like the one shown in Figure 4.17. These data may then be used in conjunction with Table 5.1 to estimate the contribution of different noise sources. The detector capacitance may be measured directly using a measurement bridge or impedance analyser with built-in bias supply. Alternatively, if there is no integrated bias supply, or if this not sufficient, a set-up like the one shown in Figure B.3 may be used. The measured capacitance is then the series capacitance of those of the diode junction (C_j) and the coupling condenser (C_C), which disconnects the measurement bridge from the bias. The measured capacitance is approximately equal to the diode junction capacitance because this is in the pF region and much less than the coupling capacitance, which is in the μF region. The leakage current can be measured in several ways; the one shown in Figure B.3 is based on a battery-powered pico-ammeter floating at the bias voltage.

The second approach is to measure the (most important) noise components directly [78]. The total noise (E_E) is easily determined by measuring the line width (FWHM) of the full energy peak of γ-ray line in the detection spectrum as a function of either detector bias or peaking time. The equipment needed in addition to the bias supply is a shaping amplifier with adjustable peaking time and a pulse height analyser (PHA). In order to express the noise in terms of radiation energy a second γ-ray (or X-ray) line is required for energy calibration. In the low-energy (<40 keV) region the variable target characteristic X-ray

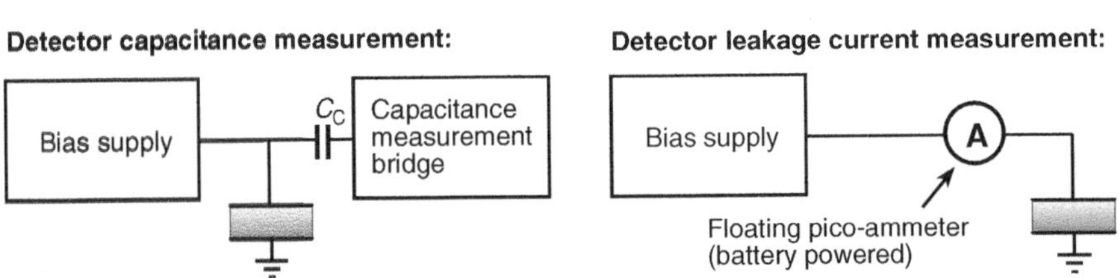

Figure B.3 Schematic diagram on measurement of the capacitance and leakage current of a radiation detector

source based on the principle shown in Figure 2.9 is very useful for this purpose. To determine the magnitude of the preamplifier noise (E_A) we also need a precision pulser and need to connect it through the preamplifier test input, as described in Section 5.1.5. The detector is then disconnected and replaced with a condenser with capacitance equal to that of the diode junction. This is a so-called cold capacitance because it generates no noise. The line width of the peak generated in the precision pulser is then measured, and because the PHA is energy calibrated we then have a measure of the preamplifier noise at this particular 'junction' capacitance. This procedure may be repeated for several capacitances so that the preamplifier noise may be found as a function of input capacitance. Provided the detector capacitance has been measured as discussed above, the preamplifier noise may also be related to the reverse bias for the particular detector. The detector noise as a function of a reverse bias may then be found using Equation (5.5):

$$E_D = \sqrt{E_E^2 - E_A^2} \tag{B.34}$$

A plot similar to that shown in Figure 5.7 (right) may then be produced. The procedure described above may also be carried out or combined with measurements as function of peaking time. By doing so the contributions of the various components of the preamplifier and detector noise may also be determined. For short peaking times delta noise components dominate and step noise contributions are negligible, and vice versa for long peaking times. This is evident from the plot in Figure 5.8. By combining several such measurements and using Equation (5.5) and the equations in Table 5.1, the contribution of all sources can be determined, or at least estimated.

Note that the system gain sometimes varies when the peaking time is changed. This may be corrected for by always using the precision pulser and by monitoring changes in its peak position (PHA channel or energy), and also when the detector is connected. The pulser peak may then be positioned with some margin above the highest full energy peak in the spectrum. If now this peak moves when the peaking time is altered, it is certainly due to changes in the system gain, and the required corrections may be applied. The precision pulser may also be used in conjunction with a γ-ray source to measure the magnitude of ballistic errors when these are present. This is determined from a plot of the gain corrected position of a full energy γ-ray line as a function of peaking time.

B.5.2 Electronics for Photodiode Read-Out of BGO Crystal

This section describes the design of a 44-channel detector and electronics system for a γ-ray fan-beam tomography scanner [84]. The energy used is the 661.6-keV line from ^{137}Cs (^{137m}Ba). All 44 channels are identical with $6 \times 6 \times 20$ mm^3 Crismatec BGO scintillation crystals attached to AME AE-994 36-mm^2 photodiodes. This calls for a careful design optimised for low noise (high SNR) because the quantum efficiency (Q_C) of this crystal is only $\approx$2.1% at 20°C, and the diode has no gain. The diode is a low-noise PIN silicon device fabricated on 280-μm-thick, 4-kΩ cm, $\langle 111 \rangle$ oriented n-type material. It depletes at about 40 V reverse bias with a junction capacitance $C_j \approx 20$ pF and a leakage current $I_l \approx 500$ pA. It is the same type of diode as presented in Section 5.1.4, except that the area is smaller. The surface of the diode is given an additional protection by a clear epoxy coating. This also serves as optical coupling and a solid attachment to the BGO crystal.

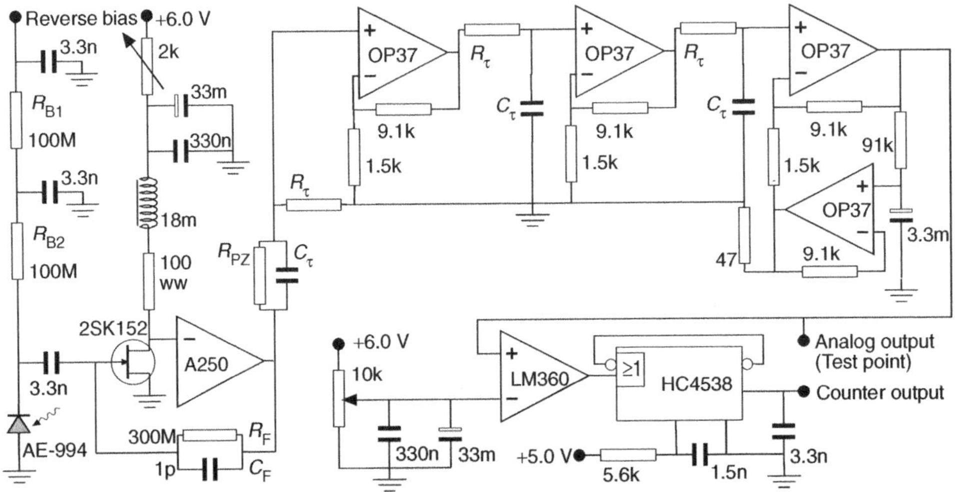

Figure B.4 Schematic diagram of the low-noise read-out electronics for a BGO-photodiode detector

TiO_2 paint was applied as a reflector material to the BGO crystals. The spectral response of the epoxy-coated diode matches very well to the emission spectrum of the BGO crystal (see Figure 4.19).

The low signal output level from the detector makes the choice of read-out electronics critical, and special care must be taken when designing the preamplifier, in particular. This has to be a low-noise, charge-integrating device with input impedance matched to the output impedance of the silicon diode. In addition, a pulse shaping amplifier is required to optimise the SNR. Finally, the filtered signal has to be fed into an SCA (single channel analyser), which then triggers the pulse counting unit.

A hybrid charge-sensitive preamplifier (Amptek A250) is selected as this uses an external field effect transistor (FET) as its input stage. Optimal impedance matching to the diode is achieved by using a Sony 2SK152 FET, which has a low input capacitance ($C_{iss} \approx 8$ pF). Next, a semi-Gaussian (CR–RC^2), unipolar shaping amplifier (based on low-noise OP37 operational amplifiers) was constructed as shown in the circuit diagram in Figure B.4. Bipolar shaping results in too low of an SNR and can consequently not be used. The SCA is composed of an comparator (LM360) followed by a monostable multivibrator (HC4538).

To minimise stray capacitance and lossy dielectrics noise, all terminals in the vicinity of the input stage, i.e. the FET gate, are lifted from the PC board and placed on Teflon stand-offs. Teflon has a very low *D*-value (see Table 5.1). Polypropene capacitors are used in the front-end circuit for the same reason. The PC board has no ground plane beneath the front-end components, again to minimise stray capacitance, and the preamplifier input is placed close to the detector to further reduce stray capacitance. The drain current of the input FET is filtered by an inductive low pass filter and adjusted to give an optimal transconductance, g_m. The diode bias is fed through a low pass filter which also separates the 44-channel detectors. The values of R_B and R_F should be very large for the purpose of noise reduction. The FET gate is, on the other hand, discharged through R_F. Hence, the average charge dissipation in the diode must also be considered when the value of R_F is determined.

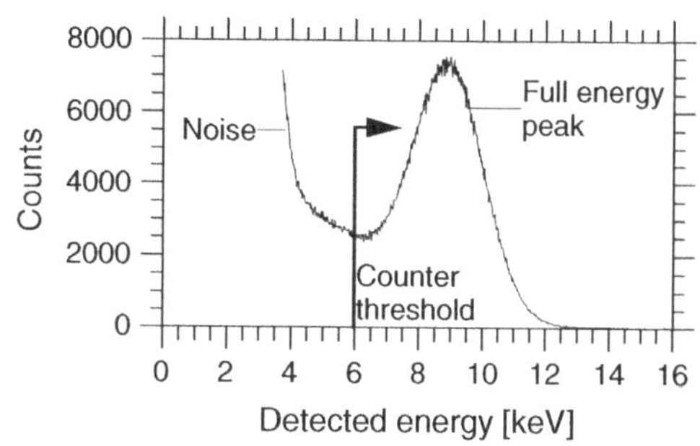

Figure B.5 Pulse height spectrum of the 36-mm^2 photodiode as read-out detector for a $6 \times 6 \times 20$-mm^3 BGO crystal exposed to 662-keV γ-radiation at 23°C. The diode is calibrated relative to direct absorption of γ-radiation, and the half-width of the 8.7-keV full energy peak is 2.8 keV

The output signal of the preamplifier is first differentiated and then integrated twice to provide a quasi-symmetric, unipolar output signal. The noise suppression characteristics of this simple filter are relatively good (see Table 5.2). The preamplifier output is connected to a CR high pass filter at the input of the shaping amplifier through a pole-zero cancellation network. This is done by adding a resistor, R_{PZ}, in parallel with the capacitor, C_τ, of the high pass filter, and adjusting R_{PZ} so that the time constants R_{PZC_τ} and $R_F C_F$ are equal. The purpose here is to eliminate the undershoot that normally will occur at the output of the shaping amplifier because of the finite time constant R_{FCF}. A DC feedback loop is added to the input of the last low pass filter to provide a stable baseline level (BLR). This is especially important to ensure correct triggering of the SCA. Separate ground-plane and power supply are used for the non-retriggerable monostable multivibrator whose output pulse has a width of about 7 μs. This means that signals from any new interactions taking place within this time are rejected and not counted. A capacitor to ground at the output is used to limit sharp transitions which disturb the sensitive front-end electronics. Finally, close attention is paid to the grounding and shielding of the detectors and the amplifiers. All the power supplies to the amplifier system are decoupled through low pass filters on each IC (integrated circuit).

The peaking time τ_0, which is twice the filter time constant τ, is the final amplifier parameter that has to be determined. The noise corner is found experimentally to be at $\tau = 1.8$ μs ($C_t = 220$ pF and $R_t = 8.2$ kΩ), which corresponds to $\tau_0 = 3.5$ μs and a total pulse width of about 15μs. A PHA spectrum was acquired and calibrated relative to direct absorption of the γ-ray energy in the diode (see Figure B.5) (Notice the better SNR compared to the 100 mm^2 otherwise identical diode whose BGO spectrum is presented in Figure 4.23.) The full energy peak is at 8.7 keV and the total noise of the system is about 2.8-keV FWHM. This means that the optimal peaking time with respect to noise, $\tau_0 = 3.5$ μs, has to be used to obtain reliable pulse counting. It is not possible to reduce this time to obtain higher count-rates, without losing the signal. Consequently, the count-rate capability of this system is limited to about 10 kc/s.

The Compton edge of 662-keV photons is 478 keV. Hence, the full energy peak and Compton edge as they appear in the diode will be about 9 and 6.5 keV, respectively. The Compton edge will then be partly buried in the noise of the system, which is about 2.8-keV FWHM. This means that the trigger level of the pulse counting unit should be set somewhere on the low-energy tail of the full energy peak (~6 keV) to avoid

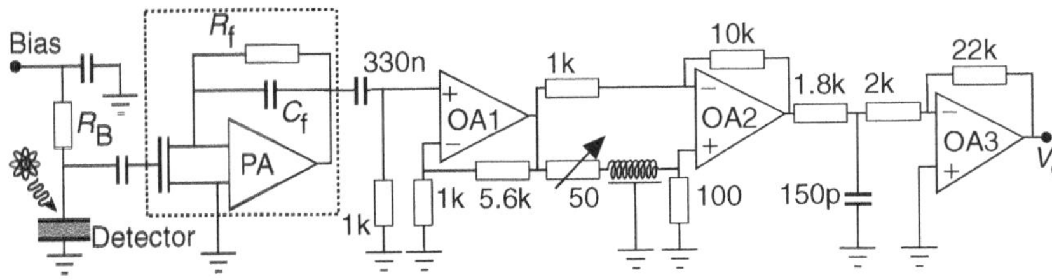

Figure B.6 Schematic diagram of the high-speed read-out electronics for a CdZnTe detector. The preamplifier (PA) is based on an eV-5093 hybrid (dotted box) with 2-MΩ feedback resistance. A Newport 11ACB15112E delay-line with 150-ns delay, and one OP37 (OA1) and two OP637 (OA2 and OA3) operational amplifiers are used in the shaping amplifier

noise triggering. The detection efficiency, i.e. the ratio of incident photons to counted photons, will then be equal to the efficiency with which full energy absorption occurs. Photoelectric absorption in 20-mm BGO gives a minimum efficiency of about 24%. Monte Carlo simulation, however, shows that the full energy efficiency will be increased to about 41% when re-absorption of Compton scattered photons is included as well.

B.5.3 High Count-Rate Electronics for a CdZnTe Detector

This section describes the design of the detector and electronics system for a 85-channel high-speed γ-ray tomograph [31, 204]. This instrument utilizes five ^{241}Am radiation γ-ray sources (59.5 keV) with five diametral detector modules each with 17 $10 \times 10 \times 2\text{-mm}^3$ $Cd_{0.9}Zn_{0.1}Te$ detectors stacked closely together in a linear array. These are eV products planar MSM-detectors with electroless gold contacts. The radiation is incident perpendicular to the electrodes, with the negative electrode as front window to optimise hole collection.

Compared to systems optimised for low noise, such as the one described in the previous section, some compromises have to be made with a high-count-rate systems. Starting with the preamplifier the value of the feedback resistor has to be reduced to avoid the preamplifier running into saturation. This will be at the cost of increased noise at high peaking times, but this is less significant because the second compromise to be made is to operate the system at shorter peaking times. The noise corner for this type of detector was found to be $\tau_C \approx 4\,\mu s$ using NIM equipment with adjustable peaking time. The maximum count-rate expected for this tomography system is about 100 kc/s. To keep the probability of pile-up low, it was therefore decided to design read-out electronics with total pulse width less than 1 μs. The feedback network for hybrid eV-5093 preamplifier was therefore prepared by selecting by $R_F = 2\,M\Omega$ and $C_F = 0.4$ pF.

A dedicated main amplifier with delay line shaping was developed as is shown schematically in Figure B.6. The preamplifier output signal is first differentiated and amplified. Both inputs of the second operational amplifier are connected to the output of the first; however, the key to the circuit's operation is that one of the input signals is delayed. The output signal of the second operational amplifier is then a short signal with square wave appearance. This signal is fed through an integrator to achieve optimal noise performance. The delay and the time constant of the integrator are chosen to obtain an amplifier peaking

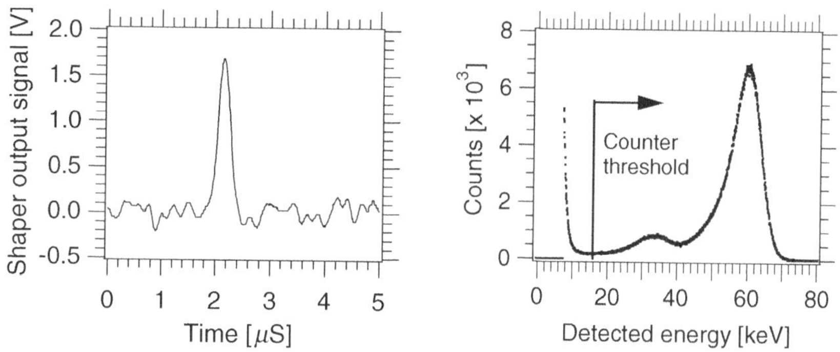

Figure B.7 Typical response to a 59.5-keV γ-ray photon at the output of the delay line amplifier as recorded by a digital oscilloscope (left). Typical PHA spectrum (right) with line width FWHM = 9.5 keV dominated by preamplifier noise (E_{sF}) because of the short peaking time ($\tau_0 = 200$ ns)

time of 200 ns. The total width of the output pulse is only about 500 ns, as can be seen from Figure B.7. Also shown in this figure is a pulse height spectrum, which clearly demonstrates the relatively poor pulse width or energy resolution at this short peaking times.

This system uses a similar comparator and one-shot configuration as the circuitry outlined in B.4, except that the pulse width of the one-shot is shorter. Also the comparator threshold is programmable using a digital-to-analogue converter for each of the 85 channels.

B.6 HALF-WIDTH CALCULATION

The distribution in the measured energy of a γ-ray emission line in a PHA spectrum is Gaussian. But as we discussed in Section 4.3.1 the FWHM is here commonly used to express the broadening of the peak. The relationship between FWHM and the standard deviation (σ) is easily derived from the expression of the Gaussian distribution:

$$p(q) = \frac{e^{[-(q-\overline{q})^2/2\sigma^2]}}{\sigma\sqrt{2\pi}} \tag{B.35}$$

Its maximum value is at $q = \overline{q}$, that is

$$p_{\max}(q) = \frac{1}{\sigma\sqrt{2\pi}} \tag{B.36}$$

At half maximum we then have

$$p(q) = \frac{e^{[-(q-\overline{q})^2/2\sigma^2]}}{\sigma\sqrt{2\pi}} = \frac{1/2}{\sigma\sqrt{2\pi}} \Rightarrow e^{[-(q-\overline{q})^2/2\sigma^2]} = \frac{1}{2} \Rightarrow (q-\overline{q})^2 = 2\sigma^2(2\ln 2) \tag{B.37}$$

Now, the full width at half maximum means that

$$\text{FWHM} = 2(q-\overline{q}) = 2\sqrt{\sigma^2(2\ln 2)} = 2\sqrt{2\ln 2}\sigma \approx 2.35\sigma \tag{B.38}$$

Appendix C: References

The URLs for all Web site references in this list are provided at the Web site for this book (www.wileyeurope.com/go/radioisotope). Most manufacturers of radiation sources, detectors, systems and other associated equipment have Web sites with relevant information regarding radioisotope gauges; so do various organisations and bodies. One has to take the usual precautions whenever taking information from the Internet: it is not necessarily quality assured, and it may be biased in favour of the Web site owner.

1. Saunders, P. *Radiation and You*, Information folder from the Comission of the European Communities, Luxembourg. ISBN 92 825 9486 6.
2. National Radiological Protection Board UK (1989). *Living with Radiation*, 4th ed., Information booklet.
3. Sutton, S. (1988). 'Inside science – radioactivity', *New Scientist*, 25 February 1988, 1.
4. Firestone, R. B., Baglin, C. M. (Ed.) and Chu, S. Y. F. (CD-ROM Ed.) (1998). *Tables of Isotopes*, 8th ed. (1998 update), Wiley, New York.
5. Masschaele, B., Dierick, M. and Vanhoorebeke, L. (2003). 'Element specific X-ray tomography at the 15 MeV electron Linac,' *Proceedings of the 3rd World Congress on Industrial Process Tomography*, Banff, Canada, 2–5 September 2003, 193.
6. Charlton, J. S. (Ed.) (1986). *Radioisotope Techniques for Problem-Solving in Industrial Plants*, Leonard Hill, London.
7. Faires, F. A. and Boswell, G. G. J. (1981). *Radioisotope Laboratory Techniques*, Butterworths, London.
8. Gardner, R. P. and Ely, R. L., Jr. (1967). *Radioisotope Measurement Applications in Engineering*, Reinhold, New York.
9. Miley G. H. (1999). 'A portable neutron/tunable X-ray source based on inertial electrostatic confinement', *Nucl. Instr. Meth.* **A422**, 16.
10. Shope, L. A., Berg, R. S., O'Neil, M. L. and Barnaby, B. E. (1981). 'Operation and life of the Zetatron: A small neutron generator for borehole logging', *IEEE Trans. Nucl. Sci.* **NS-28**(2), 1696.
11. Tertian, R. and Claisse, F. (1982). *Principles of Quantitative X-Ray Fluorescence Analysis*, Heyden & Sons, London.
12. Photkoef, USA, AIC software (1996–2000).
13. National Institute of Standards and Technology, USA, Physical reference data (ESTAR and ASTAR) web site data.

Radioisotope Gauges for Industrial Process Measurements. Geir Anton Johansen and Peter Jackson.
 ISBN 0-471-48999-9

14. Dunn, W. L. and Hutchinson, J. E. (1982). 'A nuclear technique for thin-lift gauging', *Int. J. Appl. Radiat. Isot.* **33**, 563.
15. Alfassi, Z. B. (Ed.) (1994). *Chemical Analysis by Nuclear Methods*, Wiley, Chichester.
16. Siegbahn, K. (Ed.) (1966). *Alpha-, Beta- and Gamma-Ray Spectroscopy*, Vols. 1 and 2, North-Holland, Amsterdam.
17. Storm, E. and Israel, H. I. (1967). 'Photon cross sections from 0.001 to 100 MeV for elements 1 through 100', Report LA-3753, UC-34 Physics, TID-4500, Los Alamos Scientific Laboratory, USA.
18. British Standard Institution (1966). *Recommendation for Data on Shielding from Ionizing Radiation, Part 1: Shielding from Gamma radiation*, British Standard 4094, Part 1.
19. Morris, E. E., Chilton, A. B. and Vetter, A. F. (1975). 'Tabulation and empirical representation of infinite-medium gamma-ray build-up factors for monoenergetic, point isotropic sources in water, aluminum and concrete', *Nucl. Sci. Eng.* **56**, 171.
20. Korea Atomic Energy Research Institute (Nuclear Atomic Energy Research Institute, MCNP library) web site data.
21. Lone, M. A., Leavitt, R. A. and Harrison, D. A. (1981). 'Prompt gamma rays from thermal-neutron capture', *At. Data Nucl. Data Tables* **234**(6), 511.
22. Attix, F. H. (1986). *Introduction to Radiological Physics and Radiation Dosimetry*, Wiley, New York.
23. Mladjenović, M. (1973). *Radioisotope and Radiation Physics*, Academic Press, New York.
24. International Organisation for Standardization (ISO) (1993). *International Vocabulary of Basic and General Terms in Metrology*, 2nd ed.
25. Knoll, G. F. (2000). *Radiation Detection and Measurement*, 3rd ed., Wiley, New York.
26. Dmitrenko V. V., Gratchev, V. M., Uli, S. E., Uteshev Z. M. and Vlasik, K. F. (2000). 'High-pressure xenon detectors for gamma-ray spectrometry', *Appl. Rad. Isotop.* **52**, 739.
27. Beddingfield, D. H., Beyerle, A., Russo, P. A., Ianakiev, K., Vo, D. T. and Dmitrenko, V. (2003). ' High-pressure xenon ion chambers for gamma-ray spectrometry in nuclear safeguards', *Nucl. Instr. Meth.* **A505**, 474.
28. Sauli, F. (2002). 'Micro-pattern gas detectors', *Nucl. Instr. Meth.* **A477**, 1.
29. Maia, J. M., Veloso, J. F. C. A, Morgado, R. E., dos Santos, J. M. F. and Conde, C. A. N. (2002) 'The micro-hole-and-strip plate gas detector: Experimental results', *IEEE Trans. Nucl. Sci.* **49**(3), 875.
30. Gilmore, G. and Hemingway, J. (1995). *Practical Gamma-ray Spectrometry*, Wiley, Chichester.
31. Johansen, G. A. and Åbro, E. (1996). 'A new CdZnTe detector system for low energy gamma-ray measurement', *Sensors Actuators* **A54**, 493.
32. Limousin, O. (2003). 'New trends in CdTe and CdZnTe detectors for X-ray and gamma-ray applications', *Nucl. Instr. Meth.* **A504**, 24.
33. Iwanczyk, J. S., Patt, B. E., Wang, Y. J. and Khusainov, A. Kh. (1996). 'Comparison of HgI_2, CdTe and Si (p-i-n) X-ray detectors', *Nucl. Instr. Meth.* **A380**, 186.
34. Luke, P. N. and Eissler, E. E. (1996). 'Performance of CdZnTe coplanar-grid gamma-ray detectors', *IEEE Trans. Nucl. Sci.* **43**(3), 1481.

35. Lingren, C. L., Apotovsky, B., Butler, J. F., Conwell, R. L., Doty, F. P., Friesenhahn, S. J., Oganesyan, A., Pi, B. and Zhao, S. (1998). 'Cadmium-zinc-telluride, multiple-electrode detectors achieve good energy resolution with high sensitivity at room-temperature', *IEEE Trans. Nucl. Sci.* **45**(3), 433.
36. McGregor, D. S., He, Z., Seifert, H. A., Wehe, D. K. and Rojeski, P. A. (1998). 'Single carrier type sensing with a parallel strip pseudo-Frisch-grid CdZnTe semiconductor radiation detector', *Appl. Phys. Lett.* **72**(7).
37. Parnham, K., Szeles, Cs., Lynn, K. G. and Tjossem, R. (1999). 'Performance improvement of CdZnTe detectors using modified two-terminal electrode geometry', *Proc SPIE* **3768**, 49.
38. Montemont, G., Arques, M., Veger, L. and Rustique, J. (2001). 'A capacitive Frisch grid structure for CdZnTe detectors', *IEEE Trans. Nucl. Sci.* **48**(3), 278.
39. Takahashi, T., Mitani, T., Kobayashi, Y., Kouda, M., Sato, G., Watanabe, S., Nakazawa, K., Okada, F., Ohno, R. and Mori, K. (2002). 'High-resolution Schottky CdTe diode detector', *IEEE Trans. Nucl. Sci.* **49**(3), 1297.
40. Auricchio, N., Caroli, E., Donati, A., Dusi, W., Fougeres, P., Grassi, D., Perillo, E. and Siffert, P. (2001). 'Characterization of thin back-to-back CdTe detectors', *IEEE Trans. Nucl. Sci.* **48**(4), 1028.
41. Eisen, Y. (1996). 'Current state-of-the-art industrial and research applications using room-temperature CdTe and CdZnTe solid state detectors', *Nucl. Instr. Meth.* **A380**, 431.
42. Fougeres, P., Siffert, P., Hageali, M., Koebel, J. M. and Regal, R. (1999). 'CdTe and $Cd_{1-x}Zn_xTe$ for nuclear detectors: Facts and fictions', *Nucl. Instr. Meth.* **A428**, 38.
43. Franc, J., Höschl, P., Belas, E., Grill, R., Hlídek, P., Moravec, P. and Bok, J. (1999). 'CdTe and CdZnTe crystals for room temperature gamma-ray detectors', *Nucl. Instr. Meth.* **A434**, 146.
44. Birks, J. B. (1964). *The Theory and Practice of Scintillation Counting*, Pergamon, Oxford.
45. van Eijk, C. W. E. (1997). 'Development of inorganic scintillators', *Nucl. Instr. Meth.* **A392**, 285.
46. Derenzo, S. E., Weber, M. J., Bourret-Courchesne, E. and Klintenberg, M. K. (2003). 'The quest for the ideal inorganic scintillator', *Nucl. Instr. Meth.* **A505**, 111.
47. Iwanczyk, J. S., Patt, B. E., Tull, C. R., MacDonald, L. R., Bescher, E., Robson, S. R., Mackenzie, J. D. and Hoffman, E. J. (2000). 'New LSO based scintillators', *IEEE Trans. Nucl. Sci.* **47**(6), 1781.
48. Bicron/Saint-Gobain, France, Crystals & Detectors, web site data (2003).
49. Holl, I., Lorenz, E., Mageras, G. (1988). 'A measurement of the light yield of common inorganic scintillators', *IEEE Trans. Nucl. Sci.* **35**, No. 1, 105.
50. Sakai, E. (1987). 'Recent measurements of scintillator-photodetector systems', *IEEE Trans. Nucl. Sci.*, **34**(1), 418.
51. Schotanus, P. (2003). Private communication, Scionix, Netherlands.
52. Baccaro, S., Blažek, K., de Notaristefani, F., Maly, P., Mares, J. A., Pani, R., Pellegrini, R. and Soluri, A. (1995). 'Scintillation properties of Yap:Ce', *Nucl. Instr. Meth.* **A361**, 209.

53. Johansen, G. A., Frøyen, S. and Hansen, T. E. (1997). 'Gamma-ray detection with an UV-enhanced photodiode and scintillation crystals emitting at short wavelengths', *Nucl. Instr. Meth.* **A387**, 239.
54. Lempicki, A., Randles, M. H., Wisniewski, D., Balcerzyk, M., Brecher, C. and Wojtowicz, A. J. (1995). '$LuAlO_3$:Ce and other aluminate scintillators', *IEEE Trans. Nucl. Sci.* **42**(4), 280.
55. Electron Tubes, UK, web site data.
56. Renker, D. (2002). 'Properties of avalanche photodiodes for applications in high energy physics, astrophysics and medical imaging', *Nucl. Instr. Meth.* **A486**, 164.
57. Tsoulfanidis, N. (1995). *Measurement and Detection of Radiation*, 2nd ed., Taylor & Francis, Washington, DC.
58. Hamamatsu, Japan, web site data.
59. Johansen, G. A. and Johnson, C. B. (1993). 'Operational characteristics of an electron-bombarded silicon diode photomultiplier tube', *Nucl. Instr. Meth.* **A326**, 295.
60. D'Ambrosio, C. and Leutz, H. (2003). 'Hybrid photon detectors', *Nucl. Instr. Meth.* **A501**, 463.
61. Lecomte, R., Pepin, C., Rouleau, D., Dautet, H., McIntyre, R. J., McSween, D. and Webb, P. (1999). 'Radiation detection measurements with a new 'buried junction' silicon avalanche photodiode', *Nucl. Instr. Meth.* **A423**, 92.
62. Musienko, Y., Reucroft, S. and Swain, J. (2000). 'A simple model of EG&G reverse reach-through APDs', *Nucl. Instr. Meth.* **A442**, 179.
63. Moszyński, M., Kaputsa, M., Balcerzyk, M., Wolski, I., WÍgrzecka, I. and Węgrzecki, M. (2001). 'Compariative study of avalanche photodiodes with different structures in scintillation detection', *IEEE Trans. Nucl. Sci.* **48**(4), 1205.
64. Rozsa, C., Grodsinsky, C., Penn, D., Raby, P. and Schreiner, R. (2000). 'The change of gamma equivalent energy with temperature for scintillation detector assemblies', *IEEE Nucl Sci Symp.* **N20-21**, 686.
65. Scionix, The Netherlands, web site data.
66. Rozsa, C., Dayton, R., Raby, P., Kusner, M. and Schreiner, R. (1989). ' Characteristics of scintillators for well logging to 225°C', *IEEE Nuclear Science Symposium*, San Francisco.
67. Imhof, W. L., Spear, K. A., Hamilton, J. W., Higgins, B. R., Murphy, M. J., Pronko, Wondrak, R. R., McKenzie, D. L., Rice, C. J., Gorney, D. J., Roux, D. A., Williams, R. L., Stein, J. A., Bjordal, J., Stadsnes, J., Njøten, K., Rosenberg, T. J., Lutz, L. and Detrick, D. (1995). 'The polar ionospheric X-ray imaging experiment (PIXIE)', *Space Sci. Rev.* **71**, 385.
68. Stahle, C. M., Parker, B. H., Parsons, A. M., Barbier, L. M., Barthelmy, S. D., Gehrels, N. A., Palmer, D. M., Snodgrass, S. J. and Tueller, J. (1999). 'CdZnTe and CdTe detector arrays for hard X-ray and gamma-ray astronomy', *Nucl. Instr. Meth.* **A436**, 138.
69. Centronic, Databook on Geiger Müller Tubes ISS.1.
70. Peurrung A. J. (2002). 'Recent developments in neutron detection', *Nucl. Instr. Meth.* **A487**, 123.
71. Iwanczyk, J. S., Dabrowski, A. J., Huth, G. C., Del Duca, A. and Schnepple W. (1981). 'A study of low-noise preamplifier systems for use with room temperature mercuric iodide (HgI_2) X-ray detectors', *IEEE Trans. Nucl. Sci.* **28**(1), 579.

72. Bassini, R., Boiano, C. and Pullia, A. (2002). 'A low-noise charge amplifier with fast rise time and active discharge mechanism', *IEEE Trans. Nucl. Sci.* **49**(5), 2346.
73. Nashashibi, T. (1992). 'The Pentafet and its applications in high resolution and high count rate radiation spectrometers', *Nucl. Instr. Meth.* **A332**, 551.
74. Fazzi, A. and Rehak, P. (2000). 'Performance of an X-ray spectroscopic system based on a double-gate double feedback charge preamplifier', *Nucl. Instr. Meth.* **A439**, 391.
75. Llacher, J. (1975). ' Accurate measurement of noise parameters in ultra-low noise opto-feedback spectrometer systems', *IEEE Trans. Nucl. Sci.* **NS-22**(5), 2033.
76. Llacher, J. and Meier, D. F. (1977). Excess noise in selected field-effect transistors *IEEE Trans. Nucl. Sci.* **NS-24**(1), 317.
77. Goulding, F. S. (1977). 'Some aspects of detectors and electronics for X-ray fluorescence analysis', *Nucl. Instr. Meth.* **142**, 213.
78. Radeka, V. (1968). 'State of the art of low noise amplifiers for semiconductor radiation detectors', *Int. Symp. Nucl. Electron.* **1**, Versailles 46–1.
79. Radeka, V. (1973). 'Field effect transistors for charge amplifiers', *IEEE Trans. Nucl. Sci.* **20**(1), 182.
80. Lecomte, R., Pepin, C. M., Lepage, M. D., Pratte, J.-F., Dautet, H. and Binkley, D. M. (2001). 'Performance analysis of phoswich/APD detectors and low-noise CMOS preamplifiers for high resolution PET systems', *IEEE Trans. Nucl. Sci.* **48**(3), 650.
81. Bennet, P. R., Shah, K. S., Cirignano, L. J., Klugerman, M. B., Dmitriyev, Y. N. and Squillante, M. R. (1998). 'Multi-element CdZnTe detectors for gamma-ray detection and imaging', *IEEE Trans. Nucl. Sci.* **45**(3), 417.
82. Shah, K. S., Lund, J. C., Olschner, F., Bennet, P. R., Zhang, J., Moy, L. P. and Squillante, M. R. (1994). 'Electronic noise in lead iodide X-ray detectors', *Nucl. Instr. Meth.* **A353**, 85.
83. Fiorini, C. and Lechner, P. (1998). 'Charge-sensitive preamplifier with continuous reset by means of the gate-to-drain current of the JFET integrated on the detector', *IEEE Trans. Nucl. Sci.* **45**(3), 417.
84. Johansen, G. A., Frøystein, T. and Ursin, J. R. (1993). 'A gamma detector system for tomographic imaging of porous media', *Proceedings of European Concerted Action on Process Tomography*' 93, Karlsruhe, Germany, 24–27 March, 177.
85. Radeka, V. (1988). 'Low-noise techniques in detectors', *Ann. Rev. Nucl. Part. Sci.* **38**, 217.
86. Ott, H. W. (1988). *Noise reduction Techniques in Electronic Systems*, 2nd ed., Wiley, New York.
87. Motchenbacher, C. D. and Fitchen, F. C. (1973). *Low-Noise Electronic Design*, Wiley, New York.
88. Morrison, R. (1986). *Grounding and Shielding Techniques in Instrumentation*, 3rd ed., Wiley, New York.
89. Koskelo, M. J., Koskelo, I. J. and Sielaff, B. (1999). 'Comparison of analog and digital signal processing systems using pulsers', *Nucl. Instr. Meth.* **A422**, 373.
90. Hamada, N. (2001). 'Digital signal processing: Progress over the last decade and the challenges ahead', *IEICE Trans. Fundament. Electr. Commun. Comp. Sci.* **E84A**(1), 80.

91. Jordanov, V. T., Knoll, G. F., Huber, A. C. and Pantazis, J. A. (1994). 'Digital techniques for real-time pulse shaping in radiation measurements', *Nucl. Instr. Meth.* **A353**, 261.
92. Bingham, R., Keyser, R. and Twomey, T. (2001). 'An innovative, portable MCA based on digital signal processing,' *Presented at the 23rd Brugge ESARDA Meeting on Safeguards and Nuclear Materials Management*, 8–11 May 2001.
93. Vo, D. T., Russo, P. A. and Sampson, T. E., (1998). 'Comparisons between digital gamma-ray spectrometer (DSPec) and standard nuclear instrumentation methods (NIM) systems', Los Alamos National Laboratory, LA-13393-MS, UC-706 and UC-700.
94. The European Parliament and The Council (1994). 'On the approximation of the laws of the Member States concerning equipment and protective systems intended for use in potentially explosive atmospheres', Directive 94/9/EC (also known as ATEX 95 or ATEX 100A).
95. Groh, H. (2003). *Explosion Protection*, Butterworth-Heienmann, London.
96. Eckhoff, R. K. (2003). *Dust Explosions in the Process Industries*, 3rd ed., Gulf Professional Publishing, Amsterdam.
97. Parker, D. J. and McNeill, P. A. (1996). 'Positron emission tomography for process applications', *Meas. Sci. Technol.* **7**(3), 287.
98. Parker, D. J., Forster, R. N., Fowles, P. and Takhar, P. S. (2002) 'Positron emission particle tracking using the new Birmingham positron camera', *Nucl. Instr. Meth.* **A477**, 540.
99. Nicholson, P. W. (1974). *Nuclear Electronics*, Wiley, New York.
100. International Organisation for Standardization (ISO) (1995). *Guide to the expression of uncertainty in measurement.*
101. European Co-Operation for Accreditation (1999). *Expression of the Uncertainty of Measurement in Calibration* EA-4/02. Available from EA web site.
102. Moses, W. W. (2000). 'Current trends in scintillator detectors and materials', *Nucl. Instr. Meth.* **A443**, 400.
103. Lecomte, R., Pepin, C. M., Rouleau, D., Saoudi, A., Andreaco, M. S., Casey, M., Nutt, R., Dautet, H. and Webb, P. P. (1998). 'Investigation of GSO, LSO and YSO scintillators using reverse avalanche photodiodes', *IEEE Trans. Nucl. Sci.* **45**(3), 478.
104. Johansen, G. A. (1990). *Development and Analysis of Silicon Based Detectors for Low Energy Nuclear Detectors*, PhD Thesis, University of Bergen, Norway.
105. Prescott, J. R. and Takhar, P. S. (1962). 'Resolution and line shape in scintillation counters', *IRE Trans. Nucl. Sci.* **NS-9**(3), 36.
106. Jenkins, R. (1999). *X-Ray Fluorescence Spectrometry*, 2nd ed., Wiley, New York.
107. Gardner, R. P., Metwally, W. A., and Han, X. (submitted). 'A new NaI detector arrangement for efficient detection of high energy gamma rays', *J. Radioanal. Nucl. Chem.*
108. Lee, S. H. and Gardner, R. P. (2000). 'A new G-M counter dead time model', *Appl. Rad. Isotop.* **53**, 731.
109. Baltakmens, T. (1977). 'Accuracy of absorption methods in the identification of beta emitters', *Nucl. Instr. Meth.* **142**, 535.
110. Huddleston, A. L. and Weaver, J. B. (1983). 'Dual-energy Compton-scatter densitometry', *Int. J. Appl. Radiat. Isot.* **34**(7), 997.

111. Baltakmens, T. (1970). 'A simple method for determining the maximum energy of beta emitters by absorption measurements', *Nucl. Instr. Meth.* **82**, 264.
112. Dietzel, K. I., Johansen, G. A. and McCann, H. (2001). 'X-ray fluorescence auto projection tomography (X-FAPT)', *Proceedings of the 2nd World Congress on Industrial Process Tomography*, Hannover, Germany, 29–31 August 2001, 74.
113. Watt, J. S. (1983). 'On-stream analysis of metalliferous ore slurries', *Int. J. Appl. Radiat. Isot.* **34**(1), 309.
114. Jenkins, R. and De Vries, J. L. (1970). *Practical X-Ray Spectrometry*, 2nd ed., MacMillan Press, London.
115. LaBrecque, J. J. and Parker, W. C. (1983). 'A new technique for radioisotope-excited X-ray fluorescence', *Proceedings of Conference on Applications of X-Ray Analysis*, Denver 1982, Adv. in X-ray Analys. **26**, Plenum Press, 337.
116. Alfassi, Z. B. and Chung, C. (Ed.) (1995). *Prompt Gamma Neutron Activation Analysis,* CRC Press, Boca Raton, FL.
117. Marsh, J. W., Thomas, D. J. and Bruke, M. (1995). 'High resolution measurements of neutron energy spectra from Am-Be and Am-B neutron sources', *Nucl. Instr. Meth.* **A366**, 340.
118. Hertz, K. L., Hilton, N. R., Lund, J. C. and Van Scyoc, J. M. (2003). 'Alpha-emitting radioisotopes for switchable neutron generators', *Nucl. Instr. Meth.* **A505**, 41.
119. Proctor, R., Yusuf, S., Miller, J. and Scott, C. (1999). 'Detectors for on-line prompt gamma neutron activation analysis', *Nucl. Instr. Meth.* **A422**, 933.
120. Eisler, P. L. and Huppert, P. (1979). 'A nuclear geophysical borehole logging system', *Nucl. Instr. Meth.* **158**, 579.
121. Ikuta, T., Osa, A., Taniguchi, A., Yamamoto, H. and Kawade, K. (1992). 'Portable neutron-capture γ-ray source above 3.5 MeV ^{252}Cf', *Nucl. Instr. Meth.* **A323**, 697.
122. Paul, R. L. and Lindstrom, R. M. (2000). 'Prompt-gamma activation analysis: Fundamentals and applications', *J. Radioanal. Nucl. Chem.* **243**(1), 181.
123. Holmes, J. L. and Zoller, W. H. (1992). *Prompt Gamma-Ray Spectroscopy for Process Monitoring*, American Chemical Society Symposium Series 508, ACS, New York, 229.
124. International Atomic Energy Agency (IAEA). (1990). *Guidebook on Radioisotope Tracers in Industry*, STI/DOC/10/316, IAEA, Vienna.
125. Jain, P. (2003). *Tracer Applications in Oil Field Investigations*, IAEA, Vienna.
126. Pant, H. J. (2001). 'Radiotracer applications in the chemical process industry', *Rev. Chem. Eng.* **17**(3), 165.
127. Hoffmann, A. C., Dechsiri, C., Van de Wiel, F., Ghione, A. and Paans, A. M. J. (2003). 'Tracking individual particles in a fluidised bed using a medical PET-camera', *Proceedings of the 3rd World Congress on Industrial Process Tomography*, Banff, Canada, 2–5 September 2003, 595.
128. Dechsiri, C., Van de Wiel, F., Dehling, H. G., Paans, A. M. J. and Hoffmann, A. C. (2003). 'Positron emission tomography applied to fluidisation engineering', *Proceedings of the 3rd World Congress on Industrial Process Tomography*, Banff, Canada, 2–5 September 2003, 577.
129. International Organisation for Standardization (1986). *Radionuclide Gauges – Gauges Designed for Permanent Installation*, ISO 7205.

130. The International Commission on Radiological Protection (ICRP). (1991). 'Radiation Protection – 1990 Recommendations of the ICRP', *Annals ICRP* **21**(1–3).
131. The Radiation Protection Institutes in Denmark, Finland, Iceland, Norway and Sweden (1992). *Nordic Recommendations on Radiation Protection for Radionuclide Gauges For Permanent Installation.*
132. Delaney, C. F. G. and Finch, E. C. (1992). *Radiation Detectors – Physical Principles and Applications*, Clarendon Press, Oxford.
133. Hole, E. O. (1993). 'The effect of ionising radiation: The reliability of risk estimates at low doses', *Fra Fysikkens Verden* **2**, 43. (In Norwegian) ISSN-0015-9247.
134. Hendee, W. R. and Edwards, F. M. (Ed.) (1986). *Health Effects of Exposure to Low-Level Ionising Radiation*, Medical Science Series, IOP Publishing, Bristol, UK.
135. Statens bygge- og eiendomsdirektorat. (1992). 'Depot for low and medium radioactive waste, elucidation of consequences according to the plan and building regulations (in Norwegian)', Main report.
136. Henriksen, T., Ingebretsen, F., Storruste, A. and Stranden, E. (1987). 'Radioactivity, Radiation, Health', Universitetsforlaget AS, Oslo (in Norwegian).
137. Delacroix, D., Guerre, J. P., Leblanc, P. and Hickman, C. (2002). *Radionuclide and Radiation Protection Data Handbook 2002*, 2nd ed., Nuclear Technology Publishing, Ashford, UK. Radiation Protection Dosimetry **98**, (1).
138. IAEA. (1996). *IAEA Safety Standards Series, Regulations for the Safe Transport of Radioactive Material*, Code and Safety Guides Q1-Q14, Safety Series No. 50-C/SG-Q, IAEA, Vienna.
139. European Committee for Standardisation (2000). *General Requirements for the Competence of Testing and Calibration of Laboratories*, ISO/IEC 17025.
140. Cameron, J. F. and Clayton, C. G. (1971). *Radioisotope Instruments*, Part 1, Pergamon Press, Oxford.
141. Clayton, C. G., Coleman, C. F., Mikesell, J. L., Senftle, F. E., Tanner, A. B., Charbucinski, J., Borsaru, M., Eisler, P. L., Youl, S. F., Killeen, P. G., Mwenifumbo, C. J., Prasad, A. S., Page, D., Watt, J. S., Sowerby, B. D., Cierpisz, S., Gozani, T., Fauth, G., Leininger, D., Lüdke, H., Surman, P. L., Cywicka-Jakiel, T. and £oskiewicz, J. (1986). *Gamma, X-Ray and Neutron Techniques for the Coal Industry*, STI/PUB/707, IAEA, Vienna.
142. Charbucinski, J., Watt, J. S., Castagnet, A. C., Liye, Z., Tola, F., Boutaine, J. L., Tominaga, H., Scheers, A. M., Wallace, G., Pohl, P., Hutchinson, E., Urbañski, P., Salagado, J., Carvalho, F. G., Manterigas J, Oliveira C, Conçalves, I. F., Neves, J., Cruz, C., Fedorkov, V. G. and Gardner, R. P. (2000). *Emerging New Applications of Nucleonic Control Systems in Industry*, IAEA-TECDOC-1142, IAEA, Vienna.
143. Tjugum, S.-A., Hjertaker, B. T. and Johansen, G. A. (2002). 'Multiphase flow regime identification by multibeam gamma-ray densitometry', *Meas. Sci. Technol.* **13**, 1319.
144. Hanssen, B. V. and Torkildsen, B. H. (1995). 'Status of the Framo subsea multiphase flowmeter', *Proceedings of the 13th North Sea Flow Measurement Workshop*, Lillehammer, Norway.
145. Gardner, R. P., Bean, R. H. and Ferrell, J. K. (1970). 'On the gamma-ray one-shot-collimator measurement of two-phase-flow void fractions', *Nucl. Appl. Technol.* **8**, 88.

146. Eberle, C. S., Leung, W. H., Ishii, M. and Revankar, S. T. (1994). 'Optimization of a one-shot gamma densitometer for measuring area-averaged void fractions of gas-liquid flows in narrow pipe-lines', *Meas. Sci. Technol.* **4**, 1146.
147. Amdal, J., Danielsen, H., Dykesteen, E., Flølo, D., Grendstad, J., Hide, H. O., Moestue, H. and Torkildsen, B. H. (1995). *Handbook of Multiphase Metering*, NFOGM Report No. 1, Norwegian Society for Oil and Gas Measurement, Oslo.
148. Aspelund, A., Midttveit, Ø. and Richards, A. (1996). 'Challenges in downhole multiphase measurements', *Soc. Petrol. Eng.* **SPE 35559**, 209.
149. Åbro, E. and Johansen, G. A. (1999). 'Improved void fraction determination by means of multibeam gamma-ray attenuation measurements', *Flow. Meas. Instr.* **10**, 99.
150. Åbro, E., Khoryakov, V. A., Johansen, G. A. and Kocbach, L. (1999). 'Determination of void fraction and flow regime using neural network trained on simulated data based on gamma-ray densitometry', *Meas. Sci. Technol.* **10**, 619.
151. Bishop, C. M. and James, G. D. (1993). 'Analysis of multiphase flows using dual-energy gamma densitometry and neural networks', *Nucl. Instrum. Meth.* **A327**, 580.
152. Tjugum, S.-A., Frieling, J. and Johansen, G. A. (2002). 'A compact low-energy multibeam gamma-ray densitometer for pipe-flow measurements', *Nucl. Instr. Meth.* **B197**, 301.
153. Rebgetz, M. D., Watt, J. S. and Zastawny, H. W. (1991). 'Determination of the volume fractions of oil, water and gas by dual energy gamma-ray transmission', *Nucl. Geophys.* **5**(4), 479.
154. Van Santen, H., Kolar, Z. I. and Scheers, A. M. (1995). 'Photon Energy Selection for Dual Energy γ-and/or X-ray Absorption Composition Measurements in Oil–Water–Gas Mixtures', *Nucl. Geophys.* **9**(3), 193.
155. Rafa, K. G. (1989). 'Dual-energy gamma-ray compositional measurement for oil/water/gas systems,' *Proceedings of Multiphase Flow – Technology and Consequences for Field Development*, Stavanger, Norway, 8–9 May 1989,
156. Scheers, A. M. and Letton, W. (1996). 'An oil/water/gas composition meter based on multiple energy gamma ray absorption (MEGRA) measurement', *Proceedings of the North Sea Flow Measurement Workshop*, Peebles, UK, 28–31 October 1996,
157. Bom, V. R., Clarijs, M. C., van Eijk, C. W. E., Kolar, Z. I, Frieling, J., Scheers, A. M. and Miller, G. J. (2001). 'Accuracy aspects in multiphase flow metering using X-ray transmission', *IEEE Trans. Nucl. Sci.* **48**(6), 2335.
158. Harrison, P. S., Hewitt, G. F., Parry, S. J. and Shires, G. L. (1995). 'Development and testing of the 'MIXMETER' multiphase flowmeter', *Proceedings of the 13th North Sea Flow Measurement Workshop*, Lillehammer, Norway.
159. Harrison, P. S., Parry, S. J. and Shires, G. L. (1999). 'The effects of salinity variation on dual energy multiphase flow measurements and Mixmeter homogeniser performance in high gas and high viscosity operation', *Proceedings of the 17th North Sea Flow Measurement Workshop*, Gardermoen, Norway.
160. McCoy, D. D., Warner, H. R., Jr. and Fisher, T. E. (1994). *Water Salinity Variations in the Ivishak and Sag River Reservoirs*, Prudhoe Bay Field, SPE 28577, 117.
161. Scheers, A. M. (1998). 'Multiple energy gamma ray absorption (MEGRA) techniques applied to multiphase flow metering', *Proceedings of the 4th International. Conferences on Multiphase Techinques*', London, 26–27 March, 1998.

162. Miller, G. J., Alexander, C. J., Lynch, F., Thompson, D. J., Letton, W. and Sheers, A. M. (1999). 'A high-accuracy, calibration-free multiphase meter', *Proceedings of the 17th North Sea Flow Measurement Workshop*, Gardermoen, Norway.
163. Johansen, G. A. and Jackson, P. (2000). 'Salinity independent measurement of gas volume fraction in oil/gas/water pipe flows', *Appl. Rad. Isotop.* **53**(4/5), 595.
164. Tjugum, S. A., Johansen, G. A. and Holstad, M. (2003). 'A multiple voxel model for scattered gamma radiation in pipe flow', *Meas. Sci. Technol.* **14**(10), 1777.
165. Hussein, E. M. A. and Han, P. (1995). 'Phase volume-fraction measurement in oil-water-gas flow using fast neutrons', *Nucl. Geophys.* **9**(3), 229.
166. Gay, R. R., Ohkawa, K. and Lahey, R. T., Jr. (1980). The Measurement of Local Void fraction with the Side-Scatter Gamma Technique ISA, 253. ISBN 87664-473-6.
167. Huggins, F. E. (2002). 'Overview of analytical methods for inorganic constituents in coal', *Int. J. Coal Geol.* **50**, 169.
168. Clayton, C. G. and Wormald, M. R. (1983). 'Coal analysis by nuclear methods', *Int. J. Appl. Radiat. Isot.* **34**(1), 3.
169. Sowerby, B. D. and Watt, J. S. (1990). 'Development of nuclear techniques for on-line analysis in the coal industry', *Nucl. Instr. Meth.* **A299**, 642.
170. Gravitis, V. L., Watt, J. S., Muldoon, L. J. and Cochrane, E. M. (1987). 'Long-term trial of a dual energy γ-ray transmission gauge determining the ash content of washed coking coal on a conveyer belt', *Nucl. Geophys.* **1**(2), 111.
171. Qi, X. and Yongchuan, Z. (2000). 'A novel automated separator based on dual energy gamma-rays transmission', *Meas. Sci. Technol.* **11**, 1383.
172. Sowerby, B. D. and Ngo, V. N. (1981). 'Determination of the ash content of coal using annihilation radiation', *Nucl. Instr. Meth.* **188**, 429.
173. Tominaga, H., Wada, N., Tachikawa, N., Kuramochi, Y., Horiuchi, S., Sase, Y., Amano, H., Okubo, N. and Nishikawa, H. (1983). 'Simultaneous utilization of neutrons and gamma-rays from Cf-252 for measurement of moisture and density', *Int. J. Appl. Radiat. Isotop.* **34**(1), 429.
174. Hills, A., Charlton, J. S. and Singh, G. (2002). *Radioisotope Applications for Troubleshooting and Optimising Industrial Processes*, IAEA
175. Hjertaker, B. T., Johansen, G. A. and Jackson, P. (2001). 'Level measurement and control strategies for subsea separators', *J. Electr. Imaging* **10**(3), 679.
176. Lees, R. P. (2002). 'Increasing control and accuracy in the separation process by density profiling', *Meas. Control* **35**, 164.
177. Mohammadi, H. (1981). 'Thickness gaging with scattered γ, β and X-rays', *Int. J. Appl. Radiat. Isot.* **32**, 524.
178. Gardner, R. P., Metwally, W. A. and Shehata, A. (2004). 'A semi-empirical model for a ^{90}Sr beta-particle transmission thickness gauge for aluminium alloys', *Nucl. Instr. Meth.* **B213**, 357.
179. Adaptive Technologies, USA, web site data (2003).
180. Sturm, S. P. (1999). 'On-line measurement of thin organic films on metallic sheets', *Light Metal Age* **47**, 19.
181. Hussein, E. M. A. (1989). 'Radiation scattering methods for nondestructive testing and imaging', *Int. Adv. Nondestr. Test* **14**, 301.
182. Silva, I. L. M., Lopes, R. T. and de Jesus, E. F. O. (1999). 'Tube defects inspection by using Compton gamma-rays backscattering', *Nucl. Instr. Meth.* **A422**, 957.

183. Lee, H. and Kenney, E. S. (1992). 'A new pipe wall thinning inspection system', *Nucl. Technol.* **100**, 70.
184. Zhu, P., Duvauchelle, P., Peix, G. and Babot, D. (1996). 'X-ray Compton backscattering techniques for process tomography: Imaging and characterization of materials', *Meas. Sci. Technol.* **7**, 281.
185. Jama, H. A., Hussein, E. M. A. and Lee-Sullivan, P. (1998). 'Detection of debonding in composite-aluminium joints using gamma-ray Compton scattering', *NDT & E Int.* **31**(2), 99.
186. Roach, G. J., Watt, J. S., Zastawny, H. W., Hartley, P. E. and Ellis, W. K. (1994). 'Multiphase flowmeter for oil, water and gas in pipelines based on gamma-ray transmission techniques', *Nucl. Geophys.* **8**(3), 225.
187. Watt, J. S., Zastawny, H. W., Rebgetz, M. D., Hartley, P. E. and Ellis, W. K. (1991). 'Determination of flow velocity of the liquids in oil/water/gas mixtures by dual energy gamma-ray transmission', *Nucl. Geophys.* **5**(4), 469.
188. Van Santen, H., Kolar, Z. I., Mudde, R. F. and Van den Akker, H. E. A. (1997). 'Double beam and detector γ-radiation attenuation gauge for studying bubble phenomena in gas-solid fluidised beds', *Appl. Rad. and Isotop.* **48**(10–12), 1307.
189. Hammer, E. A. and Nordtvedt, J. E. (1991). 'The application of a venturimeter to multiphase flow measurement,' *Proceedings of the Fifth Sensors and their Applications Conference*, Edinburgh, UK, 23–25 September 1991, 233.
190. Falcone, G., Hewitt, G. F., Alimonti, C. and Harrison, B. (2002). 'Multiphase flow metering: Current trends and future developments', *J. Petr. Tech.* **54**(4), 77.
191. Thorn, R., Johansen, G. A. and Hammer, E. A. (1997). 'Recent developments in three-phase flow measurement', *Meas. Sci. Technol.* **8**, 691.
192. Börjesson, J., Mattsson, S. and Alpsten, M. (1998). 'Trace element concentrations studied in vivo using X-ray fluorescence analysis', *Appl. Rad. Isotop.* **49**, No. 5/6, 437.
193. Szalóki, I., Patkó, J. and Papp, L. (1990). 'Determination of Cr and Mn in aluminium wires and sheets by XRF, NAA and AAS', *J. Radioanal. Nucl. Chem. Articles* **141**(2), 279.
194. Szabo, J. L., Simon, A. C. and Junca, R. (1994). 'Non-destructive analysis of uranium and/ or plotonium using X-ray (K or L band) fluorescence excited by sealed sources or X-ray tubes', *Nucl. Instr. Meth.* **A353**, 668.
195. Watt, J. S. (1985). 'Application of radioisotope techniques to the mineral industry,' *Proceedings of the 1985 Annual meeting of American Nuclear Society*, Boston, 9–13 June 1985, 411.
196. Shea, P., Gonzani, T. and Bozorgmanesh, H. (1990). 'A TNA explosives-detection system in airline baggage', *Nucl. Instr. Meth.* **A299**, 444.
197. Borsaru, M., Holmes, R. J. and Mathew, P. J. (1983). 'Bulk analysis using nuclear techniques', *Int. J. Appl. Radiat. Isot.* **34**(1), 397.
198. Dyakowski, T. and Jaworski, A. J. (2003). 'Non-invasive process imaging – Principles and applications of industrial process tomography', *Chem. Eng. Technol.* **26**, 6.
199. Nikitidis, M. S., Hosseini-Ashrafi, M. E., Tüzün, U. and Spyrou, N. M. (1994). 'Tomographic measurements of granular flows in gases and in liquids', KONA Powder and Particle No. 12, 53.

200. Boyer, C. and Fanget, B. (2002). 'Measurement of liquid flow distribution in trickle bed reactor of large diameter with a new gamma-ray tomographic system', *Chem. Eng. Sci.* **57**, 1079.
201. Cesareo, R. and Mascarenhas, S. (1989). 'A new tomographic device based on the detection of fluorescent X-rays', *Nucl. Instr. Meth.* **A277**, 669.
202. Stitt, H. and James, K. (2003). 'Process tomography and particle tracking: Research and commercial diagnostic tool', *Proceedings of the 3rd World Congress on Industrial Process Tomography*, Banff, Canada, 2–5 September 2003, 2.
203. Abdullah, J., Mohamad, G. H. P., Hamzah, M. A., Yusof, M. S. M., Rahman, M. F. A., Ismail, F. and Zain, R. M. (2003). Development of a portable computed tomographic scanner for on-line imaging of industrial piping systems', *Proceedings of the 5th National seminar on non-destructive testing*, Shah Alam, Malaysia 1–3 October 2003.
204. Johansen, G. A., Frøystein, T., Hjertaker, B. T. and Olsen, Ø. 'A dual sensor flow imaging tomographic system'. *Meas. Sci. Technol.* **7**(3), 297.
205. Hjertaker, B. T. (1998). 'Static characterization of a dual sensor flow imaging system', *Flow Meas. Instr.* **9**, 183.
206. Toye D. L'Homme, D. and Marchot, P. (2003). 'Perspective in data fusion between X-ray computed tomography and electrical capacitance tomography in an absorption column', *Proceedings of the 3rd World Congress on Industrial Process Tomography*, Banff, Canada, 2–5 September, 2003, 68.
207. Lange, K. and Carson, R. (1984). 'EM reconstruction algorithms for emission and transmission tomography', *J. Comput. Assist. Tomo.* **8**(2), 306.
208. Dudukovic, M. P., Dyakowski, T., Gardner, R. P., Legoupil, S. Johansen, G. A. and Thereska, J. (2002). 'Gamma industrial process tomography', IAEA Report on Consultants Meeting Raleigh, North Carolina, USA, 14–18 October 2002.
209. Frøystein, T. (1993). 'Gamma-ray flow imaging', *Proceedings of ECAPT'93* (European Concerted Action on Process Tomography), Karlsruhe, Germany, 24–27 March 1993, 338.
210. Frøystein, T. (1997). 'Flow imaging by gamma-ray tomography: Data processing and reconstruction techniques', *Proceedings of Frontiers in Industrial Process Tomography* II, Delft, 8–12 April 1997, 185.
211. Schlosser, P. A., De Vouno, A. C., Kulacki, F. A. and Munschi, P. (1980). 'Analysis of high-speed CT scanners for non-medical applications', *IEEE Trans. Nucl. Sci.* **NS-27**(1), 788.
212. De Vouno, A. C., Schlosser, P. A., Kulacki, F. A. and Munschi, P. (1980). 'Design of an isotopic CT scanner for two phase flow measurements', *IEEE Trans. Nucl. Sci.* **NS-27**(1), 814.
213. Natterer, F. (1986). *The Mathematics of Computerized Tomography*, Wiley, New York.
214. Frøystein, T. (1992). *Gamma-Ray Flow Imaging*, MSc Thesis, Department of Physics, University of Bergen, Norway.
215. Bruneau, P. R. P., Mudde, R. F., van der Hagen, T. H. J. J. (2003). 'Fast computed tomographic imaging within turbulent fluidised beds', *Proceedings of the 3rd World Congress on Industrial Process Tomography*, Banff, Canada, 2–5 September, 2003, 589.
216. Frøystein, T. (2003). Private communication, Haukeland University Hospital.

217. Hori, K., Fujimoto, T. and Kawanishi, K. (1998). 'Development of ultra-fast X-ray computed tomography scanner system', *IEEE Trans. Nucl. Sci.* **45**(4), 2089.
218. Hori, K. (2001). 'Measurement and visualization of multiphase-flow by ultra-fast X-ray CT scanner', *Proceedings of the Japanese Society for Multiphase Flow Annual Meeting*, Kitakyushu, 12–13 July 2001, 151.
219. Morton, E. J., Luggar, R. D., Key, M. J., Kundu, A., Távora, L. M. N. and Gilboy, W. B. 'Development of a high speed X-ray tomography system for multiphase flow imaging', *IEEE Trans. Nucl. Sci.* **46**(3), 380.
220. Rogers, D. W. O. and Bielajew, A. F. (1986). 'Monte Carlo techniques of electron and photon transport for dosimetry', Kase, K. R., Bjärngard, B. E. and Attix, F. H. (Eds.) *The Dosimetry of Ionising Radiation, Vol. III*, Academic Press, San Diego, CA.
221. Petler, J. S. (1990). 'Modelling the spatial response of a compensated density tool', *IEEE Trans. Nucl. Sci.* **37**(2), 954.
222. Peplow, D. E., Gardner, R. P. and Verghese, K. (1994). 'Sodium iodide detector response functions using simplified Monte Carlo simulation and principal components', *Nucl. Geophys.* **8**(3), 243.
223. Gardner, R. P. and Sood, A. (2004). 'A Monte Carlo simulation approach for generating 3 NaI detector response functions (DRFs) that accounts for non-linearity and variable flat continua *Nucl. Instr. Meth.* **B213**, 87.
224. Gardner, R. P. and Liu, L. (2000). 'Monte Carlo simulation for IRRMA', *Appl. Radiat. Isot.* **53**, 837.
225. Amersham International plc. *Instrument calibration sources* S88/85.
226. Evans, R. D. (1955). *The Atomic Nucleus*, McGraw-Hill, New York.
227. Tsipenyuk, Y. M. (1997). *Nuclear Methods in Science and Research, Fundamental and Applied Nuclear Physics Series*, Institute of Physics Publishing, Bristol, UK.
228. IAEA (1996). *IAEA Safety Standards Series, Quality Assurance for Safety in Nuclear Power Plants and other Nuclear Installations – Requirements 1996 edtion*, No. ST-1, STI/PUB/998, IAEA, Vienna.
229. Suzuki, H., Tombrello, T. A., Melcher, C. L. and Schweitzer, J. S. (1992). 'UV and gamma-ray excited luminescence of cerium-doped rare-earth oxyxorthosilicates', *Nucl. Instr. Meth.* **A320**, 263.
230. Lakatos, T., Hegyesi, G. and Kalinka, G. (1996). 'A simple pulsed drain feedback preamplifier for high resolution high rate nuclear spectroscopy', *Nucl. Instr. Meth.* **A378**, 583.
231. Dalla Betta, G. F., Manghisoni, M., Ratti, L., Re, V. and Speziali, V. (2003). 'JFET preamplifiers with different reset techniques on detector-grade high-resistivity silicon', *Nucl. Instr. Meth.* **A512**(1/2), 199.
232. Abouelwafa, M. S. A. and Kendall, E. J. M. (1980). 'The measurement of component ratios in multiphase systems using γ-ray attenuation', *J. Phys. E: Sci. Instrum.* **13**, 341.
233. Saloman, E. B., Hubell, J. H. and Scofield, J. H. (1988). 'X-ray attenuation cross sections for energies 100 eV to 100 keV and elements $Z = 1$ to $Z = 92$', *Atomic Data and Nuclear Data Tables* **38**(1).
234. Atkinson, I., Berard, M., Hanssen, B.-V. and Ségéral, G. (1999). 'New generation multiphase flowmeters from Schlumberger and Framo Engineering AS', *Proceedings of the 17th North Sea Flow Measurement Workshop*, Gardermoen, Norway.

Index

Radioisotope Gauges for Industrial Process Measurements. Geir Anton Johansen and Peter Jackson.

ISBN 0-471-48999-9

Printed and bound in the UK by
CPI Antony Rowe, Eastbourne

Printed and bound by CPI Group (UK) Ltd, Croydon, CR0 4YY
16/09/2026
14963479-0002